Philosophy of Engineering and Technology

Volume 21

More information about this series at http://www.springer.com/series/8657

Steen Hyldgaard Christensen
Christelle Didier • Andrew Jamison
Martin Meganck • Carl Mitcham
Byron Newberry
Editors

Engineering Identities, Epistemologies and Values

Engineering Education and Practice in Context, Volume 2

Springer

Editors
Steen Hyldgaard Christensen
Aalborg University
Aalborg, Denmark

Christelle Didier
Université Charles de Gaulle-Lille3
Lille, France

Andrew Jamison
Aalborg University
Aalborg, Denmark

Martin Meganck
KU Leuven
Ghent, Belgium

Carl Mitcham
Colorado School of Mines
Golden, Colorado, USA

Renmin University
Beijing, China

Byron Newberry
Baylor University
Waco, Texas, USA

ISSN 1879-7202 ISSN 1879-7210 (electronic)
Philosophy of Engineering and Technology
ISBN 978-3-319-16171-6 ISBN 978-3-319-16172-3 (eBook)
DOI 10.1007/978-3-319-16172-3

Library of Congress Control Number: 2015939348

Springer Cham Heidelberg New York Dordrecht London

Printed on acid-free paper

Springer International Publishing AG Switzerland is part of Springer Science+Business Media (www.springer.com)

Preface

And some time make the time to drive out west
Into County Clare, along the Flaggy Shore,
[T]he ocean on one side is wild
With foam and glitter, and inland among stones
The surface of a slate-grey lake....
Useless to think you'll park and capture it
More thoroughly. You are neither here nor there,
A hurry through which known and strange things pass
As big soft buffetings come at the car sideways
And catch the heart off guard and blow it open.

– Seamus Heaney, "Postscript" (1996)

We live today in a world progressively in the process of becoming an engineered artifact. We engineer not only roads and buildings but communication systems and biologies. In such a world, thinking about engineering is increasingly important – and yet incredibly difficult.

Among themselves, engineers are continuously trying to figure out what and who they are: skilled workers, project managers, applied scientists, designers, entrepreneurs, and more. Additionally, there are a host of competing interests that would enroll engineering for their purposes: military interests, nation-building interests, commercial interests, social interests, environmental interests, and more. Finally, multiple disciplines attempt to take the measure of engineers and engineering: history, sociology, philosophy, and more.

There is no simple resolution to the tensions inherent in this complexity of contextualizations for the engineered constructions in which we progressively live and move and have our being. The best we can do is take an intellectual drive through diverse intellectual landscapes, with a willingness to let what poet Seamus Heaney calls "big soft buffetings" come at us sideways, opening the mind. Open to its contexts, the mind is at once:

- More reflective in negotiating the pressures that enfold it
- Better at spanning different engineering visions and practices

- More insightful when conciliating the corporeal powers of engineering with the ethereal truths of poetry or art
- More resistant to commercial, political, and military distortions of human and professional responsibilities
- Better at constructing a more just world – one in which lives well-lived and well-examined transcend mere existence

To contribute to this opening up, not so much of the black box of what takes place behind the scenes in engineering, but of our own thinking about engineering, is the central effort of our collective reflection.

The two books we offer – *International Perspectives on Engineering Education: Engineering Education and Practice in Context. Volume 1* and *Engineering Identities, Epistemologies and Values: Engineering Education and Practice in Context. Volume 2* – are the result of an extended dialogue or bridge-building between humanists and engineers with whom we have been involved both individually and more recently as a group. Steen Hyldgaard Christensen, the editor-in-chief, studied literature and history of ideas at Aarhus University in the 1970s, and since 1987 has taught humanities for engineering and business students at what was originally a technical vocational college in Herning, Denmark (which in 1995 became the Institute of Business and Technology, and in 2006 Aarhus University). Since 2003, Christensen has been facilitating processes of collaboration between engineers, social scientists, and humanists in a series of book projects. The first, with coeditors Martin Meganck and Bernard Delahousse, was on *Profession, Culture and Communication: An Interdisciplinary Challenge to Business and Engineering* (2003); the second, with the same coeditors, was *Philosophy in Engineering* (2007); a third, again with coeditors Meganck and Delahousse, was *Engineering in Context* (2009), the precursor of the present two volumes.

Martin Meganck has a doctorate in chemical engineering and is a former Dominican friar who studied theology and currently teaches ethics for engineering students at KU Leuven in Ghent, Belgium. Bernard Delahousse was an English language scholar who taught at an engineering college in Lille, France, and served as head of the school's international office. Delahousse has retired, but participates now as coauthor of one of the chapters in Volume I. Christensen got to know them while serving as the international officer at his institution, which is now part of Aarhus University.

For each book, Christensen and his coeditors organized a gathering of potential authors. Two days of deliberations by participants lead to a table of contents, after which Christensen and his coeditors orchestrated the logistics of book production: first draft submission, final draft submission, index submission, proofreading, etc.

In 2008, Andrew Jamison was drawn into the process, as a contributor to the project that became *Engineering in Context*. But even before that book was published, Jamison, with Christensen and several other contributors to these volumes, asked the Danish Strategic Research Council to fund a four-year Program of Research on Opportunities and Challenges in Engineering Education in Denmark (PROCEED). This ambitious, interdisciplinary project took place between 2010 and 2013.

PROCEED was organized as a strategic research alliance between four universities: Aalborg University, Aarhus University (including the former engineering college in Herning), Roskilde University, and the Danish Technical University. The research was divided into five thematic projects: "Challenges and Responses in Historical Perspective," "Curriculum Design and Learning Outcomes," "Modeling and Simulation in Engineering," "Engineering Practice and Design Competence," and "Integrating Contextual Knowledge into Engineering Education" (cf. PROCEED 2010). The alliance included engineers, social scientists, philosophers, and historians; numerous chapters in these books are based on research and teaching activities that were part of the program.

Prior to the initiation of PROCEED, another project took shape that has also influenced the present two volumes. Christensen, Jamison, and Carl Mitcham teamed up to organize an interdisciplinary reflection on relationships between "engineering and development" that involved American, Chinese, and European perspectives. Christensen invited ten Europeans, Mitcham ten Americans, and Li Bocong, from the Graduate University of the Chinese Academy of Sciences (with whom Mitcham had been working since the early 1990s), ten Chinese scholars. Together these scholars met at the Colorado School of Mines in April 2010 in a workshop supported by the CSM Hennebach Program in the Humanities for an exercise in reflective, cross-cultural learning. PROCEED served as a cosponsor of the workshop by funding travel by some of the European participants.

Mitcham – a key node in the Christensen network from 2006 on – organized the CSM workshop around a series of "tutorials" designed to stimulate dialogue. Mitcham and his colleague Juan Lucena led tutorials on engineering and development from an American perspective (Mitcham for the North and Lucena, originally from Colombia, from the South); Christensen and Jamison offered a tutorial on engineering and development from a European perspective; while Li Bocong and Yanming An introduced a Chinese perspective. By the end of the meeting in Golden, CO, a table of contents was developed for a book that was eventually published in 2012 under the title, *Engineering, Development and Philosophy: American, Chinese and European Perspectives*, edited by Steen Hyldgaard Christensen, Carl Mitcham, Li Bocong, and Yanming An. The book appeared in the Springer series *Philosophy of Engineering and Technology*.

As a further contribution to the American-Chinese-European collaboration project, Li Bocong arranged another workshop on "Engineering and Sociology" in Beijing, China, in the fall of 2011. Li had long been concerned that engineering in the West was too focused on an individualistic professionalism, and he sought to stimulate reflections that would broaden the contexts of understanding in both the West and the East. It was thus in Beijing, around the pleasures of extended Chinese meals, and in a country undergoing a historically unique engineering construction, that there emerged the germ of an idea that has grown into these two volumes on *Engineering in Context*.

Another contributory linkage to these publications can be found in the European Ethics Network (EEN) from the 1990s. The EEN brought together ethicists from 40 European universities and had a broad set of objectives. One of these was creating a book series with core materials for professional ethics in the fields of biomedicine, business, press, and engineering. A kick-off conference in Barcelona, under the title "Rethinking Professional Ethics," was the starting point of a series of collaborations among ethicists involved in engineering and technology from mainly Western European countries. An immediate and tangible result was the publication of Philippe Goujon and Bertrand Hériard-Dubreuil's edited volume, *Technology and Ethics: A European Quest for Responsible Engineering* (2001). The engineering ethics team of the Catholic University of Lille (France) was the motor and the pivoting centre behind the book, and Christelle Didier and Martin Meganck were members of the editorial team. Mitcham contributed an afterword comparing American and European efforts in this area. The ethics journal *Ethical Perspectives* served for some time as the official organ of EEN and is the only *ad extra* visible remainder of that EEN period. A less visible outgrowth, however, is a continuing set of ties among ethicists in different professional fields. When the *Profession, Culture and Communication* project sought a continuation in *Philosophy in Engineering*, the ties between research groups and individual researches resulting from the EEN experience were useful in identifying new partners. The presence of Christelle Didier in the current editorial team has its basis there.

Still one more contributing stream to our collaborative effort, one that draws again on the work of Li Bocong, among others, is the 2012 Forum on Philosophy, Engineering, and Technology (fPET) held in Beijing, China. fPET-2012 was a follow-on to an earlier fPET-2010 hosted at CSM in Colorado. The fPET conferences grew out of previous workshops held in 2007 and 2008 known as the Workshops on Philosophy and Engineering (WPE). The fPET conferences, like the WPE workshops before them, have provided opportunities to bring together scholars from a variety of cultures and disciplines, all sharing a common interest in trying to better understand the human activities we call engineering, the people we call engineers, and the creations we call technology. At the latest meeting in Beijing, approximately 15 countries and 5 continents were represented. Philosophers, historians, and other humanists, along with social scientists and engineers, participated. The range of presentations included philosophical, historical, cultural, and ethical analyses of engineers, engineering, and technology. These events have proved invaluable as catalysts for ideas, scholarly exchanges, and collaborations. In fact, almost half the contributors to the present volumes have been participants in one or more of these events. Byron Newberry, another member of the current editorial team, whose background is in aerospace and mechanical engineering, served as cochair, along with Li Bocong, of the fPET-2012 meeting. Newberry also contributed to the earlier *Engineering in Context* book.

These different strands come together in the current set of two books. An international editorial kick-off workshop was initiated by Christensen and organized with the help of Louis L. Bucciarelli at MIT in May 2012. The main purpose was to define the objectives, structure, and content of the volumes. After introductory presentations by workshop host Bucciarelli, Gary Downey, and Jamison, an intensive process of discussions began. And, as the French say, *Du choc des idées jaillit la lumière*: at first confrontational ideas finally result in understanding and constructive proposals.

We hereby present the final result of a long writing and editorial process. We trust that our readers will find the work worthwhile and they may be inspired by it to do even more to think and rethink engineering contexts so as to transform engineering into a truly humanizing enterprise.

As those two books are meant to be a contribution to furthering the dialogue between engineering and philosophy in order to explore ways in which the humanities can contribute to self-development in engineering education through appreciation of the multiple contexts within which engineers increasingly work, these groups of academics are the primary audience for our books. Moreover, we believe that the very process of creating these volumes, bringing together as it has a host of scholars from a diversity of disciplinary and cultural perspectives, marks a major milestone on the path toward creating a sense of identity and shared culture, while recognizing the value of differences, and building a vibrant community of scholars dedicated to bridging the gaps between engineers, humanists, and social scientists.

However, the book is also addressing a wider academic audience and may actually function as a means to achieve greater self-understanding for both teachers in engineering disciplines and for practitioners. Educational policy makers, both on a political and an institutional level, may also find valuable matter for reflection and inspiration in this book. We believe that, not least, the process of globalization compels engineering educators to rethink and recontextualize engineering education in order to educate a better and more rounded type of engineer. We finally hope that the book may inspire students of engineering as well as students of the humanities and social sciences who are interested in the challenges and complexities that a rapidly changing and globalized world pose for higher education in general and for engineering education in particular.

Herning, Denmark — Steen Hyldgaard Christensen
Lille, France — Christelle Didier
Aalborg, Denmark — Andrew Jamison
Ghent, Belgium — Martin Meganck
Golden, Colorado, USA — Carl Mitcham
Waco, Texas, USA — Byron Newberry
1 October 2014

Acknowledgment

The editors would like to express our heartfelt gratitude to the two anonymous reviewers who provided thorough assessments of our two volumes, respectively. The comments, suggestions, and criticisms provided by these two scholars were both detailed and insightful. As a result of their feedback, we added new material on topics that deserved more attention (particularly with respect to issues of gender, race, and class), made significant improvements to several chapters, reorganized some of the chapters for better coherence and flow, and have tightened up some of our introductory sections. Our manuscript has been made stronger due to the care and diligence of these reviewers.

Contents

Author Biographies

Steen Hyldgaard Christensen M.A. in Scandinavian Language and Literature and the History of Ideas, Aarhus University. Ph.D. in Educational Studies, Aalborg University. Senior lecturer at Aarhus University, School of Business and Social Sciences, Herning, Denmark, until 2014. From 2014, adjunct associate professor at Aalborg University, Denmark. He has initiated six big international inter- and meta-disciplinary research projects on engineering including PROCEED and coordinated five of them. He has acted in roles of editor-in-chief and coauthor of four books: *Profession, Culture, and Communication: An Interdisciplinary Challenge to Business and Engineering* (Institute of Business Administration and Technology Press 2003); *Philosophy in Engineering* (Academica 2007); *Engineering in Context* (Academica 2009); and *Engineering, Development and Philosophy: American, Chinese, and European Perspectives* (Springer 2012). Besides, he has coauthored *A Hybrid Imagination: Science and Technology in Cultural Perspective* (Morgan & Claypool Publishers 2011) together with Andrew Jamison and Lars Botin. In addition, he has published a number of articles on engineering epistemology, culture, and education. Current research interest includes academic drift in engineering education and structural dynamics in higher education.

Christelle Didier B.S. in Electrochemistry Engineering, M.A. in Education, Ph.D. in Sociology from Ecole des Hautes Etudes en Sciences Sociales (EHESS), Paris. From 1993 to 2013, Assistant Professor, Lille University, France, Ethics Department. Assistant Professor, Charles de Gaulle University of Lille, Education Department. Member of CIREL (EA 4354). Coauthor of *Ethique industrielle* (DeBoeck, Brussels, 1998) and author of *Penser l'éthique des ingénieurs* (PUF, Paris 2008) and *Les ingénieurs et l'éthique: Pour un regard sociologique* (Hermes 2008). She has published many articles on ethics and social responsibility in the engineering profession and education and on the engineering profession's values (from interviews and extensive surveys). Her research areas are engineering ethics and values, including historical, cultural, and gender perspective, sustainable development and corporate social responsibility, social responsibility.

Andrew Jamison B.A. in History and Science from Harvard University, Ph.D. in Theory of Science from University of Gothenburg (Göteborg). Docent in Theory of Science, University of Gothenburg. Professor of Technology, Environment, and Society at Aalborg University. Coordinator of Program of Research on Opportunities and Challenges in Engineering Education in Denmark (PROCEED), 2010–2013, and author, most recently, of *The Making of Green Knowledge: Environmental Politics and Cultural Transformation* (Cambridge 2001), *Hubris and Hybrids: A Cultural History of Technology and Science*, with Mikael Hård (Routledge 2005), *A Hybrid Imagination: Science and Technology in Cultural Perspective* (Morgan & Claypool Publishers 2011) together with Steen Hyldgaard Christensen and Lars Botin, and *The Making of Green Engineers: Sustainable Development and the Hybrid Imagination* (Morgan and Claypool 2013).

Martin Meganck M.Sc. in Chemical Engineering from Ghent University, Ph.D. in Chemical Engineering, and M.A. Moral Theology both from KU Leuven. Lecturer in Philosophy and Ethics in the Faculty of Engineering Technology, researcher at the Center for Science, Technology, and Ethics, and the Theological Faculty of the KU Leuven. Teaching areas include Philosophy of Science, Philosophy of Technology, and Professional and Business Ethics. He was coauthor and coeditor of *Philosophy in Engineering* (Academica 2007) and *Engineering in Context* (Academica 2009).

Carl Mitcham B.A. and M.A. in Philosophy from University of Colorado, Boulder. Ph.D. in Philosophy from Fordham University. Professor of Philosophy of Science and Technology, Renmin University of China, Beijing; Liberal Arts and International Studies, Colorado School of Mines, Golden, Colorado. Scholarly contributions have been directed toward the philosophy and ethics of science, technology, engineering, and medicine and to science, technology, and society (STS) studies. Teaching areas include ethics, STS, and science and technology policy.

Byron Newberry B.S. in Aerospace Engineering, University of Alabama. M.S. in Aerospace Engineering and Ph.D. in Engineering Mechanics, both from Iowa State University. Professor of Mechanical Engineering, Baylor University. Baylor Fellow for teaching. Professional Engineer (PE) Texas, USA. Research interests include engineering design, engineering ethics, and philosophy of engineering and technology. Aircraft structural engineering consultant. Executive board member, National Institute for Engineering Ethics. Editor of the Springer *Philosophy of Engineering and Technology* book series.

General Introduction
The Engineering-Context Nexus
A Perennial Discourse

Steen Hyldgaard Christensen, Christelle Didier, Andrew Jamison, Martin Meganck, Carl Mitcham, and Byron Newberry

In 1982, Barry Barnes and David Edge published *Science in Context: Readings in the Sociology of Science*, which significantly influenced the sociology of science. The volume collected 18 previously published articles from the 20-year period 1961 to 1981 – articles almost exclusively by social scientists – to promote reflection on relationships between the subculture of science and the wider culture that surrounds it. Although the editors did not present it as such, the program for understanding

S.H. Christensen
Department of Development & Planning, Aalborg University,
Vestre Havnepromenade 5, Aalborg 9000, Denmark
e-mail: steenhc@plan.aau.dk

C. Didier
Département des sciences de l'éducation UFR DECCID,
Université Charles de Gaulle-Lille3, 3 Rue du Barreau,
Villeneuve-d'Ascq, Lille F-59650, France
e-mail: christelle.didier@univ-lille3.fr

A. Jamison
Department of Development and Planning, Aalborg University,
Erik Dahlbergsgatan 22, Malmö S-211 48, Sweden
e-mail: andy@plan.aau.dk

M. Meganck
KU Leuven, Faculty of Engineering Technology, Technologiecampus Ghent,
Gebroeders De Smetstraat 1, Ghent B-9000, Belgium
e-mail: martin.meganck@kuleuven.be

C. Mitcham
Colorado School of Mines, Golden, Colorado, USA

Renmin University, Beijing, China
e-mail: cmitcham@mines.edu

B. Newberry
Department of Mechanical Engineering, Baylor University,
One Bear Place 97356, Waco, TX 76798, USA
e-mail: Byron_Newberry@Baylor.edu

"science in context" can be read as responding to the challenge of C.P. Snow's 1959 "two cultures" lecture, which identified a debilitating split between scientific and literary intellectuals. For Snow, there really were two cultures that approached the world from antagonistic perspectives. For the social scientists collected by Barnes and Edge, however, scientific culture is always part of culture in a more expansive sense. The two cultures are really one, and science needs to be understood precisely as an aspect of what it may indeed partially oppose.

In the spirit of that earlier title, the present two companion volumes focus on *Engineering Education and Practice in Context* (EEPiC, read as "epic"). This project differs, however, not only in its concern with engineering instead of science but also in being composed of more than 40 original articles contributed by a much more interdisciplinary group: social scientists, yes, but also engineers, philosophers, historians, and even scholars from the fields of classics, communication, and film studies. Additionally, among the more than 60 contributors are representatives from 16 countries on the 6 inhabited continents. The volumes direct attention to four primary contexts of engineering: formal education, the design process, workplace and institutional experience, and civil society. Yet like Barnes and Edge, these new volumes postulate an integral if sometimes contentious relationship between engineering cultures and their larger cultural contexts.

Comparing work on science with the present work on engineering, there emerges what may be termed a contextualization-decontextualization paradox. Scientists qua scientists think of their work as decontextualized and, therefore, have trouble recognizing the ways in which it is also contextualized. Engineers qua engineers think of their work as contextual and, therefore, tend to overlook the ways in which it is decontextualized. Scientists, for example, see formulas such as $F = ma$ and $E = mc^2$ as independent of context and universally true, failing to appreciate their knowledge production can reflect particular cultures (as, in these cases, a mathematical rhetoric enacted in distinctive social institutions). By contrast, engineers engage with contexts in which they deploy those same formulas in particular projects. But it is precisely because they think of themselves as so context dependent and sensitive that engineers also so often presume they can go into any situation and provide appropriate solutions; they often too readily believe all their solutions are inherently contextual, even when this fails to be the case. The existence of such a paradox suggests the need to use the *Science in Context* project as defined by Barnes and Edge as a foil with which to exploit difference.

Beyond Science in Context

The science in context argument is in an important respect nihilistic. The significance of natural science, which the sociology of science aims to disclose, is that natural science has no special significance. Its reputed claims to significance are unmasked, demythologized, and demystified. The sociological argument, as succinctly summarized by Barnes and Edge, is that "There is no way in which [natural scientific] expertise can be guaranteed by reference to reason rather than habitual inference, nature rather than culture" (p. 11). Natural science is a social institution like any other; it rests on purely social foundations and its reasons are no more privileged than those of politics, economics, or the military.

Yet as Barnes and Edge also admit, "to conceive of expert knowledge solely in terms of advocacy" is to ignore the normative question concerning which advocates are most credible or authoritative. The normative question is not one that can be "reduced to a matter of what beliefs are immediately expedient, or immediately relevant to vested interests" (p. 10). Among natural scientists and nonscientists alike, the problem of credibility has customarily been resolved by granting natural science a measure of rational authority – although a rational authority that social scientific analysis questions.

The social scientific analysis of science in context is nevertheless faced with three problems. First, social science is not generally granted the same social recognition as natural science – that is, as the astronomical, physical, geological, and biological sciences. So its claims with regard to the natural sciences often carry little weight. It is not clear what influence the analysis of science in context can ever really have.

Second, even if the social sciences were magically to acquire social prestige and power, it is not clear how more careful and detailed sociological studies – which are repeatedly recommended by Barnes, Edge, and others, in order to give a better understanding of what really happens with science – would escape the acidic analysis that they apply to the natural sciences. That is, the sociological analysis addressed to the natural sciences would seem necessarily to apply as well to the social sciences. The social sciences, too, would have to be conceived as social constructions.

As a result, third, the social sciences can "offer no obvious solutions to the normative problems involved in the evaluation of [scientific] expertise" (p. 12). It is not just that the normative question is, as Barnes and Edge later claim, "of no sociological interest" (p. 194); normativity is not an issue that it is even possible in principle for sociology to address. The sociology of science reveals science to be without distinctive authority and thus at the mercy of political, economic, and military powers – powers that are not troubled, in their real-world exercise of power, by any alleged lack of authoritative rationality. This is what Barnes and Edge refer to as "the tragedy of the expert" (p. 237). Experts can never deploy the methods of expertise, which exist within a community of experts, to legitimate such expertise to the wider public. "If science itself is called into question, then the scientific expert can only retire gracefully" (p. 234). Scientific experts appear dependent on irrational acceptance by the public, with an irrationality that can at most and only on occasion be meliorated by programs of public participation – although Barnes and Edge acknowledge the "power" present in science, especially as reflected by the close linkages of science "with 'the higher levels' of government and industry" (p. 248).

The science in context project is thus fraught with implications the engineering in context project seeks as much as possible to avoid. To this end, we offer three observations. First, by way of a brief historicophilosophical gloss, note that while the idea of the social construction of science can be manifest among scholars without serious immediate harm, the idea has been applied elsewhere with quite harmful results. Insofar as the administration of U.S. President George W. Bush refused, when making decisions about how best to reduce teenage pregnancy, respond to climate change, and the invasion of Iraq to grant any privileged status to scientific knowledge, both natural and social, he adopted a social constructivist stance. To the realist objection that one needs to respect reality, one of Bush's senior advisers is reported simply to have replied, "When we act, we create our own reality" (Suskind 2004). Such application and its results surely provide a good reason to revisit the normative question and defend the rationality of engineering as well as of science.

Second, and more positively, as if offering a means for addressing the normative question, our engineering in context project, is inherently more interdisciplinary. It involves not just sociologists and historians but also engineers and philosophers – along with scholars in the further reaches of the humanities and the social sciences. Indeed, while science in context sought to broaden the reach of science, the broadening went no further than to describe science as not just a "source of knowledge and competence [but as] a repository of theories, findings, procedures and techniques which it makes generally available both directly, via expert intervention and consultation, and indirectly, via its interaction with technology and with specialized institutions in the economic and political structure" (p. 2). What is lacking is recognition of science as a font of social, ethical, and even environmental problems.

To recognize science or engineering as a source of problems – especially environmental problems – is not to deny that it can also contribute to solutions or, better, responses. Indeed, to adopt and adapt the naturalistic pragmatism of John Dewey and to recognize something as a problem is implicitly to imagine a better state of affairs. For Dewey, engineering is ultimately and properly subordinate to the enhancement of life and the qualitative enlargement of human experience. Insofar as science and its sibling engineering fail to accord with this transcendent end – an end that is subject to continuous reimagination and reinstitutionalization in culture – it calls forth its own reconceptualization, regulation, or delimitation along with parallel and complementary extensions and expansions.

It is precisely this that best functions as our own context for the study of engineering. We are studying engineering not simply to promote sociological understanding but in pursuit of better engagement between engineering and society – and the better education of engineers. Moreover, although to some degree a socially constructed or contingent end, it is an end for which we are willing and able to develop rational arguments. Only insofar as we can give good reasons for such ends – not just insofar as such ends are popularly accepted – should we wish to defend and built upon or toward them.

Thus the EEPiC project includes a strongly reflexive element. In the Barnes and Edge volume, for instance, there was no discussion of the meaning of context. By contrast, our two volumes both explicitly and implicitly address different meanings of context. On the explicit side, some chapters grapple overtly with the issue of context, whether trying to elucidate its meaning, to highlight its importance, or in at least one case to reject it. On the implicit side, ideas about contexts were built in via the selection of authors and topics, along with the organization of the volume sections. For example, while Barnes and Edge relied heavily on the problematic concept of culture, for which it assumes an anthropological meaning (see p. 193), here the question of culture is itself placed in context by the presence of contributions from multiple cultures and cultural perspectives, not to mention disciplines and disciplinary perspectives. In addition, chapters in the two volumes are organized in sections designed to explore particular contextual facets, whether historical, ideological, or institutional.

We should note, however, that it is not the objective of these volumes to definitively demarcate the meaning of context in engineering. For our purposes, context is not an end-in-itself but rather a means to an end. In the spirit of further reflexivity, the contingent but nonetheless rationally defensible (and inherently normative) end of the engineering in context project is to foster a better understanding of and engagement with engineering. This engagement will be intentionally provocative and argue for an end that is not explicitly given but implicitly found embedded within it: the transcendence of engineering, what has been called postengineering (see Mitcham 2009).

Remaining for the present in the European tradition, there exists a long-standing or sedimented distinction between liberal and professional education. From the perspective of liberal studies, the contrast is one between education and training, even vocational or technical training. From the perspective of professional studies, the contrast is between useless discussion or mere theory and useful or practical learning. It seems clear that engineering education accords primarily with professional or practical studies. Yet this is not to deny its possible involvement with liberal or even useless studies. We need to move beyond simple dependence on engineering. We must not become so effective at and engrossed with engineering that we forget that engineering is not everything. We need to exercise again the classical humanities disciplines of self-moderation.

Two Volumes and Their Complementarities

In summary, in relation to science in context, which it references as an ancestor, the two EEPiC volumes aim to be more interdisciplinary and original, more critical and reflexive, and more openly normative. Taken as a whole, this collection of original scholarly work is unique in its broad, multidisciplinary consideration of the changing character of engineering education and engineering practice in and from the perspective of multiple contexts.

Volume 1 on engineering education includes analyses of the history, structure, and ideologies of engineering education, challenges and critical perspectives, along with discussions of new pathways in 25 contributions by 50 authors from engineering, social sciences, and humanities. Key overlapping questions examine such issues as:

- What are the different approaches to engineering education?
- Are differences competitive or complementary?
- What special challenges are emerging for engineering from concerns for sustainable community development, energy ethics, sustainability, and demands for innovative design?

- What new efforts are being made to reform engineering education from the perspectives of design, engineering education research, and case-based learning?
- What is the role of the social sciences and the humanities in engineering education?

The chapters of Volume 1 are grouped into four sections, roughly following a see-judge-act logic. Part I historically frames engineering education in the United States, Western Europe, and a selection of locations elsewhere (India, Brazil, Slavic Europe). What appears initially simply descriptive is interwoven with a reflexive/interpretative layer. Part II groups a series of more fundamental reflections on the hidden and overt ideologies in engineering and engineering education. Parts III and IV collect contributions on experiences and approaches for reform and innovations in engineering education.

In Part I, the institutional history and evolution of engineering education in different geographical/cultural contexts is the carrying canvas. Regional, cultural, and historically bound aspects form one approach. Although these historiographical descriptions focus on regional and cultural differences, some common themes emerge. One is "academic drift": vocational-oriented training programs tend to be swept into more academic structures, inducing changes in professional profile and educational culture. A shift of focus from local toward more global perspectives can also be observed throughout the contributions. Insertion in the global economy seems to induce more pragmatic and neoliberal entrepreneurial tendencies in engineering education.

Part II shifts from institutional history to the asking of critical questions regarding theory and practice in engineering education. Like all institutionalized programs of education, engineering schools explicitly or implicitly assume and promote beliefs about how engineers should behave, not just in technical terms but in their social relationships. As previous scholars have noted, there are deeply ingrained ideas in the American context about positive relationships between engineering and business. The chapters in this section invite consideration of some alternative perspectives by calling attention to how engineering education functions differently in China and how the engineering-business nexus may not be experienced as unquestionably rational by members of nondominate social groups.

The framework of Part III extends an exploration of the limitations of received ideologies in engineering education by considering specific cases in the emergence of alternative futures. Hence the majority of chapters in Part III contribute to the construction of a counter-hegemonic discourse or "heterotopia," to use a term of Baillie et al. (2012). Some themes that come into view are engineering mindsets that get in the way of engineers seeing social justice, social justice in the context of global energy consumption and use, critique of the prevailing "weed out" culture in undergraduate programs as an impediment to diversity, developing a hybrid imagination in prospective engineering students, and questioning the ideology and codes of knowledge behind the dominant construction of the epistemological core in engineering education and more.

The chapters in Part IV focus on the renovation of engineering education. Different in their structures and approaches, the innovations that are discussed in this section have in common to refuse reducing education to a mere transmission of knowledge from a master to passive students. Instead, they rely on the active participation of the students and their personal experiences. Most importantly, rather than discussing which content should be added to enrich engineering education, some chapters focus on how to teach with pedagogical methods such as problem-based learning, and how to combine engineering teaching and engineering education research. Others propose a more radical transformation of engineering education through a definition of engineering not only as problem solution but also a contribution to problem definition or a new understanding of engineering knowledge, as the products of contextualized experience.

Volume 2 on engineering practice advances contextual analyses of engineering identity, epistemologies, and values in 23 contributions by more than 30 authors from engineering, social sciences, and humanities. Key overlapping questions examine such issues as:

- What does it mean to be an engineer?
- How are engineering self-understandings enacted in the professional world?
- What is the distinctive character of engineering knowledge?
- How do engineering science and engineering design interact in practice?
- What are the prominent norms of engineering?
- How do they interact with the values of efficiency or environmental sustainability?

The reflection on engineering identities in Part I fans out in the following sections: Is there anything like "engineering knowledge" (Part II)? Is there an inherent normativity in engineering, and how does it connect with the norms and values of the surrounding world (Part III)? The concluding Part IV gives a further exploration of the idea of context itself: in practice, a sharp delineation between "text" and "context" may appear difficult if not impossible. This can either lead to fundamentally questioning the very concept of context or to the vision that engineers can make their own context.

How do engineers distinguish themselves from scientists? From business people? From technologists? How do engineers define themselves professionally, and how are those professional identities uniquely shaped within particular national contexts. How do those outside of engineering perceive engineers? Is there a common unifying element between the diverse types of engineers? And how do gender-based stereotypes of and within engineering serve to limit equitable participation in the field? These are the types of questions that are grappled with by the chapters in Part I of Volume 2, in an effort to gain a clearer understanding of the *identities* of engineers. In addition, a final chapter provides a statistical overview of the scope of the engineering occupation worldwide.

Another field – expounded in the chapters in Part II – where the contextuality of engineering appears, is in the epistemology of engineering: the knowledge engineers need or use in their work cannot be clearly defined and demarcated. There are many uncertainties, as well in the available knowledge itself as in the evaluation of possible outcomes. Data may be lacking or hidden in an overload of information of indistinct relevance. And the boundaries within which engineering projects are to be solved are subject to negotiation with economic or political instances and societal groups and stakeholders of many kinds. Part II of this volume gathers reflections on engineering epistemology. What kinds of efficiencies are pursued by engineers? How do they situate themselves in the tension between pure science and design practice? And how can the many layers of engineering knowledge be reflected in modern curricula of engineering education?

In Part III, the central issue is the values that carry engineers and engineering (which is nowadays our common world) and cultural norms that are or should be at work in professional practice engineers. Some authors question the ambiguous influence of professional associations on the consideration by engineers of ethical issues. Others wonder how the culture of the engineers, the way they look at the world, shapes and is shaped by their relationship with the world of politics. Still others discuss the influence of social values on the attitudes of engineers and those of economic and political issues on how the problems they are asked to solve are formulated.

Do engineers create their own contexts or are they created by contexts? The authors in Part IV, the final section of Volume 2, all take explicit aim at the notion of context. Aptly titled "Competing Contexts in Engineering," the chapters present contrasting views of what context might mean or even how important the concept might be. One author argues that engineers create their own contexts. Another argues that the very idea of context is too static and should be abandoned in favor of more dynamic ways of characterizing engineering. Other chapters seek useful ways to differentiate context, whether by scale (from the micro to the macro) or by vantage point (internal versus external to the engineering activity). A final chapter explores the challenge faced by engineering practitioners with respect to reflexively incorporating an understanding of context in their work.

Contexts, Challenges, and Paths to Transformation

The notion of context in engineering education and practice is an object of heated debate. On the one hand, claims are made that context is an artificial construct, reifying a distinction between context and content and producing the sense of an inside and an outside. On the other hand, claims are made that the distinction between technical context and social context (a) reflects real tensions in engineering education and practice, (b) is constantly being re-negotiated, and, most importantly, (c) the outcome of such negotiations has real world consequences. Positions that adopt the context approach often focus on social justice, and more broadly empirical studies of engineering students' engagement with context, have been reflected in a number of path breaking works. Among these are: Cindy Atman and colleagues (1996, 2008), Caroline Baillie (2006), Donna Riley (2008), Baillie and colleagues (2011, 2012), and Juan Lucena (2013). Baillie et al. (2012), Most recently Bill Williams, José Figueiredo, and James Trevelyan in a collection on *Engineering Practice in a Global Context* (2014) have made another significant contribution.

The position taken here is that context matters and has practical consequences. The relevance of context is related to at least three different meanings of context and to inherent tensions that result:

- The embedding of institutions of engineering education into higher education systems,
- The breadth of problem scoping in engineering problem solving
- Contextual knowledge

Context, however, is an inherently dialectical concept, since contextualizing is itself dependent on definitions of what are perceived to be the relevant boundaries regarding both the education and the practice of engineers. Contextualizing unfolds its inherent dialectics in the terrain between "is" and "ought," fact and value. In this way, the quest for a recontextualizing of engineering education and practice inevitably is a value-laden enterprise and thus not without a certain degree of controversy. It is concerned with both what engineering "is" and what it "ought" to be. Ultimately a greater awareness and understanding of context should result in better preparation of engineers to render those contexts visible in their work, and consequently enable them to contribute to more socially robust and responsible endeavors.

When thinking about how far context can influence engineering and engineering education, one rapidly discovers challenges or even crises that can be roughly categorized into a number of ideal typical arguments:

- The captivity argument
- The cultural change argument
- The identity crisis argument
- The weak profession argument
- The convergence argument

This list of arguments, most of which have been developed in one form or another over recent years, should be understood as neither complete nor definitive, although it provides a useful point of departure for anyone interested in understanding and innovating with respect to engineering and engineering education. Despite overlaps between these arguments, the merit of distinguishing them is that each emphasizes a specific aspect of engineering and/or engineering education that poses challenges – and opportunities – for the engineering profession.

In many chapters of these two volumes, the ideas and analyses aim to further identify, characterize, and explicate one or more of these challenges. Other chapters, drawing on such analyses, propose responses in hopes of transforming engineering and engineering education in ways that will sustain the profession as a vital, constructive, and responsive social institution. A brief summary of relevant arguments follows.

The *captivity argument* is that the engineering profession, in regard to both education and practice, has been locked in a number of social and intellectual captivities that may be interpreted as a "fundamental usurpation of the intellectual and social dimensions of engineering as an autonomous discipline" (Goldman 1991,

p. 121). An "intellectual captivity" consists of engineering being considered subordinated to science. Engineering education requires students to master large doses of mathematics and physical sciences. Engineers in turn tend to believe that science and engineering are objective and able to exclude human values from influencing the esoteric work taking place in engineering disciplines. Engineers become overly concerned with order and certainty and adverse to ambiguity. Issues of meaning and social impact are marginalized because scientific methodology, the structure of hypothesis, proof, validation, publication, and critique are embedded in a scientific culture to which engineers find themselves attached. A "social captivity" lies with engineering practice being subordinated to a managerial agenda driven by economics and the market. Engineers exercise their power only within that mandate, which raises questions about the idea of engineers as the primary agents of technological change. According to Johnston et al. (1996), the result has been a serious limitation in engineers' capacity to examine the social meanings and effects of their work and to self-consciously reflect on their practices and professional identities.

Captivity arguments surface throughout these volumes. For example, in Volume I, Chap. 1, Atsushi Akera and Bruce Seely provide a historical account of the American system of engineering education. In it they highlight the rise to dominance of the *engineering science* paradigm, as well as the influences of "neoliberal economic doctrine." Similarly, in Volume II, Chap. 10, Stig Andur Pedersen delves into the intellectual tensions between science and engineering. Other chapters present ideas for moving beyond such intellectual and social captivities. For example, Tony Marjoram argues in Volume I, Chap. 16, for a problem-based, as opposed to science-based, education, with an emphasis on addressing human and social development goals. And in Volume II, Chap. 17, Carl Mitcham and Wang Nan advocate an expansion of engineering ethics into the political arena, so that "taking a global perspective on investing in a new technological innovation, for instance, would involve going beyond economics to include assessments of multiple risks and benefits at the social and environmental levels."

The *cultural change argument* concerns an alleged lack of diversity in engineering. In one version of this argument, feminist research criticizes the social norms of engineering culture as overly masculine. How could female students feel attracted to engineering faculties that are not only demographically dominated by men but also culturally emphasizing of male interests? Research has shown that male students go for engineering because they like to tinker; the choice of female students seems more inspired by a general interest in mathematics and physics. Even without giving in to the caricature of the pragmatic and performance-oriented male vs. the more caring and relation-oriented woman, bridging these "two cultures" is far from evident. But this is only one aspect of the cultural change argument. In Volume I, Chap. 8, Amy Slaton describes the "less-than–democratic character" of engineering and other science, technology, engineering, and mathematics (STEM) occupations and the weak influence of many inclusive efforts made in the United States to address diversity issues (gender issues, but also social diversity). Wendy Faulkner in Volume II, Chap. 2, highlights how gender operates alongside professional and organizational to produce engineering culture and proposes to disseminate "heterogeneous" images of engineering in order to create space for a more diverse range of people.

In a context where the global engineering competency becomes "a problem of engaging people from different cultures" (Downey et al. 2006), another aspect of cultural change has to do with cross cultural and globalization issues. In Volume I, Chap. 7, Qin Zhu and Brent Jesiek highlight the need to develop a better understanding of the history and cultural context of engineering education and profession in other countries and regions. They propose three key intellectual concepts enabling understanding Chinese culture: Confucianism, Marxism, and pragmatism.

A further aspect of cultural change involves preparing engineers to deal with environmental issues. In Volume II, Chap. 13, Christelle Didier and Kristoff Talin highlight French engineers' attitudes toward the environment and how they differ from those of their fellow citizens; "ecoskepticism" is the norm even among the younger generation of engineers. In Volume II, Chap. 15, Jen Schneider, Abraham Tidwell, and Savannah Fitzwater describe the tremendous difficulty of reforming nuclear science and engineering education in the United States to better integrate environmental issues. Encouraged by physics and engineering educators, student skepticism toward climate change research constitutes a cultural value and contributes to constructing an "insular culture." Rather than simply objecting to their opinions, the authors invite nuclear engineers to make their voices better heard at the "table of discussion."

The *identity crisis argument* has several manifestations, ranging from how engineering is understood – or misunderstood – by the public, to uncertainties in the roles engineers play, or will continue to play in the future, in technology development. The latter issue, for example, was developed forcefully by Rosalind Williams (2002). In a reflection that grew out of her service as Dean for Undergraduate Education and Student Affairs at MIT, she analyzes how a division of labor has eroded the identity of the engineering profession.

> What engineers are being asked to learn keeps expanding along with the scope and complexity of the hybrid world. Engineering has evolved into an open-ended Profession of Everything in a world where technology shades into society, into art, and into management, with no strong institutions to define an overarching mission. All the forces that are pulling engineering in different directions – toward science, toward the market, toward design, toward systems, towards socialization – add logs to the curricular jam. (Williams 2002, p. 70)

The challenge for engineering education is complex: it can lead to cramming more and more into the curriculum. It can lead to hyper-specialization, with a set of narrowly defined skills and competencies for preestablished jobs. But this contrasts with future demands for "educating active, rigorous and flexible individuals, rather than skilled workers for pre-established jobs." For Williams, the curricular response should be a convergence between the technological and liberal arts, educating the engineering student both for life and flexible employment.

> Only a hybrid educational environment will … prepare students for handling … life in a hybrid world. Students need to be prepared for life in a world where technological, scientific, humanistic, and the social issues are all mixed together. Such mixing will not take place if students have to decide from the outset that they are attending an "engineering school" as opposed to a "non-engineering school." (Williams 2003, p. 4)

Elements of the identity crisis argument are apparent in many chapters here. Byron Newberry, in Volume II, Chap. 1, discusses what he terms the *dialectics of identity*, which is created by ambiguities in the understanding who engineers are and what they do, ambiguities that exist both internally (engineers' self-identity) and externally (engineers as viewed by others). A detailed example of ambiguous self-identity is provided, for example, in Volume II, Chap. 3, where Mike Murphy, Shannon Chance, and Eddie Conlon present empirical results of engineering students' self-conceptions. Looking toward engineering's future Andrew Jamison, Niels Mejlgaard, and Jette Egelund Holgaard, in Volume I, Chap. 14, reimagine engineering by advocating development of what they call a *hybrid identity*:

> Fostering hybridity or a hybrid imagination involves a mixing of scientific education and training in technical skills with an appreciation of the broader cultural implications of science and technology in general and one's own role as an engineer, in particular.

The *weak profession argument* deals with the professional status of engineers. Mitcham (2009) distinguished between *strong* and *weak* professions. According to his argument, strong professions (such as medicine and law) rest on the formulations of ideals that are well embedded in the professional curriculum and practice. Weak professions (such as military and business) either lack such ideals or only weakly include the relevant specialized knowledge in a professional curriculum and practice. Somewhat provocatively he argues that engineering has more in common with weak than with strong professions.

This overlaps with the captivity argument in that engineers themselves may see their job as executing what others have decided: clients or patrons, sponsors, government, the market; decisions about the ultimate end-use of engineering work seem removed from engineers themselves. Seeing engineering as a weak profession is nevertheless at odds with the aspiration to have "engineers who will assume leadership positions from which they can serve as positive influences in the making of public policy and in the administration of government and industry" (National Academy of Engineering 2004). There is a call for engineers who would not just be technocrats, but public intellectuals, who would accompany society in dealing with a technological culture, and

> show to a broad array of audiences – politicians, engineers, scientists, and the general public – that science and technology are value laden, that all aspects of modern culture are infused with science and technology, that science and technology do play key roles in keeping society together, and that they are equally central in all events that threaten its stability. It is therefore necessary that science and technology, in their explicit and implicit forms, be subject to political debate. (Bijker 2003, p. 444)

This argument can be seen as part of the choices university education has to make in general, and not only for engineering. Will universities be training camps for professionals, under a regime run by "academic capitalism and managerialism" (Slaughter and Leslie 1997)? Or should universities be places of intellectual critique and cultural citizenship?

Especially in the second volume of this diptych, several chapters deal with the disputed professional status of engineering, either as part of a main line of discussion or at least as an aside. It is part of Newberry's consideration of the "dialectics of engineering." The "engineering-label" covers a wide range of specializations and occupational activities, and the boundaries between professionals and other educational backgrounds are blurred. This makes it difficult for engineers to gather in one recognizable group and to speak with an authoritative voice, even concerning topics that are within their realms of competence. Michael Davis has a long record of publications on professionalism and engineering. In Volume II, Chap. 4, he enters into discussion with some comments and objections his publications have raised and deals with methodological and conceptual misunderstandings that blur the vision of engineering as a profession. Martin Meganck in Volume II, Chap. 12, questions why a professionalism label should be important at all and discusses whether a professionalism-based ethics cannot be reduced to principles of ordinary morality.

Finally, the educational consequences of the above-mentioned arguments are related to a *convergence argument*, which focuses on relatively recent evolutions in higher education across many countries. Democratization of education, homogenization (e.g., through the Bologna process in Europe), political decisions, and the application of new management styles seem to lead to an academic drift – or convergence in mission – in and of nonuniversity institutions, and vocational drift in universities or institutions similar to universities. For engineering, some fear that this will lead to a gradual loss of the practice-oriented nature of engineering. Curricula will become more theoretical. Teaching staff will be evaluated more on their research activity than on their teaching or contacts with industry. The blurring of boundaries between "noble" and "less noble" institutions is a tendency that seems to occur spontaneously and organically; yet it solicits further fundamental reflection.

In Volume I, Chap. 2, Steen Hyldgaard Christensen and Newberry zoom in on major differences between and dynamics of change in European and American higher education. They examine two European examples of academic and research drift in nonuniversity institutions – Irish *Institutes of Technology* (IoTs) and Dutch *Hogescholen* (HBOs) – and three American examples – a public technical institute (Southern Polytechnic State University in Georgia), a state teacher's college (Western Kentucky University), and a sectarian liberal arts university (Baylor University). They argue that convergence in mission between universities and former vocationally oriented designated teaching institutions both in Europe and the United States are likely to create a number of tensions and dilemmas as well as winners and losers. Shifting emphases in engineering degree programs from teaching based on practical experience derived from engineering work to research-informed and research-led education creates crisis for many faculty members whose values and identities embody the core of a teaching culture. Many of these practically experienced teachers are likely to be one obvious group of losers in this process of institutional transformation.

Bernard Delahousse and Wilhelm Bomke in Volume I, Chap. 3, further substantiate the convergence argument in presenting a comparative study of two more profession-oriented institutions in Europe – the French *Instituts Universitaires de Technologie* (IUTs) and the German *Fachhochschulen* (FHs). In their study, the focus is on the historical evolution of the two types of institution in terms of degree of autonomy, creation or adaptation of curricula, pedagogical methods, student standing, personnel status, and research opportunities. The two institutions have a number of traits in common: a strong focus on teaching rather than research, fixed curricula oriented toward practice including internships, close links with companies, academic staff recruitment, a particular stand with regard to universities, insistence on graduate operational skills, and more. The authors argue that academic drift should be regarded as a natural and irreversible process: "natural" because it interacts with the inevitable evolutions of society in its economic, political, social, cultural, and technological dimensions; and "irreversible" as it constitutes a never-ending trajectory. Generally transformations take place in moments of opportunity provided by external state, public, private, or transnational agencies. Yet the void after the transformation of institutions may need filling by a new type of short-cycle institution and the process can go on once again.

Conclusion

These two EEPiC volumes thus aim to stimulate critical reflection on the past, the present, and the future of engineering in both education and practice. They offer no final answers or even a well-formed methodology. Instead, their programmatic character invites readers themselves to reflect on the engineering-context nexus and contribute their own insights to a perennial discourse – a discourse that can help us all, engineers and nonengineers alike, live more consciously and carefully in our increasingly engineered world.

With regard to issues addressed in Volume 1, engineering education in all its dimensions – histories and structures, ideologies, reforms, and innovations – can be expected to be continuing subjects for empirical research and critical reflection. More empirical research on the institutional contexts of engineering education with respect to ongoing institutional transformations both locally and globally will be a priority. Given the increased blurring of boundaries between university and nonuniversity engineering educational programs, there are ongoing needs to explore what does and does not work under what conditions to achieve diverse goals. A related issue for research and reflection is the engineering-business nexus, a multilayered relationship with implications for both engineering education and practice. More systematic empirical research along the lines of Cindy Atman and colleagues (1996 and 2008) on student engagement would also be important.

With regard to Volume 2, engineering practice as reflected in identifies, epistemologies, and values calls as well for further research and reflection. Here recent (and no doubt future) analyses of the normativity in engineering and technology are (and will become more) relevant; see, for example, the work by Ibo van de Poel and Peter Kroes (2006) and Sergei Gepshtein (2009). Additionally, the relationship between engineering, social sciences, and humanities has implications not only for education but for engineering identity, knowledge, and ethics. The need to integrate these three perspectives on engineering practice has been pointed toward by Mitcham (2014) as well as many others trying to assess large-scale social problems that have emerged in conjunction with the engineering transformations of human ways of life (see, e.g., Mike Hulme 2014).

More generally, our introduction began by referencing and criticizing a previous "science in culture" project. But we should also acknowledge the extent to which this project has received its own criticism in the science studies field. One extension of the science in context program argued that since there is no reason to grant scientific expertise any special cognitive privilege, everyone is justified in claiming expertise. In an insightful response to this developmental trajectory, Harry Collins argues at length that although everyone may be some kind of expert, we are not all scientific experts "because we do not [all] belong to the scientific community and we do not necessarily make our judgments from the platform of the norms and aspirations that drive that community" (2014, p. 131). For Collins, "If we start to believe we are all scientific experts, society will change: it will be those with the power to enforce their ideas or those with the most media appeal who will make our truths, according to whatever set of interests they are pursuing" (ibid.).

Adopting Collins' framework, we can note that there has been little temptation for any social critic to argue that "we are all engineers now." Additionally, despite Snow's blurring of any science-engineering distinction in his famous two-culture argument, engineering intellectuals are probably something different than either scientific or literary intellectuals. At the same time, there is some sense in which even literary intellectuals would have to admit their dependency on engineers much more than on scientists. This is the case, first, insofar as engineering is conceived as attempting to satisfy human needs and, second, insofar as engineering has been argued by engineers themselves to be a more refined form of that making and using that permeates all human activities (see, e.g., Koen 2003). To the extent that either of these theses is even partially true, it is all the more incumbent on us to struggle to examine engineering in context.

References

Atman, C. J., & Nair, J. (1996). Engineering in context: An empirical study of freshmen students' conceptual frameworks. *Journal of Engineering Education, 85*(4), 317–326.

Atman, C. J., Kilgore, D., Yasuhara, K., & Morozov, A. (2008). Considering context over time: Emerging findings from a longitudinal study of engineering students. *Proceedings of the Research in Engineering Education Symposium*, Davos, Switzerland, 7–10 July 2008.

Baillie, C., Pawley, A. L., & Riley, D. (2011). *Engineering and social justice: In the university and beyond*. West Lafayette: Purdue University Press.

Baillie, C., Kabo, J., & Reader, J. (2012). *Heterotopia. Alternative pathways to social justice*. Hants: Zero Books, John Hunt Publishing Ltd.

Barnes, B., & Edge, D. (1982). *Science in context. Readings in the sociology of science*. Cambridge, MA: MIT Press.

Bijker, W. (2003). The need for public intellectuals. A space for STS. Pre-presidential address, annual meeting 2001, Cambridge MA. *Science, Technology & Human Values, 28*(4), 443–450.

Collins, H. (2014). *Are we all scientific experts now?* Cambridge, UK: Polity Press.

Downey, G. L., Lucena, J. C., Moskal, B. M., Parkhurst, R., Bigley, T., Hays, C., Jesiek, B. K., Kelly, L., Miller, J., Ruff, S., Lehr, J. L., & Nichols-Belo, A. (2006). The globally competent engineer: Working effectively with people who define problems differently. *Journal of Engineering Education, 95*(2), 1–17.

Gepshtein, S. (2009). Closing the gap between ideal and real behavior: Scientific vs. engineering approaches to normativity. *Philosophical Psychology, 22*(1), 61–75.

Goldman, S. L. (1991). The social captivity of engineering. In P. T. Durbin (Ed.), *Critical perspectives on nonacademic science and engineering* (pp. 121–145). London: Research in Technology Studies Series, Lehigh University Press.

Hulme, M. (2014). *Can science fix climate change?* Cambridge, UK: Polity Press.

Johnston, S., Lee, A., & McGregor, H. (1996). Engineering as captive discourse. *Society for Philosophy and Technology, 1*, 3–4. http://scholar.lib.vt.edu/ejournals/SPT/v1n3n4/Johnston.html.

Koen, B. V. (2003). *Discussion of the method: Conducting the engineer's approach to problem solving*. New York: Oxford University Press.

Lucena, J. (Ed.). (2013). *Engineering education for social justice: Critical explorations and opportunities*. Dordrecht: Springer.

Mitcham, C. (2009a). A historico-ethical perspective on engineering education: From use and convenience to policy engagement. *Engineering Studies, 1*(1), 35–53.

Mitcham, C. (2009b). A philosophical inadequacy of engineering. *The Monist, 92*(3), 339–356.

Mitcham, C. (2014). The true grand challenge for engineering: Self-knowledge. *Issues in Science and Technology, 31*(1), 19–22.

Riley, D. (2008). *Engineering and social justice*. San Rafael: Morgan & Claypool Publishers.

Slaughter, S., & Leslie, L. L. (1997). *Academic capitalism: Politics, policies, and the entrepreneurial university*. Baltimore: Johns Hopkins University Press.

Suskind, R. (2004). Faith, certainty and the presidency of George W. Bush. *New York Times Magazine*, October 17, 2004.

Van de Poel, I., & Kroes, P. (2006). Technology and normativity. *Techne, 10*(1), 1–6.

Williams, R. H. (2002). *Retooling. A historian confronts technological change*. Cambridge, MA: MIT Press.

Williams, R. H. (2003). Education for the profession formerly known as engineering. *The Chronicle of Higher Education, 49*, 20. 24 January.

Williams, B., Figueiredo, J., & Trevelyan, J. (2014). *Engineering practice in a global context: Understanding the technical and the social*. Leiden: CRC Press.

Part I
Engineering Identities

Introduction

Tony Marjoram and Mike Murphy

This first section deals with engineering identity and identities in the context of engineering practice. Engineering identity links to epistemology and values because engineering is a knowledge-based profession, operating in professional and social contexts, where public perception is a significant factor in identification and identity formation, not to mention interest and enrolment. Interest in identity and identity theory developed in the 1970s, with various approaches and perspectives on identity and the identification of identities. Approaches from philosophy, sociology, social psychology, anthropology, and related disciplines reflect their disciplinary and sub-disciplinary viewpoints. These range from the empirical to more philosophical, conceptual and normative perspectives; some are complementary, others are contested (e.g., the role of internal and external factors). Perspectives also relate to notions of the self as an individual, at the existential and personal level, and in terms of membership of larger social and cultural groups (e.g. gender, class and race), and associated sub-cultures, beliefs, norms and values, education, employment and professional backgrounds, and affiliations.

Interest in identity developed with reference particularly to education and work, which have changed significantly since the 1970s. This is especially the case over

---

T. Marjoram
UNESCO Centre for Problem Based Learning in Engineering Science and Sustainability, Department of Development and Planning, Faculty of Engineering and Science, Aalborg University, Aalborg, Denmark

M. Murphy
Technological University for Dublin Programme Team, Dublin Institute of Technology, 143 Lower Rathmines Road, Dublin 6, Ireland
e-mail: mike.murphy@dit.ie

the last decade, with increasing emphasis on the knowledge economy, and economic growth in a time of increasing globalization and competition. While engineers have identities related to and shaped by their engineering education and practice, they also have and relate to other identities. Further, engineers have different engineering identities by discipline (e.g., civil, mechanical, etc.), and by work area (e.g., academic, consulting, public service, etc.). It is useful to note that not all engineering educators are engineers – with many having backgrounds in education and the social sciences (sociology, psychology, economics and history, for example). Also, engineers themselves have other lives and identities, as partners, parents, grandparents, hobbyists, collectors, volunteers, sportspeople, etc. These identities may relate and overlap, or they may be separate and differentiated, and many may lead different lives within the same head-space (Tomlinson 2013).

The transition from education to the world of work and career has also changed. Both education and employment are of vital importance in identity development and formation of engineers. This is especially true within the broader French context of the term, which relates both to education and the development of an identity as an engineer in terms of professional behavior and values, and personal sense of self. The scope and mechanisms regarding the development and transformation of identity, or identities, is debated (for example, what identity is, how it is formed and transformed), although there is general concurrence on the importance of identity as a framework for meaning and experience, especially in the social context. A discourse on engineering education and employability has also developed over the last 50 years, with reference to university-industry partnership and, most recently, professional attributes and competencies. This is illustrated, for example, in the context of the Washington Accord and International Engineering Alliance – where 5 of the 12 Washington Accord graduate attribute profiles relate to engineering/technical competencies, five relate to more social skills (including applying contextual knowledge to assess societal issues, understanding environmental and sustainability issues and understanding and applying ethical principles). One attribute is managerial and one is educational (lifelong learning). Education may encourage employment, opportunity and mobility, or deny it (social reproduction theory), and education may promote employment flexibility, but also competition, uncertainty and insecurity (human capital approach). This reflects the debate over liberal versus utilitarian education: of education for life contrasted against education for a job. In this context, it may be possible to "have one's cake and eat it too," with project- and problem-based learning, both of which promote life-long learning and employability. Different approaches and perspectives on identity, and reflections on the identity of engineers and others, are discussed in the chapters in this part. The six chapters, each by leading specialists in their fields, are linked by an internal-external dialogue, as reflected in the discussions of dialectics, gender, the role of design in defining an engineer, how engineering as a profession developed, the development of professional ethics and the need for better statistical data with which to identify engineering and related internal and external needs.

Identity and identity theory may be seen as overly abstract, especially when applied to the diverse world of engineering, as several contributors to this part observe. Identity also relates to identification with, and if the object of identification is vague, so will be any association. Since almost all aspects of our lives are mediated by technology, and technology is arguably researched, designed, developed, manufactured, and maintained by engineers, it follows that engineers and engineering thus play a central role in the identity formation of all of us. In terms of common usage, 'identity' may also be associated with personality, celebrity, or status. However, as engineering has become more diverse, complex and corporatized, it has become less understandable, little more than a service sector with the designers of the technology we use every day invisible within the corporations that employ them.

With identity related to status, the loss of public identity is mirrored in the declining status of the individual engineer, perhaps more particularly in developed countries. For example, in the past, in countries such as Australia and Ireland, many towns and boroughs had a municipal engineer responsible for most aspects of infrastructure. Today these roles have often been subsumed into the roles of town or county managers, with few municipal engineers remaining. Similarly, few companies still have a Chief Engineer – one of the few places that retain such a role is in teams participating in Formula 1 motor sports. With fewer engineers in high profile public engineering positions, those that transition into management often downplay – or at least see no reason to emphasise – their engineering pedigree. This general decline of the visibility and status of engineering reflects a limited understanding of engineering. Another reason for the lack of engineering identification relates to the generally poor quality of information and statistics on engineering, especially at the international level. Engineering is part of "science and technology" in OECD data, which is not disaggregated between science and engineering, or scientists and engineers.

Science studies and the sociology of science developed in the 1960s and 1970s, itself a Kuhnian paradigm shift in thought, but these new studies neglected engineering, and have been critiqued in the "science wars" literature for postmodernist knowledge relativism (Sokal and Bricmont 1998). Given the importance of engineering in social and economic development, there is a need for a new paradigm of knowledge in engineering studies and epistemology. What is engineering? What is an engineer? What do engineers do? How are engineers best educated? How can we more effectively promote awareness and understanding of engineering by non-engineers in the public and policy realms? The development of engineering studies is thus long overdue, but is beginning to be addressed through academic networks such as the International Network for Engineering Studies (INES) coordinated by Gary Downy. This also begs the question as to why engineering was ignored or overlooked by 'science studies' over the last half century, when engineering, or at least the 'engineering sciences', is widely understood to be closely related to science. For example, at UNESCO, the UN organisation officially responsible for science and engineering, which initially constituted the major part of the science programme in the 1950s and 1960s (UNESCO Engineering Report 2010). The answer lies in the diversity, complexity, ubiquity, and also opacity of engineering (a handful of core topics and around 40 sub-disciplines), and in simplistic and misleading models of science, engineering, and innovation, especially the linear model of innovation in which engineering and technology follow-on directly from underlying science.

With the subordination of engineering to science in many countries, science often looks down on engineering more as a trade than a rigorous discipline. Science is identified with people in white coats working in research labs, even though most scientists work in companies, and fewer in frontier research. Engineers practice in the public and private sectors, as academics, as researchers and as consultants, and are often professionally licenced or chartered. In this sense, engineers are more akin to medical doctors than scientists, albeit with multiple identities. Science is often regarded and reported as the basis of innovation, whereas more innovations derive from engineering, and 'rocket science' is more engineering than science. Indeed it can be argued with considerable justification that the science in rocket science is easy, but rocket engineering is hard. Old notions die hard, and although UNESCO was created in 1945 at a conference in the Institute of Civil Engineering in London, and engineering was initially a major part of its science activity, the most recent World Conference on Science in 1999 (on the main theme of "Harnessing Science to Society") totally ignored engineering despite the fact that 'harnessing science to society' could be used as a definition of engineering. The subordinate perception of engineering vis-à-vis science is captured well in the old aphorism that technological successes are most often referred to as "scientific achievements" while technological failures are usually called "engineering disasters".

Several contributors refer to this subordination of engineering to science and managerialism. Further, the heterogeneous culture, identity, and professional status of engineering all make it hard to pin down engineers and engineering. Adding to the discordancy, there may also be a schism between various branches and levels of engineering. At many universities, mechanical, civil, electrical, and chemical engineering students lead separate lives with little or no interaction. At other universities there is a form of professional isolation between engineering and technology students. Student engineers have even less contact with social science and humanities undergrads. One wonders if such divisions among engineers and others partly underlie commonly reported weaknesses in engineering graduate competences in communications and team working skills. It is yet unanswered how changes to standardised engineering degrees and postgraduate specialisation will promote understanding, at least between engineers themselves.

In Chap. 1, Byron Newberry examines the question of perceived contradictions or tensions within engineering, particularly in the internal-external context, with reference to the concept of dialectics. This is a central question in the philosophy of engineering, in terms of understanding and interest in promoting change and the transformation of engineering, especially engineering education. Newberry examines the dialectics of scope and scale (the closer one looks, the vaguer it gets, like Heisenberg's Uncertainty Principle), the dialectics of identity and status (engineering is increasingly ubiquitous, yet less understood), the dialectics of purpose (between ideals and realities), and the dialectics of method (contrasting the internal technical focus of engineering with external context and purpose). The need to understand the dialectic between increasing social, economic and cultural dependence on engineering and technology, and the decreasing social, economic and cultural understanding of engineering and technology is emphasised. The position and role of engineers in this process, and how this affects the identity, status, purpose and methods of engineering is explored. While engineering may be difficult to study, creative tension is an essential part of engineering, like innovation, and this chapter reinforces the will to proceed.

Engineering, like most of science, is largely identified as male dominated and masculine, characterised by "boys and toys", and gender is another aspect of the internal-external dialogue in engineering. The gender neutrality of technology has also been questioned, and indeed the gendered nature of science and engineering itself (Wajcman 1991, 2009). Reality is complex and nuanced, however, as Wendy Faulkner explores in the context of gender-troubled engineering identities in Chap. 2. Faulkner is a leading commentator and the gender issue reflects wider tensions within engineering, engineering identity, and the connection of engineering to the outside world. This chapter examines the tension within engineering between the nuts and bolts technical approach and its wider social heterogeneous context: the hard/soft, masculine/feminine dualistic character and identities of engineering. The people, management, social, economic, and cultural context in which engineers work are also noted. Faulkner observes that many engineers associate themselves with a technical identity following the male/masculine character of engineering, which disadvantages women in becoming "real" engineers, concluding with a call for the broad church that is engineering to promote a more heterogeneous image, and to build upon this diversity to enhance the participation of women in engineering.

Many comments on engineering education reflect the early nineteenth century model of mathematical, science, and theory-based pedagogy, developed by Humboldt in Germany and then in the *grandes écoles* in France, partly to distinguish the emerging scientific approach to engineering from other trades-based approaches. There are increasing calls for change to this model, and that what is needed today is to combine theory and practice in a problem- and project-based learning approach. Indeed this was the original model of Wilhelm von Humboldt in establishing the University of Berlin. The tension between technology focus and broader curriculum issues will likely continue as an essential feature of both engineering and engineering education. The socialisation and identity formation of an engineer generally begins at college or university, unless one has family, friends, or significant others who are engineers. This assumes, of course, that 'doing' an engineering degree 'makes' one an engineer. The decision to study engineering is also significant, if one accepts that becoming an engineer may be as much about nature as nurture; see for example the wonderful animated sketch of the young Dilbert in "The Knack". Identity formation may differ, and may be stronger or weaker, between engineering specialities, and between levels – engineer, engineering technologist, and technician. Chapter 3 by Mike Murphy, Shannon Chance, and Eddie Conlon is of central and practical importance in the identity formation of young engineers. The chapter explores the questions *who is an engineer?* and *what makes an engineer?*, with reference to a study of identity formation among two groups of final year students, in engineering and engineering technology, at the Dublin Institute of Technology. The authors begin with an informative review of the concept of engineering identity, noting the importance of engineering institutions and educators in shaping identity. The authors found strong identity development in engineering students, compared to engineering technologists, and that the inclusion of design in the engineering degree was the key to differentiating between and forming this strong sense of engineering identity among engineering students. That said, it was the engineering technologists who described themselves as more prepared for work in the real world. It is also interesting to observe that the role of design in the process of identity formation may also account for differences in identity formation in those branches of engineering with more, or less, design focus and practical approach, and consequently the need to include a problem – and project – based approach in engineering degrees.

The term 'technology' has changed significantly over the last decade – and in the media is now often used to refer particularly to information technology or IT. Methodological issues in the definition of engineering and what it is to be an engineer, how these terms, and 'technology', have changed over time, and what gets studied as engineering are covered in historical detail by Michael Davis in Chap. 4. Davis, an astute writer on the topic, observes that engineering is a function, discipline, occupation and profession (to which may be added vocation, mindset, and other descriptors). Such methodological issues are nontrivial for the development of engineering studies. Davis traces the development of engineering through military and civil engineering, professional occupations, and engineering institutions, as proponents sought to identify engineering as a discipline akin to science (viz the development of the "engineering sciences"), a gentlemanly profession, distinct from its links to and trades-based roots. Davis recapitulates previous discussion of such issues, and disposes of critical comment as relating more to pedagogy than epistemology in Popperian refutational fashion. The chapter concludes with a discussion of associated professional ethics and ethical codes.

Engineering ethics and identities are examined in detail in Chap. 5 by Gary Downey, Juan Lucena, and Carl Mitcham, with reference to the United States, Japan, France, and Germany. In each case the identities reflect local context and trajectories: professional unity and autonomy in the United States; the development of corporate culture in Japan, and the notion of "ie" or "households"; the *grandes écoles* tradition in France and the ideal of an elite public servant; and the ideal of *Bildung* or humanistic education in Germany. Many of these trajectories are changing and converging with globalisation – such as the decline of the "company family" and rise of hiring/firing contract employment in Japan, where it is also interesting to reflect on the issues of whistleblowing after the Fukushima nuclear disaster and TEPCO's prior knowledge of problems (see, for example, Cooke 2009). On the international stage, it is also useful to note that the World Federation of Engineering Organisations, of which most national engineering organisations are members, developed its first Model Code of Ethics in 1986, with the latest version covering integrity, practise, leadership, and protecting the natural and built environments. The Washington Accord international accreditation agreement also includes ethics as 1 of 12 graduate and professional attributes and competencies.

Before one can identify *with*, first there is the need to identify. The concluding Chap. 6 by Tony Marjoram looks at the international statistics and indicators currently available on engineering, science, and technology, with particular reference to the OECD data relating to human resources, research and development, and innovation. The OECD provides the main standards and collection of data on science, technology, and innovation. The lack of disaggregation in the OECD data between science and engineering, and between scientists and engineers, is a major drawback, emphasizing the need for better indicators and metrics on engineering. Better indicators would also promote the understanding of engineering identity, epistemology, and changing modes of knowledge production, application, and dissemination, models of engineering education and innovation, and the broader development of engineering studies, policy, and planning. To borrow a phrase from gender indicators on science and engineering, from the point of view of evidence-based policy, where there are no data there is no visibility and where there is no visibility there is no priority.

References

Cooke, S. (2009). *In mortal hands: A cautionary history of the nuclear age.* London: Bloomsbury Publishing.

Sokal, A., & Bricmont, J. (1998). *Intellectual impostures,* London: Profile Books.

Tomlinson, M. (2013), *Education, work and identity: Themes and perspectives,* London: Bloomsbury.

UNESCO Report. (2010). *Engineering: Issues, challenges and opportunities for development.*

Wajcman, J. (1991). Feminism confronts technology. *Cambridge Journal of Economics.*

Wajcman, J. (2009). Feminist theories of technology. *Cambridge Journal of Economics, 34*(1), 149–152.

Tony Marjoram Guest Professor at Aalborg University, Honorary Fellow at Melbourne University and *Manchester Institute of Innovation Research.* Responsible for the Engineering Program in the Division of Basic and Engineering Sciences at UNESCO from 2001 to 2011. Ph.D. from the University of Melbourne on Technology and Development in the Pacific Islands, M.Sc. on Science and Technology Policy from Manchester University, B.Sc. Hons in Mechanical Engineering from UMIST. Conceived, edited and produced the UNESCO Report *Engineering: Issues, Challenges and Opportunities for Development* in 2010 – the first UNESCO Report on engineering. Worked for UNESCO for 18 years, first at the Regional Office for Southeast Asia and the Pacific in Jakarta. Prior to UNESCO was a Senior Research Fellow at the International Development Technologies Centre of the University of Melbourne, Senior Development Fellow at the University of the South Pacific, and Research Fellow at Manchester University. Published over 50 papers, books, articles and reports.

Mike Murphy H. Dip. from Dublin Institute of Technology, B.Sc. (Eng) in Electrical Engineering, Trinity College Dublin, M.Sc. and Ph.D. in Electrical Engineering, Stevens Institute of Technology. Fellow of Engineers Ireland. Member of the IEEE. He worked in Bell Labs and later at Bell Communications Research before returning to academia in 2002. Director of Dublin Institute of Technology and Dean of the College of Engineering & Built Environment. Member of SEFI Administrative Council, and Member of European Engineering Deans Council and current President. Area of special interest is engineering and technology education.

Chapter 1
The Dialectics of Engineering

Byron Newberry

Abstract Dialectics of engineering are here defined as tensions pulling the engineering enterprise in opposite directions simultaneously, or ways in which engineering seems to be at odds with itself or with our perceptions of it. We offer several examples of such dialectics in the belief that they represent some of the key issues upon which any deeper understanding of engineering hinges. The introduction highlights an initial dialectic of scope that is encountered when it comes to studying the activity of engineering – that the closer it is scrutinized, the less well-defined engineering seems to become. The following section features dialectics concerned with engineering's identity. These include the enigma of engineering's simultaneous ubiquity and obscurity in society, the question of engineering's status as a distinct profession, and the tensions between the technical and organizational roles of engineers. Next, dialectics of engineering's purpose are highlighted, including a comparison of engineering ideals with practical realities, and an outline of engineering's equivocal contribution to societal understanding of technology. Finally, a dialectic of method is presented which contrasts the inward-focused nature of engineering methods with the outward-focused nature of engineering's purposes.

Keywords Engineering profession • Engineering method • Engineering management • Technological literacy • Public perception

B. Newberry (✉)
Department of Mechanical Engineering, Baylor University,
One Bear Place 97356, Waco, TX 76798, USA
e-mail: Byron_Newberry@Baylor.edu

S.H. Christensen et al. (eds.), *Engineering Identities, Epistemologies and Values*, Philosophy of Engineering and Technology 21,
DOI 10.1007/978-3-319-16172-3_1

Introduction

This book is predicated on the assumption that *engineering* is an activity that ought to be studied and understood, in no small part because of its critical role in the creation of modern technology, technology which we know has transformational power for both the natural and social structures of our world. But engineering is not some external agent exerting its influence from the outside. It is endogenous to the world upon which it acts. It is an emergent process that coevolves with, and is inseparable from its medium. And this makes the study of engineering both fascinating and complex. It is fascinating because of engineering's fundamental provenance in human nature, expressed succinctly in the title of Henry Petroski's book (1992), *To Engineer is Human*, and in the words of Billy Vaughn Koen (2003), who writes, "[T]he engineering method is coterminous with any reasonable definition of the human species." To truly understand engineering, therefore, is to understand something essential about humanity.

But this essentiality is also what complicates the study of engineering. When viewed as a most basic characteristic of human nature, the activity of engineering cannot be easily excised and examined in isolation from the larger ecology of human action. Like all ecological elements, it is inextricably coupled with its surroundings. At a very high level we might be able to create serviceable definitions of what it means to be an engineer, or to describe the products of engineering and the reasons for their creation. Such definitions of *engineering* are abstractions that we use to aggregate particular aspects of human activity for purposes of conceptual manipulation. But as we begin to dig to deeper levels of understanding, we get the feeling that the more we learn about engineering, the less plainly we can demarcate it. The more we study its causes and effects, the less clear are the distinctions between them. This is because, to use the words of Levins and Lewontin (1985), "abstraction becomes destructive when the abstract is reified and when the historical process of abstraction is forgotten, so that the abstract descriptions are taken for descriptions of the actual." That is, if we investigate *engineering* as if it were an actual ontological entity, we are destined to be unsatisfied with the result. Thus we have a *dialectic* tension in the study of engineering – a fundamental antagonism in which our examination of the contents of the engineering box, so to speak, dissolves the box and intermixes the contents with its surroundings.

Engineering is characterized by such dialectic tensions at many levels. They exist in overarching questions, such as those concerned with engineering's ultimate goals. They also exist in more narrow questions, such as those concerned with the technical methods engineering employs. Our aim in this chapter is to illuminate some of these dialectics – these ways in which engineering appears to be at odds with itself, or with our perceptions of it. We do this in the belief that grappling with them is central to refining our larger understanding of engineering. The list of dialectics discussed herein is not exhaustive, nor are they all necessarily unique to engineering. But when taken collectively, we hope these examples will shed some useful light when it comes to understanding *engineering in context*.

Dialectics of Identity

Omnipresence Contra Invisibility

I have elsewhere used the *Shoemaker's Elves* fairy tale as a metaphor for engineers (Newberry 2007). Like the elves' role in the making of the shoemaker's shoes, engineers play an instrumental role in the design and production of nearly all the artifacts of life in a modern industrial society. A citizen of such a society is hard-pressed to touch or use anything during a typical day that is not either itself engineered (e.g., an appliance), or else was made available via engineered systems (e.g., a tomato purchased at the grocery store). And like the elves, the engineer's role in the existence and availability of many of these artifacts remains largely a mystery. Samuel Florman (1996) calls engineering the "anonymous profession." By this he literally means that modern engineers, unlike their historical counterparts – Henry Ford, say, or Thomas Edison – are rarely famous; that is, the public does not generally know the names of the engineers who have designed and developed the latest technologies that are so prominent in our lives. Rather, technologies are seen as the products of teams of nameless, faceless engineers. This observation is echoed in an article titled, "The Invisible Engineer," in which Gary Downey et al. (1989) write that in the twentieth century, "Engineers lost their visibility as individuals and became instead corporate men buried within organizations."

Engineers are also anonymous in another, perhaps more significant, sense. When surveyed, people often correctly associate engineers with the production of certain iconic technologies – such as vehicles, bridges, spacecraft, computers, and electronics. But those same surveys indicate that people do not know much about what engineers actually do. Nor do people necessarily realize engineering's role with respect to the existence of the vast remainder of artifacts and products that are less iconically technological, such as paper clips or toothbrushes. So not only are engineers anonymous as individuals, but what they collectively do at work, and how that translates into the resulting technologies, products, and goods, is also largely unclear. "Our culture's lack of attention to the artifacts and people of engineering," writes David Goldberg (2006), "causes it to misunderstand engineering education, engineering practice, and engineers themselves in important ways." We might say that engineers and engineering are largely opaque to the public. Like the proverbial black box, raw materials go into engineering and artifacts come out, but no one really knows what goes on inside. "Though ours is an age of technology," writes Petroski (1992), "the essence of what engineering is and what engineers do is not common knowledge." This lack of knowledge is true even of many nascent engineers. As an engineering professor, I meet frequently with high school students and their parents. These are students who have generally made up their minds to study engineering and are primarily trying to decide which school to attend. You would think that the decision to pursue a particular career would be based on a good understanding of what that work entails, but surprisingly one of the questions I am most frequently asked is some version of, "What is it that engineers really do?"

This lack of clarity about engineering work, however, is not limited to the general public. Sociologist Robert Zussman (1985) spent time shadowing engineers at work in order to help fill a void he perceived in the academic studies of the engineering profession. "Although there is broad convergence around the idea that what engineers do is analytically important," writes Zussman, "there is little consensus as to what they actually do…" And Downey et al. (1989) highlight the lack of definitive scholarship on the modern engineering profession, not only from sociologists, but also from historians and philosophers.

Unified Profession Contra Diverse Occupations

In many countries, including mine (the USA), engineers widely consider themselves to be part of a *profession*. When an occupation is considered a profession, it connotes a somewhat privileged status in society. Professions are thought of as doing work critical to the well-being of society, work that requires a high level of education and expertise, and work which is worthy of a measure of prestige. Further, professions are often regulated to ensure that only those qualified are permitted to practice, and that the practice is governed by both procedural and ethical rules aimed at protecting the interests of the society which the profession serves. These rules, in turn, are the purview of the professionals themselves; that is, because of their expertise, professionals are allowed the autonomy of specifying their own constraints. As a result, professionals tend to see themselves as having obligations to society that may transcend the necessities of their particular jobs. In short, a profession comprises practitioners of a discipline having a formal, shared set of qualifications, ideals, and obligations.

But engineering's status as a profession is complicated. Much of that complication is due to the staggering diversity of engineering disciplines and occupations. This is compounded by differences worldwide in educational criteria and regulatory constraints. For example, if we examined the work of a biomedical engineering researcher, a civil engineering construction manager, and an electronics sales engineer, we would likely find little in common between their technical knowledge, their daily work activities, their work environments, their work objectives, or their employer types. If they are employed in the USA, for example, only one of the three is likely to be professionally licensed. Requirements for professional licensure for engineers vary by country, ranging from the non-existent to the strict. In the USA, certain types of engineering work require licensure, but most do not, resulting in approximately 80 % of engineers being unlicensed. Most unlicensed engineers employed by industry in the USA have accredited academic engineering degrees, which assures some minimal educational commonalities. Yet in the USA, as well as in many other countries, companies are free to employ people in jobs titled "engineer" without regard to academic credentials. In France (Didier 1999), "almost half of the people working as engineers in corporations are not graduate engineers, but self-taught. The practice of an engineering profession is neither controlled nor regulated by French law." It is also likely that in some companies employees with neither the title "engineer," nor any formal engineering education, perform work substantially indistinguishable from that of engineers. In a related vein, at my own university there is a professor of electrical and computer engineering who holds no degrees in engineering – his educational background is in physics. Yet his theoretical knowledge and practical experience with electronics allow him to hold an engineering title and to teach engineering. And quite likely he could qualify for many engineering jobs within the electronics industry.

Among other differences, the three engineers used in the example above would probably not share membership in a common professional organization – in fact, one or two of them may not belong to any such organization at all, despite the fact that engineering organizations proliferate. Florman (1987) highlights the consistent failure of any efforts to establish organizational unity across engineering disciplines. The reason, he concludes, is because the engineering community does not have "any discrete message." The goals, interests, and concerns that both drive and constrain engineering are as diverse as the underlying technical subject matter. Rosalind Williams (2003) writes, "Engineering has evolved into an open-ended Profession of Everything in a world where technology shades into science, art, and management, with no strong institutions to define an overarching mission." This divergence of engineering disciplines and the cross-boundary diffusion between engineering and non-engineering fields occur both in engineering as a whole as well as within individual disciplines. In an article aptly titled "Electrical Engineering's Identity Crisis: When does a Vast and Vital Profession Become Unrecognizably Diffuse," Paul Wallich (2004) discusses electrical engineering's rapid divergence in many directions, some of which blur the boundaries with other fields such as biology, physics, or computer science. Wallich's article seeks, *with little success*, to identify the common thread that binds together the diversity of people who call themselves electrical engineers – to define what it means to be one. Given this trouble establishing what it means to be an electrical engineer, it is no wonder that it may be difficult, if not impossible, to identify a coherent nucleus of attributes that is general enough to apply to all engineers *yet* specific enough to unequivocally differentiate them as a distinct professional group (at least in a more meaningful sense than as a "profession of everything").

Technoscientist Contra Businessperson

In his book *The Ancestor's Tale*, biologist Richard Dawkins (2004) tells of salamanders that inhabit the mountains that ring California's Central Valley. The salamanders, it is believed, migrated (over time) south from the northern end of the valley, following the two mountain chains that line its east and west sides – the valley itself is inhospitable to the creatures. By the time the salamanders arrived at the southern end of the valley, where the mountains rejoin, the east–west divergence had resulted in the evolution of two separate species that do not recognize each other and will not interbreed. The interesting point, however, is that if we start with one of the species at the southern end, follow its path back north along one side of the valley, and then proceed down the other side of the valley back south again, we will find a continuum of interbreeding salamanders bounded on either end by the two distinct southern species. Thus, while these two species appear clearly demarcated when viewed in isolation from their context, there is no such clear demarcation when the context is restored – that is, there is no way to determine where one species ends and the other begins.

I recount this story as a metaphor for engineering. Engineers are employed at all levels of responsibility within organizations, from the lowest level technical work to, in some cases, corporate CEO. A rank and file engineer doing basic technical tasks will interface seamlessly with his or her team leader. The team leader has some supervisory responsibility for several engineers, but will also be intimate with the technical details of their work. That team leader will also interface seamlessly in the other direction with a department head, say. The department head manages the budgets and schedules of several teams, in addition to overseeing the programmatic and technical objectives of the department. The department head in turn reports to a program manager, and so forth up the line. At each level both the business and technical aspects may be equally important – the main difference is in the level of abstraction with which they are engaged. The low level engineer will certainly be cognizant of the importance of budgets, deadlines, and other business objectives, but will typically engage them only in fairly abstract ways. The technical issues, on the other hand, are very concrete for that engineer. At a program manager level, the business issues are engaged much more concretely, while the technical issues are engaged at a much higher level of abstraction.

It is tempting to conclude that someone who has risen from the engineering ranks to the CEO level of a big corporation has ceased to function as an engineer and is now employed in some other capacity. We might say the CEO has become a completely separate species from a low-level engineer. But as with the chain of salamanders, it is difficult to say where along the continuum between the two one crosses the line from being an engineer to being something else. Or rather than crossing a clear line of separation, perhaps one goes from being a whole engineer to being a fractional engineer, with the fraction gradually decreasing as non-technical responsibilities accrue. Or, does the attempt to separate the technical from the business miss the mark altogether with respect to characterizing engineering? Is Goldberg (2006) correct when he writes, "The businessperson who says that engineering is 'mere technology applied to the needs of business' could more accurately be told that modern business is merely the application of the engineering method to the design of commerce?"

These questions highlight a long-standing tension between engineers' technical and organization roles. Is engineering primarily about providing technical expertise or is it about accomplishing business or societal objectives with respect to technology? Are engineers technicians or technocrats? Are they labor or management? The answers are not easily forthcoming. In fact, this tension extends into the realm of engineering education. In the USA, for example, engineering education has long been heavily invested in the technical and scientific elements of the curriculum. But this emphasis is constantly challenged by voices from the engineering community that would prefer to see more attention given to developing engineers' organizational leadership skills and business acumen (e.g., NAE 2004; Duderstadt 2008). This tension has only been heightened in recent years in the context of engineering's adjustments to globalization.

Dialectics of Purpose

Ideals Contra Realities

The mission statement of the Institute of Electrical and Electronics Engineers (IEEE), one of the world's largest engineering societies, states, "IEEE's core purpose is to foster technological innovation and excellence *for the benefit of humanity*" (IEEE 2008). Likewise, the mission statement of Cal Tech, one of the world's leading engineering education and research institutions, states, "The mission of the California Institute of Technology is to expand human knowledge and *benefit society* through research integrated with education" (Cal Tech 2008). This same overarching idealism – pursuing the benefit of humanity/society – is common rhetoric for engineering organizations and institutions worldwide. This is not surprising coming from a discipline that aims to be a profession in the fullest sense. But is there substance to these claims? Against them is the reality that the benefits proffered by engineering accomplishments are almost always attended by some measure of undesirable side effects or unintended consequences. We might even be able to name some products of engineering that have not worked to any reasonable public benefit at all. In many cases, benefits may be distributed inequitably, accruing to one group at the expense of another. For reasons such as these, public opposition can and does arise in response to various engineering projects or products. It is also important to note that the majority of engineering work is carried out by private firms aiming most explicitly at financial success, not humanity's benefit. Some see engineers as being captive to these private interests, and thus limited in their ability to make good on any overarching professional ideals (Goldman 1991; Noble 1977).

So how might engineers reconcile tension between their overarching ideals and a somewhat different reality? One way is in the interpretation of what it means to *benefit humanity*. "No one claims technology is omnipotent or omnibenevolent," writes Sunny Auyang (2004). In engineering, "the underlying philosophy is usually utilitarian." Benefit, therefore, comes to mean *on balance*. We can accept a dose of negative in return for generating a greater dose of positive. Of course, as with all utilitarian thinking, the crux of the matter lies in how we choose to define positives and negatives, and in what (often incommensurable) values we assign them. Tradeoffs have always been a staple in engineering, but they are conceived most often in terms of balancing quantifiable technical parameters within the details of a design. But how do engineers ensure favorable tradeoffs at the far more consequential – yet often far less tangible – level of wide-ranging societal benefit/detriment? If their highest ideal is societal benefit, and if societal benefit is a complex and contested concept, then we might imagine engineers being heavily engaged in discourse about it. But in the view of many observers, such is not the case. "If Socrates' suggestion that the 'unexamined life is not worth living' still holds," writes Langdon Winner (1986), "it is news to most engineers." Winner softens that acerbic statement (slightly) by noting that there are exceptions. But his broad contention that engineers as a group are largely unreflective about their work remains intact, and it is shared by others. Richard Devon (2004), for example, writes that "it is still easier for engineers to understand a lot about how a technology works as a technology, while having a limited understanding of its possible uses and its social and environmental impacts."

It can be argued whether or not engineers are really as unreflective as these claims would have it. But even if they are, does being unreflective mean that engineers are indifferent, or that their mission statements are simply fodder for lip service? Not necessarily. "Every engineer I have ever met," writes Florman (1987), "has been satisfied that his work contributes to the communal well-being, though admittedly, I had never given much thought to why this should be so." Florman is suggesting an axiomatic presumption, perhaps instinctive to many engineers, that their work is organically beneficent, at least in a utilitarian sense. Engineers tend toward pragmatism, a belief in progress, an action orientation, and, of course, an affinity for technology. "Engineering is an inherently constructive profession," writes Goldberg (2006), "attempting to make a better world through change." Many engineers may tend to view engineering work as operating like a Smithian *invisible hand* that inexorably promotes the collective benefit. From that viewpoint, reflection may seem largely superfluous – what is important is action. Of course, whether engineers realize it or not, the belief that technological progress will invariably work, in the manner of an invisible hand, for the collective good, is a disputed claim. So even if engineers might be absolved of any general indifference towards the broader implications of their work, allegations that their views are too limited may still be levied. Whatever the case, it is clear that significant tensions exist between engineering's ideals and the realities of engineering practice and technological development. This dialectic is a fundamental dynamic that must be grappled with in any quest to understand engineering in context.

Technological Understanding Contra Technological Concealment

It is a widely accepted premise that the public needs to be more technologically literate for a variety of reasons that will benefit both individuals and society. For individuals, technological literacy will enhance the potential to acquire technology related jobs (Barus 1989), to make wise consumer choices, and to participate in public discourse about the pros and cons of technologies. Society benefits from having a skilled workforce capable of sustaining technological industries, as well as from having a citizenry capable of making informed contributions to public policy (Wacker 1991).

If there is a need for the increased technological literacy of people in our society, then it would seem patently obvious that engineers could and should play a vital role in helping to fulfill that need. After all, engineers are collectively the group most intimate with the workings of technologies. Toward that end, many engineering professional organizations have become active in seeking ways to promote technological literacy. The IEEE, just to give one example, recently launched an initiative called "Technological Literacy Matters!" Aside from the societal benefits, the engineering profession has more selfish motives for promoting technological literacy. The profession is generally aware of its own public invisibility, which has led it to undertake efforts, such as the annual National Engineers Week in the USA, to enhance understanding of engineering and technology. This also aids in filling the educational pipeline with young people having the interest and preparation to pursue careers in engineering – a crucial issue for the profession.

But with respect to the causes of technological illiteracy, an influential NAE/NRC report (Pearson and Young 2002) makes the following statement: "Most modern technologies are designed so users do not have to know how they work in order to operate them." The key word is *designed*. If technologies are black boxes which people learn to use without any real understanding, and if that is a leading cause of technological illiteracy, then engineers are in some sense the architects of that illiteracy. Albert Borgmann (1984) coined the term *device paradigm* to describe much of modern technology. The device paradigm suggests that modern technologies are designed specifically to enhance the ends of the technology (such as ease of communication in the case of the telephone) while removing the means from view as much as possible. "The concealment of the machinery and the disburdening character of the device go hand in hand… A commodity is truly available when it can be enjoyed as a mere end, unencumbered by means." Because of the powerful marketability of *ends unencumbered by means*, engineers have been proficient and prodigious in making the concealment of means a reality, and they do so as an explicit design goal.

We are all familiar with the term *user-friendly*, which we apply to a technological product that is easy to use and to understand. But when we say easy to understand, we do not mean it is easy to understand the underlying technological principles. Rather, we mean it is easy to understand how to get it to do what we want it to do, and this often is purposely divorced from any knowledge of those underlying technological principles. In fact, the design trend is toward technologies that will do what we want them to do with less and less explicit input or manipulation on our part. Take for example the ideas of Donald Norman (1998), a proponent of human-centered computer technology. He writes, "Today's technology imposes itself on us, making demands on our time and diminishing our control over our lives. Of all the technologies, perhaps the most disruptive for individuals is the personal computer. The computer is really an infrastructure, even though today we treat it as the end object. Infrastructures should be invisible: and that is exactly what this book recommends: A user-centered, human-centered humane technology where today's personal computer has disappeared into invisibility."

Norman echoes Borgmann in pointing out that technology imposes a cognitive burden on us, one which we are generally happy to relieve if possible via user-friendly, invisible technologies. But whereas Borgmann – a philosopher – views that trend with uneasiness, Norman – an engineer – celebrates it as a worthy objective. The more invisible the technology, the less the user has to know about it, and the more successful the designer. Such design goals can stem from a positive desire to enhance the user's experience and productivity. On the other hand, sometimes such goals are couched more negatively as palliatives for users' ignorance. An article in the EETimes (Wallace 2006), an industry newspaper for electronics engineers, states that technology "wants – and needs – to become transparent, if not completely invisible to today's techless, clueless consumer." The article refers to such designs as *invisible facilitation*, which it says "is rapidly emerging as the design rule of the day." The article implies that the *techless, clueless consumer* – i.e., technologically illiterate consumer – is a problem to be solved. But the solution strategy in this case is not to educate consumers about technology, but rather to increasingly design technology to cater to consumers' low level of technological knowledge. This notion of designing to compensate for users' ignorance is illustrated, for example, in Inagaki's (2004) discussion of automation for transportation technologies: "[I]n cases of non-professional operators, such as private car drivers, it would not be sensible to assume that their levels of knowledge and skill are high. Their understanding of machine functionalities can be incomplete, or even incorrect."

The intentional design for concealment of means is pervasive. In addition to terms such as *user-friendly* and *black box*, other familiar terms which convey the notion of usability without understanding – and which are pursued during design as desirable things – include *plug-and-play*, *turnkey system*, *human-centered design*, or *user-centered design*. The great irony – the key to this dialectic – is that even as engineers recognize the need for, and work to promote technological literacy, in the context of their actual work they are caught in a spiral that works against that objective. The more engineers make their designs user-friendly, the less users need to know about the underlying technology. But the less users know about the underlying technology, the more they demand increased user-friendliness. And so on.

A Dialectic of Method

The General Contra the Specific

In his *Metaphysics*, Aristotle (2001) writes, "Actions and productions are all concerned with the individual; for the physician does not cure man, except in an incidental way, but Callias or Socrates or some other called by some such individual name, who happens to be a man." That is, while physicians may value a general ideal (health) or work toward a global objective (curing the sick), their actions are always local and specific. Engineering's general ideals perhaps include such things as progress, efficiency, or improvement in the quality of human life. And global objectives might include such items as providing energy, transportation, communications, and the like. In practice, however, engineering concentrates on the local and the specific. Localization manifests itself in two primary ways in engineering, one circumstantial and the other methodological.

Circumstantial localization – or particularity – exists by virtue of the fact that solutions to engineering problems are always local and never universal. Engineering concentrates "on what is possible in narrow localities of the universe and definitely not everywhere" (Jarvie 1966). For example, if engineers design a suitable drinking water distribution system for a small town, they have not solved the problem of drinking water distribution for all people everywhere. The solution for the one town is particular; it depends on the particularities of, among many other things, the nature and quality of the local water source, the geography of the locale, the size of the town, the size of the town's budget, the engineers' inherent preferences for some materials and techniques over others, and the capabilities of local construction firms. This is not to suggest that there is no universal engineering knowledge. Certainly, much of the knowledge and reasoning that went into the design of the one drinking water system can also be applied to the design of other such systems. Nevertheless, the application of that engineering knowledge is always "concentrated on local conditions and their transformation," conditions "which might be absolutely unique" (Poser 1998). To take another example, the engineer does not solve the problem of communications, except in an incidental way, but rather solves the problem of communicating a specific type of information, at a specified rate and fidelity, between specific types of points. In fact, engineering cannot address generalized or abstracted problems. The fundamental object of engineering is to meet a set of specifications. And as the very word implies, specifications define the concrete and particular manifestation of a problem.

Hand-in-hand with this circumstantial localization of the problems with which engineering is concerned, engineering practice invariably attacks those problems by engaging in methodological localization, a form of reductionism. Not only is it the reality that each engineering problem is unique, but at all levels within the solution process, from overall system analysis, to the minute detailing of individual components, forms of reduction prevail. In order to cope with real world complexity and uncertainty, engineers invariably isolate, subdivide, and simplify. This reduction is what allows engineers to be successful. Carl Mitcham (1997) writes, "[I]t is not only permissible to ignore complex subtleties, but better to do so." In Larry Bucciarelli's (1994) analysis of engineering design, this notion of methodological localization surfaces time and again. "Object world stories," he writes, in reference to the domain of thought, actions, and artifacts comprising design, "work better with fewer elements; abstraction and reduction go hand in hand in this business. Sparseness characterizes a good, workable model." Reductionism, he concludes, "is the essence of technique within object worlds." Similar observations also appear in Walter Vincenti's (1990) account of engineering. He writes, for example, "Such successive division resolves the airplane problem into smaller manageable subproblems, each of which can be attacked in semi-isolation."

By their very nature, both types of localization help engender a mindset that is restrictive rather than expansive, exclusive rather than inclusive, convergent rather than divergent. Earlier we mentioned that engineers are sometimes accused of being unreflective with respect to the broader implications of their work. It should be no wonder that people who are constantly engaged in the solution of concrete, particular, and finite problems, and who habituate themselves to solution methods that discretize and simplify those problems, are not as a rule always and instinctively cognizant of the more abstract and potentially generalized effects of those solutions. We might posit that when engineers visualize an overall problem as a collection of relatively independent subproblems, each of which has been simplified and idealized, they nonetheless believe that they are manipulating actual components of external reality. This opens the engineer to a criticism, articulated for example by Larry Hickman (2001). When resolving a complex problem into component parts for the convenience of achieving a solution, it is a mistake, according to the criticism, to view those parts as somehow unique and absolute, existing independently of the process that led to the parts being identified and isolated – i.e., it is a mistake to view the use of a particular taxonomy of parts as somehow logically inevitable and necessary. This criticism is echoed by Levins and Lewontin (1985), who suggest that this biases solutions by favoring problems that are amenable to being reduced in the preferred ways.

The danger lies in the potential to foster a belief that engineering methodology follows a rigidly deterministic and logical path, rather than recognizing the biases, contingencies, and subjective decisions that skew the process toward the expedient achievement of specific, narrow objectives. As Koen (2003) suggests, engineering solutions always provide the right answers, just not always to the same questions that were initially asked. In other words, the way in which engineering problems are parsed in the solution process can serve to alter the problem itself. This has ramifications for the previously-discussed dialectic of engineering's ideals versus the realities of engineering practice and technological development, which this dialectic tension – and sometimes disconnect – between engineering's globalized objectives and its localized methods can exacerbate.

Conclusion

In sum, the core thesis presented here is that in one sense engineering can never be finally understood because it is neither discreet nor static. Nonetheless, engineering can be usefully investigated and those investigations can broaden our understanding – not just of engineering but of humanity in general – and we suggest that taking a dialectical approach can be beneficial. In making their case for the use of dialectical thought in biology, Levins and Lewontin (1985) write, "Things change because of the actions of opposing forces on them, and things are the way they are because of the temporary balance of opposing forces." Thus, they conclude, biological study advances with the investigation of these dialectical tensions. Likewise, we posit that many of the key entry points for our investigations of engineering are precisely at such points of dialectical tension. It is important, for example, to understand the dynamic arising from the tension between the increasing dependence of society upon engineering and technology and the simultaneously decreasing understanding by society of that same engineering and technology. It is important, for example, to understand the dynamic arising from differences between what engineers say they are trying to accomplish, or think they are trying to accomplish, and what they actually do accomplish, or what we perceive them to have done. In this chapter we have not even attempted a full assessment of these or any of the dialectics mentioned – those concerned with our study of engineering, those concerned with engineering's identity, those concerned with engineering's purpose, and those concerned with engineering's methods – space has permitted only the briefest of discussions (and no discussion of others that might be identified – "performance/capability contra risk" is one that comes easily to mind). But hopefully this chapter has served to frame some of these key conundrums that challenge our understanding of engineering in context. The subsequent chapters of this book will explore engineering in more detail from a variety of angles. In the process, many will encounter and grapple with various aspects of one or more of these dialectics.

References

Aristotle. (2001). *The basic works of Aristotle*. New York: The Modern Library.

Auyang, S. Y. (2004). *Engineering: An endless frontier*. Cambridge, MA: Harvard University Press.

Barus, C. (1989). On technological literacy. In *Delicate balance: Technics, culture and consequences* (pp. 292–298). Torrance: Los Angeles Chapter IEEE SSIT.

Borgmann, A. (1984). *Technology and the character of contemporary life*. Chicago: University of Chicago Press.

Bucciarelli, L. (1994). *Designing engineers*. Cambridge, MA: MIT Press.

Cal Tech (California Institute of Technology). (2008). *Mission statement*. Available at: http://www.caltech.edu/at-a-glance/. Accessed 20 Nov 2008.

Dawkins, R. (2004). *The ancestor's tale*. Boston: Houghton Mifflin.

Devon, R. (2004). Towards a social ethics of technology: A research prospect. *Techné, 8*, 1.

Didier, C. (1999). Why there are no engineering ethics in France: A historical interpretation. *The OEC 1999 international conference on ethics in engineering and computer science*. Available at: http://www.onlineethics.org/cms/.aspx. Accessed 18 Nov 2008.

Downey, G. L., Donovan, A., & Elliot, T. J. (1989). The invisible engineer: How engineering ceased to be a problem in science and technology studies. *Knowledge and Society: Studies in the Sociology of Science Past and Present, 8*, 189–216.

Duderstadt, J. (2008). *Engineering for a changing world. A roadmap to the future of engineering practice, research, and education*. The millennium project. Ann Arbor: University of Michigan. http://milproj.ummu.umich.edu/publications/EngFlex_report/download/EngFlex%20Report.pdf. Accessed 3 Dec 2008.

Florman, S. (1987). *The civilized engineer*. New York: St. Martins.

Florman, S. (1996). *The introspective engineer*. New York: St. Martins.

Goldberg, D. E. (2006). *The entrepreneurial engineer*. Hoboken: Wiley.

Goldman, S. L. (1991). The social captivity of engineering. In P. T. Durbin (Ed.), *Critical perspectives on nonacademic science and engineering* (Research in technology studies series, pp. 121–145). London: Lehigh University Press.
Hickman, L. (2001). *Philosophical tools for technological culture: Putting pragmatism to work.* Bloomington: University of Indiana Press.
IEEE. (2008). *IEEE mission and vision.* Available at: http://www.ieee.org/web/aboutus/visionmission.html. Accessed 20 Nov 2008.
Inagaki, T. (2004). Design of human–machine interactions in light of domain-dependence of human-centered automation. *Cognition, Technology, and Work, 8*(3), 161–167.
Jarvie, I. C. (1966). The social character of technological problems: Comments on Skolimowski's paper. *Technology and Culture, 7*(3), 384–390.
Koen, B. V. (2003). *Discussion of the method: Conducting the engineer's approach to problem solving*. New York: Oxford University Press.
Levins, R., & Lewontin, R. (1985). *The dialectical biologist*. Cambridge, MA: Harvard University Press.
Mitcham, C. (1997). Engineering design research and social responsibility. In K. S. Shrader-Frechette & L. Westra (Eds.), *Technology and values*. Lanham: Rowman & Littlefield.
NAE. (2004). *The engineer of 2020: Visions of engineering in the new century*. Washington, DC: The National Academies Press.
Newberry, B. (2007). Are engineers instrumentalists? *Technology in Society, 29*(1), 107–119.
Noble, D. F. (1977). *America by design – Science, technology and the rise of corporate capitalism.* New York: Alfred Knopf.
Norman, D. (1998). *The invisible computer: Why good products can fail, the personal computer is so complex, and information appliances are the solution*. Cambridge, MA: MIT Press.
Pearson, G., & Young, T. (Eds.). (2002). *Technically speaking: Why all Americans need to know more about technology*. Washington, DC: National Academies Press.
Petroski, H. (1992). *To engineer is human: The role of failure in successful design*. New York: Vintage.
Poser, H. (1998). On structural differences between science and engineering. *Technè, 4*(2), 81–93.
Vincenti, W. G. (1990). *What engineers know and how they know it: Analytical studies from aeronautical history*. Baltimore: Johns Hopkins University Press.
Wacker, G. (1991). Societal technological literacy is essential for sustainability. ISTAS '91: Preparing for a sustainable society. Post Conference Edition. *Proceedings of the 1991 international symposium on technology and society*, Ryerson Polytechnic Institute, Toronto, Canada, June 21–22, 1991, pp. 311–317.
Wallace, R. (2006). It's design, not technology: A warmer, fuzzier engineering realm is emerging, one that puts a priority on humanity rather than geek trip. *EETimes Online*, 27 Feb 2006. Available at: www.eetimes.com
Wallich, P. (2004). Electrical engineering's identity crisis. *IEEE Spectrum, 41*(11), 45–51.
Williams, R. (2003). Education for the profession formerly known as engineering. *Chronicle of Higher Education, 49*(20), B12.
Winner, L. (1986). *The whale and the reactor*. Chicago: University of Chicago Press.
Zussman, R. (1985). *Mechanics of the middle class: Work and politics among American engineers.* Berkeley: University of California Press.

Byron Newberry B.S. in Aerospace Engineering, University of Alabama. M.S. in Aerospace Engineering and Ph.D. in Engineering Mechanics, both from Iowa State University. Professor of Mechanical Engineering, Baylor University. Baylor Fellow for teaching. Professional Engineer (PE) Texas, USA. Research interests: engineering design, engineering ethics, philosophy of engineering and technology. Aircraft structural engineering consultant. Executive board member, National Institute for Engineering Ethics. Editor of the Springer *Philosophy of Engineering and Technology* book series.

Chapter 2
'Nuts and Bolts and People' Gender Troubled Engineering Identities

Wendy Faulkner

Abstract How and where boundaries are drawn between 'the technical' and 'the social' in engineering identities and practices is a central concern for feminist technology studies, given the strong marking of sociality as feminine and technology as masculine. I explore these themes, drawing on ethnographic observations of building design engineering. This is a profoundly heterogeneous and networked engineering practice, which entails troubled boundaries and identities for the individuals involved – evident in interactions between engineers and architects, and amongst engineers, around management and design. There are complex gender tensions, as well as professional tensions, at work here. I conclude that engineers cleave to technicist engineering identities in part because they converge with (and perform) available masculinities, and that women's (perceived and felt) membership as 'real' engineers is likely to be more fragile than men's. Improving the representation of women in engineering requires foregrounding and celebrating heterogeneity in genders as well as engineering.

Keywords Engineer identities • Heterogeneity • Technical/social dualism • Gender

Introduction

In conversation with a friend who has been an engineer for some 40 years, I discovered he had worked in quite different sectors and technologies, from toy manufacturing to road bridge maintenance. He explained, 'It's all engineering really – all nuts and bolts.' Then he paused for a minute and added, as if to correct himself, 'Well, nuts and bolts and people'.

Engineers have two types of stories about what constitutes 'real' engineering: in sociological terms, one is *technicist*, the other *heterogeneous*. For instance, engineers commonly report that their biggest surprise when they started their first

W. Faulkner (✉)
34 Queens Crescent, Aberfoyle, Stirling FK8 3UP, Scotland
e-mail: wendyfaulkner34l@btinternet.com

S.H. Christensen et al. (eds.), *Engineering Identities, Epistemologies and Values*, Philosophy of Engineering and Technology 21,
DOI 10.1007/978-3-319-16172-3_2

engineering job after graduating was how little time is spent on 'real' engineering, by which they typically mean 'calculations and drawings'. The emphasis on calculations is hardly surprising. The core of university-based engineering education is a mathematical approach to analytical problem solving in which problems are 'reduced' to their physical properties and social complexity is pared away (Bucciarelli 1994, p. 108). This training stands in stark contrast to the huge importance of 'social' expertise in engineering jobs, which engineers soon learn is actually vital to their work. Some, like my friend, come to view these aspects of the job as the more challenging and rewarding; others cleave to a 'nuts and bolts' identity. But virtually all the engineers I have met oscillate between or straddle, not always comfortably, technicist and heterogeneous engineering identities.

Social studies of engineering problematize the deep technical/social dualism at the heart of engineering identities and practice. As numerous case studies have demonstrated, the knowledge mobilised in engineering practice is never 'just technical' with 'the social' bolted on (e.g. Bucciarelli 1994; Downey 1998; Vinck 2003). Rather, these two dimensions are in a very practical sense *inseparable* – hence the unhyphenated term 'sociotechnical' (Hughes 1986) – and the boundaries drawn between them are inevitably arbitrary. As Downey and Lucena conclude, engineers live 'on the constructed social boundaries between science and society and between labor and capital' (1995, p. 167). But is this the full story? In this chapter, I seek to 'write gender in' to these accounts.[1] How and where boundaries are drawn between 'the technical' and 'the social' in engineering identities and practices is a central concern for feminist technology studies, given the strong marking of sociality as feminine and technology as masculine, and the continued numerical dominance of men in engineering in most disciplines and most Western countries (Faulkner 2000b). Yet the connection is rarely made in the otherwise interesting literature on engineering identities.

This chapter seeks to redress this gap. It draws on ethnographic fieldwork in two UK offices of a building design engineering consultancy company. This involved job shadowing six engineers over the course of 5 weeks, two of whom – Karen and Fraser – I followed for over a week each, offering many opportunities for extended conversation. During this fieldwork, I was able to observe closely the routine office-based practices of some 20 engineers, plus several meetings with external partners.

The design of buildings is a networked and staged process involving a heterogeneous array of partners – engineers (various disciplines), architects, clients, cost consultants, building contractors, suppliers – each of which is vital but none of which could do the job on their own. Because of the complexity and scale of any major building, and the diverse expertise required, control is a major pre-occupation and relations can become very political. The engineering expertise required in building design is itself heterogeneous. Much of it is specific and can only be built up on the job: an appreciation of particular client/user requirements; knowledge of relevant products, regulations, etc; networks of contacts; and above all, a cumulative body of

[1] Other intersections of identities have been addressed, including intersections of national identities and engineering identities: e.g., Meiksins and Smith (1996), Downey and Lucena (1997 and 2004).

experience of 'what works and what doesn't'. In addition, building design engineering demands considerable people skills – precisely because the design process necessitates sophisticated management and 'conversations' between the diverse partners.

The body of this chapter examines some of the troubled engineering identities and *boundary work* which flow from this heterogeneity.[2] It looks at boundaries between engineers and architects, and at boundaries amongst engineers around design and management. In both cases, the troubled identities reflect very real and rather intractable professional and organizational dynamics; but they also reflect very real and rather complex gender dynamics. These dynamics help explain both the persistence of technicist engineering identities and the tensions between these and heterogeneous ones. First, I open with some background on the framing of the study and on how I understand gender.

Genders in/of Engineering

The fieldwork presented here was part of a larger study, 'Genders in/of Engineering', which sought to examine gender dynamics within engineering practices, cultures and identities.[3] The study was posited on the conviction that we need to know more about the men and masculinities in engineering if we are to understand better the continuing poor representation of women in engineering. By using ethnographic methods, it addressed the premise that the retention and progression of women engineers is impaired not only because of well-rehearsed structural barriers (e.g., lack of flexible work practices), but also because of more 'taken-for-granted' gender dynamics not always evident to participants. In particular, the study has identified a number of subtle dynamics by which people come to *belong* (or not) in engineering communities of practice (e.g. Faulkner 2009a).

In line with much feminist technology studies, I understand gender and engineering as co-constructed or *co-produced*.[4] This often operates symbolically. Thus, the technical/social dualism does not necessarily map on to actual people and practices – which are typically diverse – yet it performs gender work. For instance, the 'nerd' stereotype is of men who are passionate about technology but rather a-social. The fact that these two poles of the dualism are posited as mutually exclusive – to be technical is to be not-social – is one of the ways in which engineering appears 'gender inauthentic'[5] for women, given the strong association of women/femininities

[2] Gieryn's (1995) concept of *boundary work* has been helpful in illuminating the constructed nature of boundaries around many areas of science or technology. The key analytical point is that how and where boundaries are drawn at any one time and place is often consequential.

[3] Faulkner (2000a) indicates the framing of this larger study. In total 66 engineers were interviewed and/or observed; where not attributed, later claims are derived from this wider fieldwork.

[4] This framework has been elaborated and reflected on in: Lerman et al. 1997; Lohan 2000; Wajcman 2000; and Faulkner 2001.

[5] The concept of 'gender in/authenticity' is elaborated in Faulkner 2009b.

with caring about people. I take it as one of the valuable contributions of well grounded social scientific research that it can reveal the extent of mismatches between such stereotypes and actual people or practices, and so serve to destabilize stereotyped assumptions. For example, my fieldwork revealed no evidence to support the common assumption that women engineers have better people skills than men engineers, which is an important but largely un-trumpeted challenge to the 'to be technical is to be not-social' stereotype of engineers.

I have coined the term *gender in/authenticity* to capture the normative pressures of 'the way things are' – pressures that lead people to expect the gender norm (in this case, the man engineer) and to notice when they see an exception (the woman engineer). There is nothing remarkable about a man choosing to be an engineer, while the reactions of outsiders are a constant reminder that being a woman engineer marks them out as unusual. I must stress that I use the term gender in/authenticity in a non-essentializing way. The term is not meant to imply that 'the way things are' can never change: far from it. Much of the evidence in the larger study profoundly challenges the presumed non-congruence of gender and engineering identities for women. The point is that gender in/authenticity issues are consequential; they *perform gender work*. Thus, the perceived gender inauthenticity of the woman engineer means that women engineers face in/visibility problems (cf Tonso 2007) which men engineers never experience: they tend to be highly visible as women but not as engineers, so have routinely to (re)establish their engineering credentials (Faulkner 2009b).

Troubled Boundaries Between Engineers and Architects

Without exception, the building design engineers I met distinguish the professional orientation and interests of engineers and architects around a dualized boundary: architects want a building that 'looks good', while engineers want a building that 'works'. The distinction drawn – between 'design' and technology – is misleading. In practice, there is considerable overlap between the two communities in terms of what they actually do and know. There has to be. The form of the building has to accommodate all its functions, including the building services of power and water supply, air quality, etc. So, building design engineers and architects acquire what Harry Collins and Rob Evans (2002) call *interactive expertise* in each other's specialisms in order to collaborate effectively. They also share important subjectivities: they both derive huge pride and satisfaction on seeing the finished building, and enjoy talking about other publicly visible buildings.

Mechanical engineer Karen illustrates many of these points. Unusually, her degree combined architecture and engineering. She describes architecture as being 'more design than sums' and says she might have become an architect but 'I felt more of an engineer. I was a bit too practical for architecture … I need more to justify a space than "it's the right aesthetic" – it has to fulfil its function, it has to make people comfortable, it has to use the appropriate amount of energy, etc.' Karen has a particular interest in low energy and sustainable building design. She asserted her belief that 'we engineers understand more about it than architects' with some humour at a meeting to prepare a bid to design sustainable offices. The engineers and architects present chat about the 'green' gherkin-shaped building in London. Karen asks 'Is there any assessment of the Atrium and how it will work? [since] *air doesn't do that!*'. Eilidh, the other engineer present, explains that the architects behind this building 'love arrows: blue one for cold air and red ones for hot ... *They think they can change the laws of physics!*'. Karen joins in, lamenting the marketing of 'stupid ideas that don't work'. Perhaps sensing that the architects present don't really understand the problem with arrows, she then explains, grinning, "They behave as if you can make air do what you want it to do! [But] cold air pushes hot air up. Hot air doesn't rise – *it's a myth!* It's displaced by cold air, which is denser and needed to drive it. In a room full of hot air there is no air movement." She laughs openly because, like many engineers, she identifies strongly with the apparent certainty which flows from their reliance on science; she relishes the fact that such expertise distinguishes her from non-engineers.[6]

The technicist professional identity Karen is expressing here is associated both with science and with a kind of practical materiality – something I encountered repeatedly amongst men and women engineers. Engineers' educational grounding in mathematics and science enables them to claim an identity in the material and (mostly) predictable phenomena governed by 'laws of nature', backed up by a faith in cause and effect reasoning (see also Mellström 1995). And this same materiality and scientificity enables them to claim, as the central contribution of engineering design, that it creates technologies that 'do the job'.[7] This is a very empowering identity, in the very literal sense that buildings are empowering: they enable users to do things. This is why engineers in all sectors celebrate the visible outcomes of their work (Florman 1976; Hacker 1989, 1990). And it is why engineers' practical and scientific expertise *feels* empowering to them, when contrasted with a lack of such expertise in others.

The certainty and materiality associated with science and technology can also be very powerful *symbolically* – with significant gender connotations, at least historically. As feminist scholars have demonstrated (e.g. Merchant 1980; Noble 1991), achieving control and domination over nature was a central plank in the Baconian project – and a central justification, at the time, for excluding women from that project. Similarly, Ruth Oldenziel (1999) has demonstrated that the strong association of engineering with industrial technology (machines), with science and with corporate might, served to code engineering as heavily masculine during the period of its professionalization. In short, the establishment of both science and engineering involved the emergence of new versions of what Bob Connell terms *hegemonic masculinity* (1987, 1995). The 'mastery of nature' remains a powerful emblem of

---

[6] Humour ridiculing the lack of 'technical' knowledge amongst others is a common feature of engineering communities (Hacker 1990, Chap. 4; Mellström 1995, Chap. 5).

[7] I am not suggesting that engineering guarantees certainty. The point is that engineers see their role as seeking to reduce uncertainty to acceptable levels, and that the palpable successes of modern technologies in achieving this gives them comfort (Kleif and Faulkner 2003).

technology, both within engineering (e.g. Florman 1976, pp. 121–26) and in wider culture (e.g. Caputi 1988).

Elsewhere I have suggested that engineers' shared pride and pleasure in the technologies they build can be read as a vicarious identification with the power of technologies, perhaps even a kind of *symbolic compensation* for a felt lack of power in other aspects of their lives (Faulkner 2000a). It has been suggested that this might explain the particular appeal of engineering to men – to the degree that performances of masculinity 'demand' a sense of mastery over something (Edwards 1996), and that men 'have a problem' with interpersonal relationships (Hacker 1990, Chap. 4). Tine Kleif and I (2003) found this hypothesis to hold for some men but by no means all. Many women engineers I have encountered also like science and maths 'because there's always a right answer', and many also get excited by 'big bits of kit!'. The fact that the theme of power resonates with hegemonic versions of masculinity does not prevent women engineers from enjoying the felt power of built technologies as much as men.

To recap The need for 'conversations' between specialists in a networked design process creates contradictory impulses about what counts as 'real' engineering. On the one hand, all the partners have to be able to 'meet in the middle' in order to collaborate. On the other hand, engineers have an occupational interest in foregrounding the 'core' scientific and technical expertise which only they, as engineers, can bring to the design process. I have suggested that there are other subjective dynamics at play here too – engineering identities strongly tied up with the actual and felt power of built technologies, and with the apparently certainty afforded by their use of mathematics and science. Whilst these subjectivities are strongly associated historically and symbolically with available masculinities, they are no longer confined to men.

Boundary Spanning

In practice, some engineers are more proactive than others in their relations with architects. Karen enjoys working at the interface between engineering and architecture, and says her job 'is as much about people and relationships as sizing stuff'. In the same spirit, Karen is often animated about 'the people aspect' of building services engineering, including the very real difficulties associated with engaging the end user in making the building 'work'. As she commented during the preparation of the sustainable office bid:

> They [the client] need to think about the control system. Sustainability ends when you put people in! You need to train staff to ensure that the building is operating correctly. We mustn't leave once it's built. … If you don't get buy in, the buildings won't operate properly, and it will overheat. You probably need some automatic features [but] if it's all automatic, they'll also complain. You need people to like being there. …. For example, you can introduce digital displays in the building about water and energy use – so people know. It increases awareness and ownership. … It's all about people: designing buildings people can use!

Eilidh also works at the engineering-architecture boundary. Later in the meeting, when Karen has been running through some of the requirements for water use and heating, she says to the architects present, 'It's useful to have this discussion in front of [the client] to show we know the subject, to show we're not soft engineers who can't deliver!' Both women make frequent use of the hard/soft dualism. Eilidh comments on the landscape architect's concept of a 'pavilion': 'It can sound very soft and not very commercial'. There is a clear value hierarchy in these quotes: hard is associated with being effective commercially, with the 'nuts and bolts' of engineering and with 'being able to deliver'; soft is associated with 'aesthetics', with the people aspects of design and, perhaps, with idealism in relation to sustainability. The symbolic gendering of this 'hard-soft' dualism is fairly self-evident (see Faulkner 2000b). What I want to emphasize here is how Eilidh and Karen are building a space in which the importance of *both* 'hard' and 'soft' issues – and the need for *both* 'hard' and 'soft' expertise – is acknowledged.

Troubled Boundaries Amongst Engineers

The identity work performed by engineers is wrought with tensions and contradictions. Karen is clearly a 'nuts and bolts and people' type of engineer. Yet she sees maths, science and practical technology as central to her engineering identity. As noted earlier, the mathematical competence so emphasized in engineering education is often all but absent in engineering practices. I was treated to many playful asides about this – like, 'See, I did some sums there!' and, after 2 days of shadowing, 'This is the first calculation you've seen me do!'. I feel such jokes indicate a realistic irony, even wistfulness, about the inadequacy of the technicist version of engineering. They serve simultaneously to challenge and reproduce an image of engineering that is at odds with the actual work. The loss of a technicist identity is a readily recognized lament.

These tensions beg several questions: What do individual engineers feel about the mismatch between the actual heterogeneity of their work and the technicist focus of their education? How do they position themselves in relation to the (implied) technical-social scale? And what are the implications of different positionings in terms of 'getting the job done', career progression and perceived membership as 'real' engineers? The cases Karen and Fraser provide interesting insights.

Karen

Karen's joint degree has not held back her career as an engineer. Five years after graduating, she was responsible for the design of the mechanical building services in a major iconic building, incorporating many principles of sustainability. She subsequently won a prestigious national prize for this work, became chartered and was promoted to a level where she could bring in new business, undertake concept design and run projects unaided. Like many others, however, she feels some ambivalence about leaving behind the more 'back room' work of detailed design:

> There are weeks when I feel I've done no engineering at all. The person I am now is a project manager/design manager. … Every now and then I get a craving to do some sums. It used to bother me more. The feeling of 'not producing anything' has made me unhappy at times.

Karen juxtaposes the 'up front' and 'backroom' roles in a way that echoes the technical/social dualism. She has a sense that her new roles are less 'real' engineering, perhaps because they are further away from the materiality of 'producing' things. Nonetheless, having established her engineering credentials, she now feels she has earned the right to concentrate on the more up front work, which she feels her personality is suited to. She believes people should be allowed to concentrate on the jobs where their strengths and interests lie. Unfortunately for Karen, this view is not shared by her manager, Tom, although he esteems her highly:

> Tom used the term 'captain and cabin boy' when I joined – i.e., we all have to do a bit of everything, from basic stuff through to management stuff. … Problem is, I don't really want to be the cabin boy anymore – again been there, done that – and I've worked very hard to progress to a point where I don't have to do that role anymore. … I'm more than happy to do the concept design and get things kicked off and then run the job, but the thought of spending the next several years tied to my desk detailing and personally putting tender packages together fills me with dread. … I definitely see my future as a project/design manager and not sure that I can do this within [the company].

Five months later, Karen left the company in which she has had such a brilliant early career, for a job in project management – a move which, though still in mechanical building services, she sees as leaving engineering.

It would be wrong to view Karen's story as a tragedy. For Karen, this will probably prove a good career move, and her obvious talent is not being lost to the design of new buildings. What her story does illustrate, however, is how perceptions of what counts as 'real' engineering can have a material bearing on who is and is not deemed to *belong* in engineering – and thus, on who gets to stay and progress. In part, as Karen rightly perceived, she did not 'fit' because the business model of the regional office where she now works (the captain and cabin boy) differed from that in the head office she previously worked (where post-chartered engineers are *expected* to move into managerial roles and detailed design is conducted almost exclusively by junior engineers). In part, however, I suspect that Karen's fragile membership in the regional office was also due to the culture and ethic of her colleagues, many of whom appear to celebrate a 'practical', 'nuts and bolts' version of engineering. There were gasps of astonishment once when Karen admitted that she'd never 'sized' a gas pipe – *'You've got this far and never sized a gas pipe?!'*.[8]

[8] Sizing here refers to the calculation needed to establish what diameter of pipe is needed for a particular purpose. Karen asked how to size a gas pipe because, as it happens, she'd never had to do it for gas before. She acknowledged their astonishment in her reply: 'I know, but how do you do it?'.

In this setting, it seems likely that some of Karen's colleagues were unimpressed by her disdain for practical and backroom jobs.

Fraser

Fraser is more in line with the culture and ethic of the regional office. Like Karen, he works in mechanical engineering building services and is in his early 30s. He also has a demonstrated talent for management and up front roles. He is currently project managing the company's design of building services for a major office development. This means he plays a pivotal role between the dozen or so company engineers doing the detailed design and the wider network of partners in the project. It is the first time Fraser has done so much 'people management' and financial control. In his own time, he has developed a detailed plan for the design process – breaking down the jobs into tasks, with milestones and estimates of the number of drawings required for each. From this he worked up a spreadsheet of the hours per month needed from everyone on the job. These two documents are bound together with selected drawings marked up to show areas of flexibility. The whole document is half an inch thick. Fraser gave copies to all staff on the project, 'so they own it and know where their work fits in and have personal targets'.

This very heterogeneous reality of Fraser's work does not sit entirely comfortably with him, however. Coming out of an on-site design team meeting one day, he expressed deep disdain for the role of the contractor manager who chaired the meeting. When I commented on the man's ability to 'keep it all in his head', Fraser's immediate and pained response was 'But that's *all* he does is manage!' There is a similar feel to a later comment: 'They [the contractors] will never get blamed because all they do is management contracting; the subbies [the subcontractors] do the work.' By implication, then, the *real* work is designing and building, not managing. So for Fraser, there is a tension between design and management, where for Karen it is between backroom and up front roles. But both of them experience the move into management as a move *away* from engineering. Fraser laments that he now gets to do less and less engineering (i.e., design), and frequently voices the heartfelt view that engineering should attract the same kudos and pay as management.

Science and technology are both part of Fraser's engineering identity. The science connection surfaces in the way he dualizes 'facts and politics'. Time and again he finds himself having to operate politically, but he is clearly more comfortable when he can 'stick to the facts'. Fraser presents the 'technological' part of his engineering expertise identity in terms of a focus on the design work. For example, in a telephone conversation with a contractor to whom the company is bidding in order to pair up for a major hospital project, Fraser says they need to talk 'with the people responsible for managing and delivering the thing as well as the nuts and bolts'. He then suggests 'an informal meeting with everyone chipping in ... That's what I like. I'm more of a nuts and bolts person, than sitting talking about the thing. It's all about delivery at the end of the day'.

Design and Management

The story of this bid is interesting for what is illustrates about the troubled boundary between management and design.[9] There are two internal teleconferences – with three directors and two senior engineers, Fraser and Peter, from different offices of the company – to brainstorm their strategy for the bid. Everybody present recognizes that management and design need to be *integrated* if the design is to be 'delivered'. Yet the distinction between management and design runs throughout the preparatory discussions, with 'delivery' emerging as an ambivalent boundary term.

The management challenges in the hospital project are considerable. But whilst the team know they must have a convincing story to tell about this, time and again they come back to the need to demonstrate their 'design depth' – especially because the people they have to persuade in this bid are contractors. Tom emphasizes this: 'At some point, we will talk about design and delivery, and we will want depth in the meeting. … They will talk nuts and bolts. They'll want to know what your [waving to Fraser and Peter] duct drawings look like.' Peter has extensive experience of project management but, like Fraser, cleaves to a 'nuts and bolts' engineering identity – perhaps because he comes from a contracting background. He is asked to lead this side of the bid. Unlike Karen, Peter does not relish the role. He replies: 'Good designers don't necessarily do well up front. … I'm not necessarily the man for the job. I'm not comfortable with strangers. My confidence is in my technical ability'. Seeking to persuade him, one of the directors then suggests, 'I could be the project director, delivering some up front bullshit, alongside Peter as the bid manager'.

In some ways, the relationship between the directors and senior engineers is similar to that between engineers and architects. The two must be able to work effectively together, but without losing their respective strengths – the directors, their 'up front bullshit', business experience and networks of contacts; and the senior engineers their day-to-day, 'hands on' control and knowledge of design projects. The directors would find it hard to 'talk engineering' in specific detail with the contractors; they need the two senior engineers to be 'nuts and bolts' people in the context of the bid. But if the company *is* to deliver the eventual hospital design, Fraser and Peter will need to be what they in fact are – 'nuts and bolts and people' engineers – at which point their staff, and not they, are cast as the 'nuts and bolts' engineers.

We see here the fluidity of management-design boundaries within engineering. In both cases, engineers are attributed a technicist engineering identity in contrast to colleagues senior to them, while the directors and the senior engineers are managers, albeit with somewhat different management roles. At the same time, they are both still 'doing engineering' in these management roles. For example, Tom routinely reviews the designs of his staff and makes presentations of their work to architects or clients. In such ways, while engineers need interactive expertise in

[9] Notice how 'design' is used in relation to engineering when the contrast is with management, but in relation to architecture when it is contrasted with engineering.

relation to architecture, they use what Collins and Evans call *referred expertise* as managers.[10]

Gender Trouble Around 'Real' Engineering

Karen and Fraser have much in common beyond their shared discipline and age. Both are relatively senior, respected by their peers and managers alike; both are hardworking and ambitious; both do a lot of up front and managerial work and have good people skills; both have engineering identities rooted in science and technology; and both lament the loss of 'real' engineering work to some degree. The main difference between them is that Fraser is still trying to hold on to some of the 'nuts and bolts' work and has a strong sense of this as central to his engineering identity, whereas Karen is moving away from the 'nuts and bolts' of design and doesn't foreground this in her engineering identity. And in this particular company, Karen has had to leave in order to continue doing the type of engineering work she enjoys, whereas Fraser is likely to stay and progress through the ranks.

In drawing a comparison between the two, I do *not* wish to imply that Karen is typical of women engineers and Fraser of men engineers: plenty of men engineers happily gravitate away from backroom, design roles and plenty of women engineers prefer these roles. Rather, I see the cases of Karen and Fraser as illustrating *how gender symbols co-produce,* alongside professional drivers, *engineering identities.*

Most obviously, the 'nuts and bolts' identity paraded by Fraser and others takes its marker from hands-on work with technology; it is modelled on the technician engineer, virtually none of which are women. This identity therefore resonates with a working class 'muscular masculinity'. Its blue collar associations may be a particular draw for engineers in the UK, where professional engineering attracts more working class entrants than in other countries (Whalley 1986), many coming in through apprenticeships. In addition, the blue collar associations are especially prominent in relation to building contractors, who generally have a stronger working class presence and culture than does design engineering. Yet, even in countries where fewer engineers come from blue collar backgrounds, it seems common for men engineers to celebrate a 'nuts and bolts' identity. In their extensive study of engineers in the USA, Judith McIlwee and Gregg Robinson (1992) found that men engineers often engage in 'ritualistic displays of hands-on technical competence' even when the job does not require this competence.

So, the traditional association of men and engineering tools still marks professional engineering as masculine, which makes 'nuts and bolts' feel 'manly'. This does reflect a real, if diminishing, gender difference. 'Tinkering' with car engines and the like has long been a typical route into engineering for men (e.g., Mellström

[10]Thus: 'to manage a scientific project at a technical level requires, not contributory expertise to the sciences in question but *the experience* of contributory expertise in some related science' (Collins and Evans 2002, p. 257: emphasis original).

1995). Although a growing proportion of those now entering engineering do not come from a tinkering background, and although some women opt for hands-on work, still considerably more men than women engineers have been socialized into a hands-on relationship with technology. As many women engineers testify, this can seriously undermine their confidence and their sense of belonging, especially when they first enter engineering degrees.

The term ‘practical’ seems to me very gender-troubled in this context. As we have seen, both women and men engineers celebrate a ‘practical’ engineering identity – practical in the sense that as engineers they come up with solutions that ‘get the job done’. Yet many of the women engineers I have met tell me, unprompted, that they are ‘not practical’ – practical in this context meaning that they do not have a strong background or interest in ‘hands-on’ aspects of engineering.

Significant though the ‘hands-on’ theme certainly is, the gendering of engineering identities is rather more complex than this, on a number of counts. For a start, women and men engineers both foreground technicist engineering identities, and science is an important marker of these identities for women and men alike. I sense that most women engineers foreground science more than ‘nuts and bolts’ in their engineering identities. This is not terribly surprising. The gender norms surrounding science are less strong these days than those surrounding ‘nuts and bolts’ technology, in the obvious sense that there are vastly more women scientists than women technician engineers. Yet, the strong emphasis on practical materiality – of designing things that work – is shared by all engineers. This is a unifying theme of both the ‘nuts and bolts’ and the ‘laws of physics’ versions of technicist engineering identities – and so cuts across the heavy masculine coding of the former.

Another source of gender complexity is that the two versions of ‘real’ engineering with which I opened this chapter are associated with two very available versions of masculinity. Where the technicist engineering identity takes its marker from science and technology, the heterogeneous identity takes its marker from corporate authority and business. It is modelled on the senior manager or entrepreneur, of which relatively few are women. Like engineering, senior management is a materially powerful role, but here the power wielded is a money power or organizational power rather than a physical power. A man engineer who moves into management may lose his credentials as a ‘nuts and bolts’ engineer, and unsettle the blue collar associations, but he does not lose his credentials as a man. If anything, he gains in this regard, since the authority wielded by managers, and the money made in business, are widely applauded markers of achievement in men (Connell 1987, 1995) – what Michel Kimmel (1994) calls *marketplace manhood.*

Why, then, does Fraser parade a technicist engineering identity even when his job is so heterogeneous? Why is he so reluctant to embrace an identity more consistent with his growing management role? Many oilfield engineers I studied also voice disdain for ‘collar and tie’ men. Two of them independently told me they dislike the career model that moves engineers from being specialists to generalists in management. Like Fraser, their gender identity is closely tied up with technology. If their ambitions could be met by staying in more narrowly technical roles, they would probably not opt to go into management. However, as well as being ambitious, all three men get excited by (feel vicarious pleasure in) the 'money power' of the businesses they work for, which is precisely what management gets them closer to. So they are torn between identifying with technology and getting on in engineering, between the power of technology and the power of the corporation.

Of course, such ambivalence is not unique to the engineering profession. People in many walks of life have to move progressively into management and away from their original specialist skills if they want to progress their careers. These are organizational drivers. But I believe a further, gender dynamic may be operating here – namely, that the gender symbolism surrounding management is itself somewhat ambivalent. There are two, readily gendered dualisms operating here: hard/soft and technical/social. Note that the people skills required for management are widely referred to by engineers as 'soft' skills, in contrast to the 'hard' skills required for engineering. But management is also an arena of 'hard' commercial reality – readily cast as hardnosed, hard hitting and so on – as earlier quotes from Eilidh and Karen remind us. The gender connotations are clear. Management and business is likely to feel, and be perceived as, more 'masculine' (and more gender authentic for men engineers), to the degree that these roles carry real authority over others and/or deal with profit and loss aspects of running the business. Management and business is likely to feel, and be perceived as, more feminine (and more gender authentic for women engineers) to the degree that these roles draw heavily on interpersonal skills.

Where 'the technical' and 'the social' are gendered and presumed to be mutually exclusive, the technical/social dualism similarly creates tensions for men engineers doing or contemplating management roles. It means that identifying with 'the technical' (masculinity) means distancing oneself from 'the social' (femininity) – or at least playing down its importance, as Fraser does in relation to management. It also explains why management roles are portrayed as 'just' social by many women and men engineers. For men whose gender and engineering identity is tied up with technology, a move into management potentially undermines both their masculine and their professional identities.

The technical/social dualism also creates tensions for women engineers. On the one hand, it means that moving out of narrowly technical roles is likely to feel, and be perceived as, more gender authentic for them than for men. On the other hand, it means that those women who move away from the more narrowly technical aspects of engineering are in greater risk of losing their membership as 'real' engineers than are men who make the same move. Two older women engineers told me that women engineers who become senior managers are more likely to stop calling themselves engineers than are men who make the same move. It seems the gender authenticity issue never quite goes away for women in occupations dominated by men. Significantly, the tendency for women engineers to be invisible as engineers (cf Tonso 2007) can serve to reinforce a technicist identity amongst women engineers, many of whom choose to stay on the 'technical' side. After all, engineering generally attracts women who 'love technology' and all women engineers *per force* make a huge investment in becoming and belonging as engineers.

Evidence on the types of management jobs women and men engineers end up in reveals an interaction of the gendering of these two dualisms – hard/soft and technical/social – in a pattern Mike Savage calls 'Women's expertise and men's authority' (1992). It seems engineering is typical of other occupations in that men disproportionately occupy positions of power and authority where they are involved in high level line management and the control of organizational resources, whilst women are disproportionately in management of support roles which demand specialist expertise (e.g., in charge of IT systems).[11] Women engineers also tend to get stuck in lower level management jobs, such as project or team management, which can be dead-ends in terms of progression into more powerful and remunerative seats of management (Evetts 1993, 1996).

The upshot of all this is that Fraser's membership as a 'real' engineer is likely to remain more solid, and Karen's more fragile, as they each move progressively from design into management. And Karen's move into management is more likely to be seen as – and sadly, in the case of her recent job move, to feel like – a move away from engineering, in spite of her obvious credentials on that front. In this regard, I would conclude, we *can* see Fraser and Karen and 'typical' of their gender.

Conclusions

We can now return to our opening challenge – to 'write gender in' to accounts of heterogeneity in engineering identities. A key question is: *why do engineers so often foreground a technicist engineering identity in spite of the lived heterogeneity of their actual work?* Clearly a key professional factor is that the 'core' expertise in scientifically-based analytical problems solving which engineers get from engineering education, in their unique professional contribution in a networked design process. But there are also two critical gender factors operating here.

First, technicist engineering identities are as strong as they are in part because these identities converge with available masculinities, in at least two ways: they brings them close to a sense of hands-on technical work (even though they rarely do this themselves); and it makes them feel powerful (they make 'buildings that work'). Thus, many men engineers cleave to a technicist engineering identity because it feels consistent with versions of masculinity that are comfortable for them. Whilst most women engineers also take pleasure in and identify with the material power of the technologies they build or work with, the majority nonetheless identify more readily with the science base of engineering than with hands-on engineering.

Second, the conventional gendering of the technical-social dualism simply cannot be ignored if we are to understand the strength of technicist engineering identities – and, by this token, the continued predominance of men in engineering. The technical/social dualism makes it easier for men to identify with the 'nuts and bolts' of engineering, and casts people skills as 'soft', for women. The tendency to see 'the technical' and 'the social' as mutually exclusive is likely to reinforce some men's resistance to embracing a heterogeneous engineering identity. In any case, presenting

[11] See also Halford et al (1997) on gender segregation of management roles in other sectors.

as a 'nuts and bolts' person is rather more 'gender authentic' symbolically for a man than for a woman in our culture; just as moving away from the 'nuts and bolts' is rather more 'gender authentic' for a woman than a man. Little wonder that women's membership as 'real' engineers is often more fragile than that of men colleagues.

Notwithstanding the pull of technicist identities, engineers routinely experience contradictory impulses about how much of 'the social' is admitted in their engineering identities and in what counts as 'real' engineering. A second key question, then, is: *why are the tensions surrounding the two versions of 'real' engineering so apparently intractable, and what are they about?*

Again there are gender dynamics operating alongside professional and organizational ones. Professionally and organizationally, there is a tension between the need for engineers' 'core' expertise in maths, science and technology, and the need for them to also be able to collaborate and communicate effectively with the other partners in a networked design process. In a similar way, there is a mutually dependent but partially overlapping relationship between those engineers who do more design and those who do more management.

The gender tensions operating around technicist and heterogeneous engineering identities concern men and women engineers in different ways. For men engineers, tensions can flow from the fact that the two versions of masculinity that these engineering identities map so readily onto are *very distinct*: one associated with technology, the other with business. Although these are both in some sense hegemonic masculinities, they are not necessarily compatible for all men, as Fraser's story illustrates. For women engineers, tensions can flows from the very gender inauthenticity of the woman engineer, which means that women engineers have a constant struggle to prove that they are not only 'real engineers' but also 'real women' (Faulkner 2009a). In this context, moving away from narrowly technical roles is a case of 'damned if you do, damned if you don't'.

My central conclusion from this analysis is that engineering as a profession must find ways to foreground and celebrate heterogeneous understandings of engineering and heterogeneous engineering identities. There are two really strong reasons for this conclusion.

First, that is what engineering is! Every aspect of engineering is heterogeneous; even the most apparently technical roles have social elements inextricably within them. Moreover, *good* engineering (as in engineering which is effective) demands the thorough integration of these elements, in ways which *transcend* the normal dichotomizing ways of thinking. Witness, Eilidh's mission to integrate 'hard' and 'soft' elements in sustainable building design; and the hospital bid team's mission to integrate management and design if the hospital is to be 'delivered'. The crucial and (for some) radical challenge is to convey that all engineering is, of necessity, *both* technical and social.

Second, foregrounding and celebrating more heterogeneous images of engineering can only serve to make the profession more inclusive. Engineering encompasses a wide diversity of roles, in which the relative weight of technical and social elements (amongst other things) varies along a spectrum. Within this 'broad church', individuals tend to gravitate to roles which suit their particular skills and personality.

As we have seen, some are more comfortable with the 'up front' roles and others with the 'backroom'; some are more comfortable interacting with contractors and suppliers, and others with architects and clients. If the profession does not promote an identity for itself which welcomes this broad range of interests and aptitudes, then it will fail to attract some very valuable talent. And if the profession remains a 'mono-culture', in which only people from one spot on that spectrum really feel they belong, then it will lose some very valuable talent.

So, promoting heterogeneous images of engineering will create space for a more diverse range of people to be engineers. If such moves are to be more *gender* inclusive, however, they must also challenge the gendering of 'the social' as feminine and 'the technical' as masculine – and thus promote new 'co-constructions' of gender *and* engineering simultaneously.[12] In the words of Evelyn Fox Keller many years ago (1986), we need to learn to 'count past two'. Counting past two is about challenging the very dualisms that (re)produce women and men as necessarily different, and engineering as necessarily technical or social. As my ethnography of building design engineering demonstrates, heterogeneous engineering requires heterogeneous genders – in the sense that it requires various mixes of stereotypically masculine and feminine strengths.

Acknowledgments I acknowledge the financial support of the UK Economic and Social Research Council for the study this chapter draws on (ref: RES-000-23-0151). Heartfelt thanks are due to the many engineers who put up with me trailing them around and so generously shared their experiences with me. A longer version of this chapter appears in *Social Studies of Science* (Faulkner 2007). I am grateful to Michael Lynch for agreeing to this version being published, also to the referees and colleagues who took the time to read and comment on earlier versions.

References

Bucciarelli, L. L. (1994). *Designing engineers*. Cambridge, MA: MIT Press.

Caputi, J. (1988). Seeing elephants: The myth of phallotechnology. *Feminist Studies, 14*(3), 487–524.

Collins, H. M., & Evans, R. (2002). The third wave of science studies: Studies of expertise and experience. *Social Studies of Science, 32*(2), 235–296.

Connell, R. W. (1987). *Gender and power: Society, the person and sexual politics*. Cambridge, UK: Polity Press.

Connell, R. W. (1995). *Masculinities*. Cambridge, UK: Polity Press.

Downey, G. L. (1998). *The machine in me: An anthropologist sits among computer engineers*. New York/London: Routledge.

Downey, G. L., & Lucena, J. A. (1995). Engineering studies. In S. Jasanoff, G. E. Markle, J. C. Petersen, & T. Pinch (Eds.), *Handbook of science and technology studies* (pp. 167–188). Thousand Oaks: Sage Publications.

[12] This represents a challenging message for 'women into engineering' campaigns, many of which draw on gender stereotypes by playing down the technical content of engineering and playing up the social content. Lagesen (2007) demonstrates that playing to such stereotypes can 'miss the mark' for the young women being targeted.

Downey, G. L., & Lucena, J. A. (1997). Engineering selves. In G. L. Downey & J. Dumit (Eds.), *Cyborgs and citadels: Anthropological interventions in emerging science and technologies*. Santa Fe: SAR Press.

Downey, G. L., & Lucena, J. A. (2004). Knowledge and professional identity in engineering: Code-switching and the metrics of progress. *History and Technology, 20*(4), 393–420.

Edwards, P. N. (1996). *The closed world: Computers and the politics of discourse in Cold War America*. Cambridge, MA: MIT Press.

Evetts, J. (1993). Women and management in engineering: the glass ceiling' for women's careers'. *Women in Management Review, 8*(7), 19–25.

Evetts, J. (1996). *Gender and career in science and engineering*. London: Taylor and Francis.

Faulkner, W. (2000a). The power and the pleasure? A research agenda for "making gender stick" to engineers. *Science, Technology, & Human Values, 25*(1), 87–119.

Faulkner, W. (2000b). Dualisms, hierarchies and gender in engineering. *Social Studies of Science, 30*(5), 759–792.

Faulkner, W. (2001). The technology question in feminism: A view from feminist technology studies. *Women's Studies International Forum, 24*(1), 79–95.

Faulkner, W. (2007). "Nuts and bolts and people": Gender-troubled engineering identities. *Social Studies of Science, 37*(3), 331–356.

Faulkner, W. (2009a). Doing gender in engineering workplace cultures: Part I – observations from the field. *Engineering Studies, 1*(1), 3–18.

Faulkner, W. (2009b). Doing gender in engineering workplace cultures: Part II – gender in/authenticity and the in/visibility paradox. *Engineering Studies, 1*(3), 169–189.

Florman, S. (1976). *The existential pleasures of engineering*. New York: St Martin's.

Gieryn, T. F. (1995). Boundaries of science. In S. Jasanoff, G. E. Markle, J. C. Petersen, & T. Pinch (Eds.), *Handbook of science and technology studies* (pp. 393–441). Thousand Oaks: Sage Publications.

Hacker, S. (1989). *Pleasure, power and technology: Some tales of gender, engineering, and the cooperative workplace*. Boston: Unwin Hyman.

Hacker, S. (1990). *Doing it the hard way: Investigations of gender and technology*. Boston: Unwin Hyman.

Halford, S., Savage, M., & Witz, A. (1997). *Gender, careers and organisations*. Basingstoke: Macmillan.

Hughes, T. (1986). The seamless web: Technology, science, etcetera, etcetera. *Social Studies of Science, 16*(2), 281–292.

Keller, E. F. (1986). How gender matters, or, why It's so hard for us to count past two. In J. Harding (Ed.), *Perspectives on gender and science*. Brighton: Falmer Press; edited version in: Kirkup, G., & Keller, L. S. (Eds.). (1992). *Inventing women: Science, technology and gender* (pp. 42–49). Cambridge, UK: Polity Press.

Kimmel, M. S. (1994). Masculinity as homophobia: Fear, shame and silence in the construction of gender identity. In H. Brod & M. Kaufman (Eds.), *Theorizing masculinities* (pp. 119–141). Thousand Oaks: Sage.

Kleif, T., & Faulkner, W. (2003). "I'm no athlete [but] i can make this thing dance!" men's pleasures in technology. *Science, Technology, & Human Values, 28*(2), 296–325.

Lagesen, V. A. (2007). The strength of numbers: Strategies to include women in computer science. *Social Studies of Science, 37*(1), 67–92.

Lerman, N. E., Mohun, A. P., & Oldenziel, R. (1997). The shoulders we stand on and the view from here: Historiography and directions for research. *Technology and Culture, Special Issue on Gender Analysis and the History of Technology, 38*(1), 9–30.

Lohan, M. (2000). Constructive tensions in feminist technology studies. *Social Studies of Science, 30*(6), 895–916.

McIlwee, J. S., & Robinson, J. G. (1992). *Women in engineering: Gender, power, and workplace culture*. Albany: State University of New York Press.

Meiksins, P., & Smith, C. (Eds.). (1996). *Engineering labour: Technical workers in comparative perspective*. London: Verso.
Mellström, U. (1995). *Engineering lives: Technology, time and space in a male-centred world.* Linköping: Department of Technology and Social Change, Linköping University.
Merchant, C. (1980, 1982). *The death of nature: Women, ecology, and the scientific revolution* (2nd ed.). London: Wildwood House.
Noble, D. (1991). *A world without women: The evolution of the masculine culture of science.* New York: Knopf.
Oldenziel, R. (1999). *Making technology masculine: Men, women and modern machines in America, 1980–1945*. Amsterdam: Amsterdam University Press.
Savage, M. (1992). Women's expertise, men's authority. In M. Savage & A. Witz (Eds.), *Gender and bureaucracy*. Oxford: Blackwell.
Tonso, K. (2007). *On the outskirts of engineering: Gender, power, and engineering practice.* Rotterdam: Sense Publications.
Vinck, D. (2003). *Everyday engineering: An ethnography of design and innovation.* Cambridge, MA: MIT Press.
Wajcman, J. (2000). Reflections on gender and technology studies: In what state is the art? *Social Studies of Science, 30*(3), 447–464.
Whalley, P. (1986). *The social production of technical work: The case of British engineers.* London: MacMillan.

Wendy Faulkner B.Sc., Biology, M.Sc. and D.Phil. Science and Technology Policy Studies, all from University of Sussex. Recently retired from the Institute for the Study of Science, Technology and Innovation Studies, University of Edinburgh, where she developed the Edinburgh Masters and Doctoral Programmes in Science and Technology Studies. Her research has spanned innovation studies, especially inter-institutional knowledge flows in innovation (*Knowledge Frontiers*, 1995, with Jacqueline Senker), and feminist studies of science and technology (*Alice through the Microscope*, 1980, Brighton Women and Science Group; *Smothered by Invention*, 1985, with Erik Arnold, *Technologies of Inclusion: Gender in the Information Society,* 2011, with Knut Sørensen and Els Rommes). She now provides training to encourage and enable researchers to bring dialogic approaches to their public engagement work.

Chapter 3
Designing the Identities of Engineers

Mike Murphy, Shannon Chance, and Eddie Conlon

Abstract In 2007 Gary Downey, Juan Lucena and Carl Mitcham argued that a "key issue in ethics education for engineers concerns the relationship between the identity of the engineer and the responsibilities of engineering work". They suggested that "one methodological strategy for sorting out similarities and differences in engineers' identities is to ask the 'who' question. Who is an engineer? Or, what makes one an engineer?" (Downey et al. 2007). This chapter explores these questions of *who is an engineer* and what *makes one an engineer* by examining how engineering and engineering technology students in Dublin Institute of Technology (DIT) describe and differentiate themselves. DIT offers both 4-year engineering degrees (that are equivalent to the educational standard required for professional status) and 3-year degrees in engineering technology. Annually DIT graduates the largest combined number of engineering and engineering technology majors in the country. We present results that show that there is no distinct sense of identity for a technologist. For faculty as well as engineering students and engineering technology students, design is perceived as a key differentiating activity that separates the engineer from the engineering technologist. Paradoxically, while all students chose DIT based on its reputation and practical focus, it is engineering technology students who indicated they are prepared for the 'real world' as they near graduation. Results also show, in terms of their own responses, that engineering and engineering technology students have fairly consistent views of their education and preparation for the workforce.

Keywords Self-direction • Purpose • Engineering identity • Engineering technology • Design • Real world • Career

M. Murphy (✉)
Technological University for Dublin Programme Team,
Dublin Institute of Technology, 143 Lower Rathmines Road, Dublin 6, Ireland
e-mail: mike.murphy@dit.ie

S. Chance • E. Conlon
College of Engineering and Built Environment, Dublin Institute of Technology,
Bolton Street, Dublin 1, Ireland
e-mail: shannonchance@verizon.net; edward.conlon@dit.ie

S.H. Christensen et al. (eds.), *Engineering Identities, Epistemologies and Values*, Philosophy of Engineering and Technology 21,
DOI 10.1007/978-3-319-16172-3_3

Introduction

In the foreword to a book by Sheri Sheppard et al. on educating engineers, Lee Shulman related an anecdote in which a number of senior (i.e., final year) engineering students from a highly regarded public university were asked to characterize the engineer's place relative to other professions by answering the question "What's an engineer?" Shulman explained:

> Their response – collaboratively crafted and framed – was unforgettable: "An engineer is someone who uses math and the sciences to mess with the world – by designing and making things that people will buy and use; and once you mess with the world, you are responsible for the mess you've made" (quoted in Sheppard et al. 2009, p. ix).

Engineering education has evolved differently in various countries. In the United States, engineering education has developed two broad streams: engineering and engineering technology. There, differentiation between the two streams is generally described by way of a theoretical-versus-applied approach, with different accreditation criteria for each. Graduates are differentiated by the name of the degree they earn (bachelor of engineering as opposed to bachelor of technology). In Germany, engineers are educated in technical universities and in universities of applied sciences (*Fachhochschulen*). Their differentiation is similar to that in the USA, being along a theoretical-applied continuum, but graduates of both the universities and the universities of applied sciences earn engineering degrees.

Ireland provides an interesting example of these two typical ways of differentiating "engineering technology" from "engineering". This country has distinguished the two based on the relative levels of theory and application they offer. The separation along the theoretical-to-applied engineering continuum aligns structurally with the university-institute of technology dimension, with universities providing more theoretical "engineering" degrees and institutes of technology generally offering more applied "engineering technology" programs. But within Dublin Institute of Technology, programs at both levels are offered. Further, there is a well-established transfer route from engineering technology programs onto engineering programs.

DIT differentiates between traditional 4-year degrees in engineering (that are accredited as professional engineering programs) and 3-year degrees in engineering technology. This differentiation is made at enrolment, where engineering students are required to have earned higher college entrance exam scores than engineering technology students (based on Ireland's Leaving Certificate examination). The single largest differentiating factor between the incoming classes of engineering students (4-year cycle) and engineering technology students (3-year cycle) is in their mathematics ability at entry. To enroll in a 4-year engineering program directly from second-level school requires each student to achieve a minimum C3 grade in higher-level mathematics. To enroll in a 3-year engineering technology program requires a passing grade of D3 in lower-level mathematics. Survey responses consistently show that a significant percentage of students enroll in engineering technology at DIT because they want to become engineers but have not achieved the minimum mathematics standard in their Leaving Certificate examination. Upon completion of their 3-year program, students can apply to transfer onto the junior year (third year) of the 4-year engineering program, which they are allowed to do provided they achieve a minimum threshold grade. Consequently, approximately 50 % of graduates from engineering technology programs transfer onto engineering programs.

What we set out to examine were the similarities and differences between how the two groups of students – engineering and engineering technology – describe themselves. Do the engineering technology students see themselves as engineers, technicians, or technologists? Are there common factors in the identity of engineering students across disciplines, or are identity factors discipline-specific? In this chapter, we examine the identities of students who are about to graduate in order to understand: (1) why the students chose to study engineering in the first place, (2) how their engineering teachers see, describe, and characterize the identity of their students and future graduates; and (3) how these students see and describe themselves as engineers, or technologists. We used Ireland's accreditation standards for each of the two different degree programs as a guide to writing a number of the survey questions, because we wanted to gauge if the differences (implied in the standards) were detected by the students themselves.

What Is Identity?

Self-identity can be seen as the conception individuals develop of *who* and *what* they are (Tony Watson 1997). Identity develops in the course of interactions with others. In a sense, an individual's life can be seen as a *career* during which the person moves through different situations, interacts with others, and adjusts to achieve a sense of selfhood. Watson (1997) asserts, "self-identity is constantly in the process of being won from the social environments in which we find ourselves" (p. 129).

There are two broad dimensions to identity formation: the invented and the constructed. Identity is a social product. It is continually appropriated by individuals for themselves as well as bestowed on individuals by others (Kerry Meyers et al. 2010; Paul Thompson and David McHugh 2002). People actively construct their identities out of the materials presented during social activities and in their various roles. Individuals engage in securing identities that can provide personal stability and help in directing their activity. Identity is thus a tool people use; it helps them project images appropriate to the specific social, cultural, and work contexts they encounter. There are, however, limits to this active creation of identities. The typified self tends to be created from factors that arise in various social situations that fall into specific categories.

Watson (1997) has identified two aspects of individual identity. The first is self-concept involving such matters as self-efficacy, self-esteem, and self-confidence. According to Robin Leidner (2006), this can be derived from the experience of education or work providing the "satisfactions of feeling oneself competent to accomplish one's intentions, overcome difficulties (and) create something" (p. 436).

The second is a social identity that includes various attitudes, values, beliefs, and commitments in relation to society and social institutions. Personal values are influenced by the culture of society and the groups within it; our social and professional identities are often shaped by occupational culture.

Personal and social identities are inevitably intertwined. For example, Leidner (2006) asserts participation in an occupational culture "frequently involves an explicit reframing of self-identity as well as development of a new collective identity" (p. 436). She makes the point that in well-defined occupations, processes of initiation are explicitly intended to transform the identity of newcomers:

> Novices gain skills and a body of practical and… abstract knowledge. When socialization is successful they also learn and internalize the occupation's ideology, ethos, traditions, and norms, including criteria for judgment, craft pride, and rules for interacting among themselves and various others. (Leidner 2006, p. 436)

The literature identifies two broad approaches to identity formation (see Jan Stets and Peter Burke 2000): (1) *identity theory* which focuses on roles and the manner in which individuals (through a process of identification) come to occupy a role and incorporate the meanings and expectations associated with that role into their sense of 'self', and (2) *social identity theory* where the emphasis is on group membership and self-categorization by individuals to identify themselves as members of particular groups. In this, "Having a particular social identity means being at one with a certain group, being like others in the group, and seeing things from the group's perspective" (p. 226).

They argue for a more integrated view of the self and assert the differences between these theories are more of *emphasis* than of *kind*:

> In general one's identities are composed of the self-views that emerge from the reflexive activity of self-categorisation or identification in terms of membership in particular *groups* or *roles*… theorists in both traditions recognise that individuals view themselves in terms of meanings imparted by a structured society.… Both identification with a social category and role behaviour refer to and reaffirm social structural arrangements (Stets and Burke 2000, pp. 225–226, 232).

This is not simply an issue of personality; organizational, institutional, and situational factors play a role in shaping identity (Olga Pierrakos et al. 2010).

We can conclude that 'being' and 'doing' are both central features of one's identity. Moreover, central questions to ask in exploring engineering identities are: who is an engineer, what does an engineer do, what does performing the role of an engineer entail, and what are the responsibilities of engineers? Drawing on Michael Hogg and Deborah Terry (2000), Kevin Anderson et al. (2010) argue that engineering groups imagine archetypes that capture dependent features of group membership which are abstractions of group features: "These archetypes then show what the group values and serve to distinguish the ways of doing and thinking of one group from another" (p. 157).

The approach described above requires us to focus not just on the emerging identities of engineering graduates but also on the way that the role of the engineer is socially constructed within different societies and how that role is reproduced (or challenged) for each new generation of engineers. A key focus must be on engineering education because "formative processes in education serve as key locations for negotiating and renegotiating of the relationship between the person of the engineer and the definition and responsibilities of engineering work" (Downey et al. 2007, p. 466). In the course of obtaining education, students will develop technical and professional expertise but will also "undergo changes in their identity and self-conception of what it means to be an engineer" (Pierrakos et al. 2010). Thus, Downey et al. (2007) say, engineering educators typically bear primary responsibility for addressing and answering the question: What does it take to become a good engineer?

Of course there may be more than one answer to this question arising from national differences in the organization of engineering work and different approaches to the education of engineers. The two issues are clearly linked. Engineering, and technical work, is structured differently in different societies and the processes that reproduce the engineering and technical workforce also differ. The manner in which engineers are formed has implications for their understandings of their roles and their relationships with other groups – especially management (see Peter Meiksins and Chris Smith 1996; Chris Smith 1987). It may be the case that the professional identity of engineers is weak, as other forms of self-categorization and identification have greater significance. This is a collective issue and not just an issue for individual engineers. Such is the case in Japan, where engineers have traditionally identified with the enterprise where they are employed rather than their profession (Downey et al. 2007; Meiksins and Smith 1996). National variations in the processes for reproducing engineering work and engineers led Meiksins and Smith (1996) to conclude it may be "impossible to develop a definition of what an engineer is, or where the boundaries of engineering lie, which would apply to all industrial capitalist societies" (p. 3).

Andrew Jamison (2013) has mapped the relationship between different approaches to engineering education and different archetypes of engineering identity. He identified three broad approaches to engineering education: science-driven, market-driven and socially driven. These are related to three aspects of identity: academic, commercial and hybrid. However, in most societies these ideal types do not exist in a pure form.

Indeed the identification of deficiencies associated with the science-driven model has led to the development of a second layer of market-driven engineering education in many countries. It is aimed at the production of "more practically trained engineers" (Meiksins and Smith 1996, p. 245). The development of institutes of technology in Ireland can be seen to fit into this pattern. The *Mulcahy Report* (1967) set out the rationale for their formation in the following terms:

> We believe that the long-term function of the colleges will be to educate for trade and industry over a broad spectrum of occupations ranging from craft to professional, notably in engineering and science but also in commercial, linguistic and other specialities. They will, however, be more immediately concerned with providing courses aimed at filling gaps in the industrial manpower structure, particularly in the technician area.

This fits in with a pattern – that can be identified in a number of countries – whereby the state defines various categories of technical worker based on the abstract-practical continuum. This increases the degree of hierarchy in technical labor, so that the workforce becomes "stratified by credentials and mode of entry into the technical workforce" and a direct correspondence emerges between the "the type of qualification possessed and the engineer's position in the division of labour" (Meiksins and Smith 1996, p. 240). This leads to a more fragmented occupational community for engineering.

Cutting across issues related to the structure of the engineering workforce are debates about what characterizes a good engineer. What makes a good engineer is contested (see Matthew Wisnioski 2012). Debates in engineering education have focused on shortcomings of traditional engineers and argued for the need for "New Engineers" (Sharon Beder 1988) and, more recently, for "Green Engineers" (Jamison 2013).

Jamison (2013) voices the need for educators and professionals to conceptualize engineering as both a social and technical activity. This includes the need: (1) for the technical component of engineering to be combined with social and cultural understandings, and (2) the need for engineers to have skills and capacities other than technical proficiency. All this should be done with the aim of furthering public, rather than corporate, good. Competencies for sustainable engineering span a number of knowledge domains; they include skills such as critical and systematic thinking, the capacity to work with and integrate the perspectives of others, sustainable development values and ethics, and a wide range of interpersonal skills (Iacovos Nicolaou and Eddie Conlon 2013).

Kevin Anderson et al. (2010) interviewed engineers in six firms and noted the significance of communication skills. As one engineer told them: "Engineering is the easy part. It's the people who are difficult" (p. 162). These researchers found that engineers "walked around with an unstated equation in their head: Problem solver + team player + life-long learner" (p. 166). Despite this finding, they discovered engineers still value the technical core of engineering work: "Authentic engineering tends to be viewed as getting one's hands dirty" (169)[1] and they struggled with including non-technical elements in their definition of engineering. Yet they still believed that effective communication is intimately intertwined with engineering problem solving and that engineering cannot be done without it. They also did not see themselves as being engineers in order to contribute to the public good. "Their identity was more likely to be grounded in solving problems well – for themselves, for their team, for their organization and for their client" (p. 170). This can be explained to an extent by their understanding of the constraints, particularly fiscal

[1] Compare this with research by Llewellyn Mann et al. (2009) on engineering graduates: "Most of the participants talked about being able to fall back on their technical knowledge when they were unsure of how to proceed. Their technical knowledge became almost a safety blanket, *something that makes them sure they are an engineer*" (emphasis added).

constraints, they faced as engineers and the realization that the demands of engineering work do not always, as one engineer put it, “mesh with the romantic visions he held as an undergraduate” (p. 166).

The undergraduates studied by Meyers et al. (2010) identified three factors which define engineering: (1) ability to make competent design decisions, (2) capability to work with others, and (3) maturity to accept responsibility for one’s actions (p. 1554).

What emerges from this is that the identity of engineers can be explored through looking at: (1) how they understand their engineering work, (2) the skills and relationships they need to do that work, and (3) how they understand their responsibilities as engineers. But such an understanding must be contextualized with regard to the organization of engineering (work and education) and the archetypes of engineers that are promoted within these structures.

We know that educational institutions shape student identities both during the recruitment process and while they are studying to be engineers. For instance, Carney Strange and James Banning (2001) argue that certain types of colleges attract specific types of students. The scholars identify four general typologies of (American) colleges and four typologies of (American) college students. They describe relationships between the two sets of typologies. Where the type of institution successfully matches the ‘type’ of student who attends (e.g., the student’s interests, expectations, temperament, inclinations, and abilities) an appropriate ‘fit’ is usually achieved. In the process of finding the right fit, students typically absorb messages that colleges send out (using websites, brochures, campus tours, and the like). Students do this prior to selecting the specific college where they will enroll. This helps match their own values and personal identities to the college. Once a student arrives on campus, he or she typically accepts the values of that community and begins to internalize such messages even more deeply. However, where there is misalignment between the student’s personal values and those of the campus community, the student may become unhappy and leave. Thus, the identity of the college (and its programs) is shaped by, and helps shape, the identities of the individuals who join and maintain it.

Reed Stevens et al. (2008) and Kerry Meyers et al. (2010) have pointed to the importance of the labeling and categorization processes that take place in education. How institutions identify students as engineers has a profound effect on students’ identification of themselves as engineers (Stevens et al.). It matters “what we call students and more specifically the curricular and institutional structures that classify students within departments… as this contributes to the social portion of psychosocial identity” (Meyers et al. pp. 1555, 1558).

A further issue is that there seems to be diversity in how engineering educators understand engineering. Alice Pawley (2009) studied how engineering faculty define engineering. While common themes emerged – such as problem solving, applied science, and making things – there were a range of beliefs as to what engineering is. According to Meyers et al. (2010) “Many engineering educators are challenged to define succinctly what engineering is to students” (p. 1557).

Study Methodology

Our work in this chapter is based on a mixed-methods exploratory study that sought to address the following questions:

1. How do students nearing graduation in engineering and engineering technology identify themselves?
2. To what extent are their identities similar?
3. What differences exist between the groups in what they think they have learned and what they envision as their future roles?

Our study is situated within the constructivist paradigm. In conducting it, we have sought, by exploring points of similarity and contrast, to understand how groups of students see themselves. We began by developing research questions that aligned with constructivist beliefs that groups of people define themselves – and thus shape their own culture – collectively. Together, they develop a shared sense of reality that constitutes truth for them. In this study, we sought to identify points of shared understanding among the two groups (engineering and engineering technology students) as well as factors that distinguished the two groups from each other in the context of a DIT education. We included final year engineering and engineering technology students at DIT in the general fields of mechanical and electrical engineering.

To gain a basic understanding of relevant issues and begin to identify important factors differentiating the student groups, we conducted interviews with faculty from two countries (Ireland and the USA). We analyzed their responses qualitatively and used our findings to construct an instrument for surveying students. We pilot tested the surveys using think-aloud protocols; then we disseminated the survey to graduating students via email. Responses were analyzed using IBM's Statistical Package for Social Sciences (SPSS, version 20) to detect statistically significant differences in the ways the two groups responded. To broaden our understanding of student perception, we examined a series of surveys conducted with students entering DIT in the years 2003–2007. We also drew from preliminary results in conducting a brief case study of how both student groups tackled a design challenge.

Thus, the study reported in this chapter utilized a four-strand approach. *Strand 1* involved conducting and analyzing semi-structured interviews with seven senior lecturers who teach final year engineering or engineering technology students. *Strand 2* involved an online survey of final year students from engineering and engineering technology programs. *Strand 3* involved a review of previous surveys of incoming freshman (i.e. first year) students that included a range of questions such as why they chose to study engineering and who influenced their decisions. *Strand 4* is a case study of design approaches that differentiate the two groups of students. This case study is included because the two groups of students reported significantly different perceptions of the role of design in their work.

Strand 1: Faculty Perspectives

Consistent with the constructivist paradigm, we believe that educators are involved in the dialogue of professional identity and are not separate from it. We also wanted to address existing confusion on the topic of engineering versus engineering technology that seems evident in DIT and other institutions. We interviewed five DIT faculty members. Our objectives were: (1) to elicit how these educators describe and characterize the identity of their final year students and soon-to-be-graduates, and (2) to understand the language used by educators in describing their students. We analyzed these interviews in search of themes that could inform our interviews with students. We also interviewed two faculty members from Purdue University as external reference, and because we hope to expand this study in the future to the discussion in the United States on engineering and technology.

Findings from Faculty Interviews

Two main themes became evident in DIT faculty responses related to the question: *What is it that engineering technologists (or engineers) do?* The first theme is that faculty members see engineers and engineering technologists as generally performing different roles. The second is that the two groups also perform these roles at different depths or levels. Sample comments, illustrating this, are shown in Table 3.1.

Another key distinction is that DIT faculty see the role of the engineer as significantly bound up with design activities, and therefore the identity of the engineer aligns with becoming a designer, a creator of solutions. The role of technologist, even if it contains design elements, is not as fully invested in the design process. So, engineering technologists are involved in the more limited re-design of existing systems, whereas engineers are involved at a conceptual level (see Table 3.2).

Faculty provided a range of views on engineering and engineering technology that included seeing them as overlapping disciplines characterized by different emphases on the one hand and different depth of activities on the other. Table 3.3 below illustrates this dichotomy of views.

Table 3.1 Respective roles

Role of engineering technologists	Role of engineers
"Engineering technologists clarify, confirm, apply, test and ensure"	"Engineers are responsible for conceptual designs and mathematical constructs whereas engineering technologists flesh out these designs"
"Engineering technologists are responsible for operating, managing and supervising processes"	"Professional engineers are responsible for considered design and systematic/methodical problem solving"

Table 3.2 Design as an identifying role

Design role of engineering technologists	Design role of engineers
"Engineering technology graduates will see 'how we can make it better' rather than designing new"	"A professional engineer has the ability to do research and design at the highest level"
"Engineering technologists are involved in design of sub-stations based on the modification of existing designs"	"Engineering graduates design, test and deploy systems"

Table 3.3 Different emphases and different depth

Different emphases	Different depth
"Engineering and engineering technology are almost interchangeable terms"	"The level of application and depth of understanding are the key differences"
"Engineering Technology is more hands-on, practical focused, more applied, less theoretical, less mathematical, less analytical"	"The difference is between mastering design methods versus using technology to implement a solution"
"Engineering technology students may have fantastic applied knowledge but have no analytic skills"	"Engineering is at a superior level with respect to analysis and understanding of fundamental principles"

Table 3.4 Identity and confusion

Identity	Confusion
"Engineering technologists would see themselves as engineers"	"Students don't see the difference between engineering and engineering technology"
"Engineering Technology graduates would characterize themselves as engineers"	"Students could not yet describe themselves or their discipline"
"Recent technology graduates would describe themselves as engineers"	"Students may not be able to characterize the difference"

Important from an identity perspective, faculty noted that engineering technology graduates generally see themselves as engineers and that the students themselves are not well positioned to differentiate between engineering and engineering technology. In putting themselves in the role of their graduating final year students, the educators commented as shown in Table 3.4 above.

In summary, while understanding the curricular and academic differences in the education of the two groups, the faculty we interviewed acknowledged that (1) students about to graduate don't differentiate between engineering and engineering technology, (2) engineering technology graduates will see and identify themselves as engineers, and (3) there is a complete absence of identity as a technologist or engineering technologist. A question yet to be answered is to what extent the final year students' views of themselves have been shaped over the course of their studies at DIT by the views of their educators.

Strand 2: Student Survey

The findings for this portion of our work – the crux of our study – are gathered from students in their final year of study. We wanted information from students whose identities had been shaped (at least partially) by the educational culture they had engaged in for the preceding 3 or 4 years. Based on the results of the faculty interviews, we developed an on-line survey for final year students. This survey was tested and refined through separate 'talk aloud' sessions with four students from our target population: two from engineering technology and two from engineering programs. Of the population of 425, a total of 153 students accessed the survey, for a response rate of 36 %. What we report below as significant meets the 95 % threshold (meaning that there is less than a 5 % chance that each difference we found was random). We also assessed the qualitative responses that students submitted to open-ended questions, looking for themes.

Quantitative Findings from Student Survey

Regarding the survey, there were a number of questions where the two student groups responded in statistically different ways. Engineers ranked each of the following statements higher than technologists did:

(a) I want to use my knowledge to design and create new things.
(b) I can devise and generate new designs and solutions.
(c) My program has prepared me for a wide range of jobs after graduation.
(d) My program gave me detailed knowledge and understanding in my technical area (for example in mechanical engineering).
(e) I have focused significant efforts on developing competence in my profession.
(f) I have focused significant efforts on balancing my independence with my dependence on others.
(g) As a result of my program I can design new systems.

Engineering technology students ranked the following statement higher:

(h) I want to control and maintain equipment in an engineering environment.

The two groups responded in statistically similar ways to the items "my program has taught me how to apply my technical skills" (69 % of students), "my program has taught me how to tackle problems creatively" (45 %), and "my program has taught me how to develop/create successful new technologies" (14 %). However, the two groups responded differently to the statement that "my program has taught me how to solve problems I will face in the real world," with 66 % of technology students ticking this box, but only 47 % of engineers ranking this in their top two.

The survey included questions that were generated using Bloom's revised taxonomy (Anderson et al. 2001; Benjamin Bloom and David Krathwohl 1956). In these questions, the majority of responses were similar for both groups: 42 % of our population said that they had best mastered "to analyze things," 42 % said they had best mastered "to understand things," 39 % said "to apply knowledge," 33 % said "to evaluate things," 18 % said "to remember things," and 14 % said "to create things." However, there was a difference on the item "to apply knowledge," for which 51 % of technology students selected this as one of their top two responses. Just 31 % of engineers did the same. Although this might appear to contradict the finding above that both groups responded similarly to the statement "my program has taught me how to apply my technical skills," what is important to realize is that engineering technology students selected "apply knowledge" as most important to them in selecting their top two choices. This is consistent with the qualitative findings described below regarding how both groups aligned around either 'design' (engineers) or 'apply' (technologists).

Qualitative Findings from Student Survey

The survey asked final year students to give a reason – in one sentence – as to why they chose to study their particular program. There were clear differences in the explanations provided by the two groups. For engineers, responses generally were along the lines of the student always knowing that they wanted to be an engineer, or that they always liked analytical subjects, or that they liked the possible careers and career paths that an engineering qualification would open up. For engineering technologists, the responses tended towards how the engineering technology degree would ensure the graduate would get a good job, or that the program was practical, or that the engineering technology program itself was a follow-on to an earlier program. In DIT this earlier program almost invariably is skills-based (such as electrician training). Some specific responses are provided in Table 3.5 below.

Table 3.5 Why did you choose to study this particular program?

Sample responses of engineering technologists	Sample responses of engineers
"I wanted to move up from being an electrician and be able to work at a higher profile"	"I felt it would give me the widest range of career choices"
"many job prospects afterwards"	"I felt the degree would give me a lot of options after graduation"
"Job opportunities and an interest in machines"	"Interest in maths, physics and all things mechanical. I liked making stuff …"
"I chose this program because it was a practical program that provided skills that could be applied in the real world"	"Mechanical engineering keeps the world ticking and I wanted to be part of that background work"
	"Buildings are great; the idea of applying maths to create solutions for buildings is exciting"

It must be emphasized that there were clear overlaps in the response types with, for example, engineers saying that "there are lots of jobs in engineering" and engineering technologists saying "I always want[ed] to design buildings". But, generally, engineers saw the study of engineering as a stepping stone to a career that was aligned with an inner sense (perhaps ill-defined) of the nature of engineering work – design – that attracted them to study engineering. Technologists tended to have a more immediate horizon: the program was practical and hands-on and would lead to good job opportunities once they graduate.

We asked the final year students to describe what they wanted to do in their first job after graduating. Here, engineers generally responded along one of three themes: they wanted to work in design, or they wanted to gain experience by applying their knowledge, or they wanted to make money. One noteworthy response from an engineering student combines all three of these themes: "get as much money as possible and gain as much experience as possible in design[ing] different systems. Apply anything I've learned while in College." Table 3.6 below provides indicative responses by students to what they wanted to do in their first job.

The key action verb that differentiates the two groups is that engineers again and again brought the word 'design' into their responses: "I would love to work in a design engineering role", "develop independent design skills, learn to work creatively", "design the systems for buildings", "design to help people in any way I can", and "contribute to the skyline of a major city in the world, be involved in projects which reduce the carbon emission and energy use of the world using evidence based design, start on the road to becoming chartered, being referred to as Dr. would be nice and a healthy bank account would be an advantage."

While the engineering technologists did not exclude working in design or as part of a design team from their responses (e.g., "I want to get a graduate position in a design office"), the responses tended to be less career focused and more oriented to applying their skills (e.g., "I want to do something related to my skills", "utilize my skills and knowledge I acquired from my course"). Responses generally were open-ended but based on the knowledge and skills that they had acquired through their studies: "I am open to any type of work related to my program".

Table 3.6 What do you really want to do in your first job after graduating?

Sample responses of engineering technologists	Sample responses of engineers
"Be an engineer"	"I would like to work in a design office applying what I have done in my final year project"
"Be able to run equipment such as machinery and be able to solve their problems"	"Earn money and gain work experience"
"Plant maintainer with computer aided skills"	"Get the most experience I can in technologies that interest me"
"Get a job in a programming environment to control systems"	

Lastly, we asked the final year students themselves what differences they perceive between people who leave college as "Engineers" and those who leave as "Engineering Technologists". The key finding was that significant numbers either gave no answer or said they did not know (37 % of engineers and 52 % of technologists). A further group said there was no difference (11 % and 10 %). In total, two-thirds of technologists gave no answer, did not know or said there was no difference.

This seems to clearly align with statements by the faculty that technology students "may not be able to characterize the difference". We find further congruence with faculty views when we explore what were seen as the differences (see Table 3.7). The key differentiation was seen to focus on the issue of design. Engineers were more likely to be associated with design while technologists were seen to be more practical and involved with the implementation of designs. Engineers were also seen as better educated and having higher status.

In a memorable comment on the difficulty of completing an engineering degree, one final year engineering student said of engineering technologists: "the latter leave college with around €5,000 less p/a and about 5 % more hair!!" In line with the earlier note that engineering technologists were more job-focused than career-focused, this group noted that the job, pay, and promotion prospects were better for engineers. Both groups, when comment was made, noted the higher standing or esteem that engineers would have. Finally, and again supporting the statistical results, both groups overwhelmingly used design activity as a key differentiator between the two.

Table 3.7 What differences do you perceive, if any, between people who leave college as "Engineers" and those who leave as "Engineering Technologists"?

	Engineering technology students said	Engineering students said
None/don't Know	"Didn't know there was a difference"	"None really. Generally most people don't have a clue about the differences between them. The only people who point it out or are bothered by it are the 'engineers' and the 'technologists' themselves"
	"There are no apparent differences, they both have the same fundamental background"	
Recognition	"Engineering technologist doesn't sound as good"	"Engineers are probably more highly thought of";
	"Engineers may pursue a management role"	"Engineers have more opportunity ..... Engineering Technologist may not be able to advance beyond a certain level within their career without further study"
	"Engineers are more employable and better educated"	"You get more respect from lectures, laboratory staff and future employers"
	"Engineers will get a job before engineer technologists"	"Less opportunities for technologists"
	"Engineers are more respected"	"Engineers have more responsibility"
	"Engineers have more scope for promotion and higher salaries"	

Table 3.7 (continued)

	Engineering technology students said	Engineering students said
Educational level	"People who leave with a BE as opposed to a BEngTech have a more administrative approach to engineering"	"Engineers as a whole have learned to learn";
	"Different lines of work, different levels of degrees"	"Engineers would be of a higher educational standard";
	"Engineers have higher qualifications"	"Different level degree, almost same knowledge";
	"Engineering Technologists are more confident with practicality than the theory"	"Engineers know how and why things happen while Engineering Technologists are mostly shown how things work"
	"To me it's the fields chosen by the individual so there is no difference, both have taken the course for their intended career choice"	"Engineers have achieved a broader education in the field whereas technologists have received education in a more specific area of focus in Engineering"
Function	"Engineers would be more inclined to design and numerical analysis, where as engineering technologists would have a stronger sense of operation and maintenance ...."	"In my opinion, Engineers will leave focusing their careers on the design and evaluation of new technologies as [opposed] to engineering technologists who, in my opinion, will focus more primarily as technicians, maintaining systems, carrying out tests, evaluations, etc. that the engineers have assigned them"
	"Engineers can design things and analyze errors when building things. Engineering technologists focus more on theories rather than technicalities"	"Engineering technologists will have a more hands on job while engineers will be more design or management role"
	"Engineers create design and develop new technologies. Engineering technologists integrate existing technologies and systems"	"Engineers focus on using their knowledge to design, improve and innovate technology. Engineering Technologists use their expertise to operate and efficiently maintain technology"
	"Engineers have more responsibilities and are more involved in design whereas engineering technologists operate and carry out tasks"	

Overall, what emerges from the survey is that both groups see themselves as having different roles and functions. Engineers are more likely to be seen as designers with a careers focus. Technologists have a narrower job focus, were seen as more practical and better prepared to tackle real world problems. While these differences can be identified in the responses to the full range of survey questions it is also the case that two-thirds of technology students were unable (or did not want) to distinguish themselves from engineers. This suggests a weak social identity as engineering technologists and an inability to distinguish themselves from engineers.

Strand 3: Prior DIT Surveys of First Year Students

In considering the responses above (from faculty in Strand 1 and from final year students in Strand 2), the quantitative and qualitative differences between the two groups of students center on: (1) students' views on design as an identifying activity, (2) how they wish to use their education once they graduate, (3) their initial views on their future careers, and (4) their own development as engineers and people. To provide further insight, we contrast these responses from *final year* students with responses from an earlier DIT study of *first year* students that sought to examine why students chose to study engineering and why they chose to come to DIT for their studies.

Between 2003 and 2007, DIT conducted surveys of incoming engineering and engineering technology students in an effort to understand attraction and retention issues in engineering education. This work involved surveying students about *why* and *how* they chose an engineering-related field of study and why they selected DIT. The overall response rate was around 65 % each year the study was conducted. These data have been reported previously (Eddie Conlon 2006) but not analyzed statistically. We reviewed the findings of these prior studies. Then, we extended them by using statistics to compare the 2007 responses provided by engineering majors with those provided by technology majors. In 2007, a total of 525 students entered DIT's various engineering programs. Of these, 307 submitted responses from our target population of programs. We compared responses from the 114 engineering students with those provided by 193 engineering technology students. We wanted to better understand what motivated them to become engineers in the first place and see if there were different factors at play with the two groups.

The cohort of first year engineering and engineering technology students who commenced studying in DIT was asked to select, from a list of possible reasons, the two most important reasons they saw for choosing to study engineering. The survey results for 2007 show the percentages selected by incoming students:

- 41 % chose "I was always interested in how things work" (46 % of engineers and 39 % of technologists)
- 36 % chose "I am interested in designing things" (28 % of engineers and 41 % of technologists)
- 28 % chose "Engineering is a good career" (28 % of engineers and 27 % of technologists)
- 24 % chose "I want to build things" (21 % of engineers and 26 % of technologists)

This prioritization of response was consistent across the 5 years for which the survey was conducted. When we analyzed the 2007 response data for the two groups, we found that while 25 % of engineering majors listed "I like maths and physics" as their first or second choice, just 14 % of technologists did likewise. Engineering majors were significantly more likely to have an engineer somewhere in the family (99 % say they do as opposed to 94 % of technology students). The technology students who did have an engineer in the family were likely to have just one (60 % as opposed to 51 % of engineering majors). Significantly more of the engineers were influenced positively by an engineer (71 % as opposed to 49 % among technologists) or a mathematics teacher (58 % as opposed to 37 % of technologists).

There were also significant differences in why the two groups chose to study at this institution. "DIT courses are more practical and applied" was important to 80 % of engineers (i.e., among the student's top five choices) but just 62 % of technologists. A significantly higher number of technology students selected "I like working with computers" as one of their most important reasons for selecting the career (40 % of technologists listed it, as opposed to 25 % of engineers).

These survey responses raise two issues. Firstly, the engineers interviewed in 2007 were less likely than technologists to say they were interested in designing when they started their engineering studies, but those moving towards graduation in 2013 were more likely to see designing as a key distinguishing feature of their identity. Secondly, the engineers were more likely to say they came to DIT because the programs are more practical and applied (that at other institutions). This needs to be understood in the context of the students having a choice to study at DIT or at a university that would have a more 'theoretical' orientation.

What can be noted is that the experience of studying at DIT seems to enhance the identity of engineers as designers but leaves them less prepared than technologists to solve real world engineering problems. It might be the case that their education as engineers in DIT is less practical than initially thought. This has clear implications for DIT in attracting and retaining students as it suggests a mismatch between the expectations of students and their actual experience in DIT.

Strand 4: Case Study – Observations on How Engineering Students and Engineering Technology Students Approach a Design Problem

DIT has a design course titled "Engineering Practice and Design" (popularly known as RoboSumo), in which teams of students design and build a robot which then competes one-on-one against other student robots in a competition to locate and push the other robot from a round table. Teams are comprised of either second year electrical engineering technology students or first year electrical engineering students. Because the faculty interviews and student survey responses identified *design* as a differentiating factor between the two groups of students, we asked a colleague to describe his experience with both groups as they engage in the same design course. In the following case study, Dr. Ted Burke describes his observations of various approaches student teams take with regard to the RoboSumo design task.

Case Study: How Students Approach a Design Problem – RoboSumo

There are different design approaches that dominate within each cohort. In particular, the archetypal 'good' engineering technology team approach to design is to get 'stuck in' straight away and start building a robot. This sometimes appears rash – as though the team has completely bypassed the important step of critically analyzing a proposed design before committing to it. Based on my observations however, this criticism is often not applicable. In fact, this early building behavior should be regarded more as a 'mocking up' exercise than as an attempt to produce the final design in a single hare-brained step. By building these flawed designs, teams learn a huge amount that will inform their final design. A mock-up helps teams to build a shared understanding of design features and also to get a clearer sense of each other's strengths and weaknesses ("My team mate says he can cut and bend sheet metal, but can he really?". "My team mate swears she can get the program written by tomorrow – I suppose I'll wait and see if she can deliver.").

The fact that these teams are often perfectly happy to build the robot 'wrong' a couple of times before building it right reflects two significant factors: (1) the students' belief (which I share) that this approach (let's call it 'Build Early and Build Often' – BEBO) is a very effective way of learning; (2) The students' level of confidence in building physical things. Many of these engineering technology students have a lot of practical fabrication experience. Perhaps in the past they have found it rewarding to make physical things. As a result, many of these students probably assign a lower 'effort cost' to this approach than another student with less prior manufacturing experience would.

I would describe the archetypal weak engineering team approach as follows:

- Spend a lot of time thinking about the problem. This step typically involves a considerable amount of meditation, hand-wringing, soul-searching, and very occasionally critical analysis of proposed design features.
- Devise an ingenious, over-complicated solution, often with very fundamental design flaws (e.g., wheels attached directly to DC motors without any gearing).
- Underestimate the difficulty of building the proposed solution. By and large, our engineering RoboSumo teams of the last few years have seemed more confident with computers and less confident making physical things.
- Leave it until far too late to pull the whole thing together.
- Panic (optional).

The archetypal strong engineering team approach is actually something like the above, but with two critical differences: (a) for whatever reason, the thinking stage is much more fruitful. Terrible ideas are successfully weeded out without anybody needing to build anything. Good ideas are refined to make them more practical. Future problems are anticipated and possible solutions formulated. (b) A working prototype gets built much earlier, allowing wrinkles to be ironed out and the design (mechanical, electrical, software) to be tweaked as required. What's different here to the BEBO model is that more thinking happens before the first build, and there probably won't ever be a second build – just testing and refinement of the first prototype. Let's call this 'good engineering approach' Build Once After Thinking (BOAT).

BEBO versus BOAT
All in all, I see both approaches as very effective when done right. I suspect that good engineers will produce a good robot either way. Part of what draws some good engineer towards the BEBO approach is confidence in (or enjoyment of) building things, which I suppose is influenced to a large degree by prior experience. An engineer who is already in his or her comfort zone building things will assign a lower effort cost to mocking up design ideas to get a better feel for them. For such a person, BEBO is a reasonably painless strategy for shaping design ideas. By contrast, someone with less manufacturing experience may assign a higher effort cost to the same process since they have fewer existing skills to fall back on. Someone in this situation might be more naturally drawn to BOAT. For strong RoboSumo teams, I don't really mind which of the two approaches they use. However, for weaker teams, I'm inclined to nudge them towards BEBO, since they'll at least get a reality check early in the process about the complexity of the task (when their first prototype stinks).

Ted Burke

Discussion of Results

Sense of Purpose Arthur Chickering and Linda Reisser (1993) developed a framework for understanding the broad issues associated with identity development. These scholars theorized that college students develop their personal identities along seven primary "vectors", with development in each vector taking a unique direction and rate of speed. We had the students rank their own effort with regard to each of the following vectors:

1. Developing competence in my profession
2. Managing my feelings and emotions
3. Balancing my independence with my dependence on others
4. Developing mature relationships with others
5. Establishing my own personal identity
6. Developing a strong sense of purpose
7. Developing a sense of integrity in the way I behave

More often than technologists, the engineering students in our survey reported a focus on developing professional competence and balancing independence with dependence on others. On the other hand, technologists indicated greater focus on developing a "strong sense of purpose". These responses suggest that the engineers are more career-oriented and that technologists have focused on more general (less profession specific) aspects of their identities.

Aligning Identity with College Values First year surveys demonstrated that both sets of students had a very practical orientation as they entered DIT. This practical orientation is underlined by consistent responses across all programs as to why students chose to study at DIT. In all years the most popular response was that "DIT has a good reputation for engineering," followed by "DIT courses are more practical and applied". These findings support the outcomes of research by the IEEE (2003) in which student respondents indicated that their primary reason for doing engineering was that they "wanted to invent, build or design things".

However, the survey and the case study point to a divergence in how both groups of students see themselves being prepared for the "real world" they will shortly face. Engineering technology students were significantly more comfortable with the statement that "my program has taught me how to solve problems in the real world." The case study also highlighted that engineering students today may not be as confident making physical things (e.g., robots in the case study) and this also can generate a self-perception of not being prepared for the real world, especially if the student came to DIT expecting it to be practical and hands-on. One could also speculate that the difference in confidence in preparedness for the real world is related, in part, to the open-endedness of design: technologists see themselves as doing more deterministic work (i.e., applying concrete principles to specific situations in a prescribed fashion), whereas engineers see themselves confronted with problems which don't yet have a solution and they will be expected to find one by conjuring up a design (which might appear to them as a more daunting task).

There is a difficulty for DIT in aligning the expectations of students with the requirements for professional engineering. This difficulty is made more difficult in that the Institute has to attract and retain students of engineering and engineering technology. The latter may require a greater emphasis on the practical nature of DIT programs. But this may lead to the wrong message being conveyed to engineering students (as per Strange and Banning 2001) who may not be prepared for a program of study that may be a good deal more analytical and theoretical than they expect.

Intrinsic Motivation and Role Our surveys of incoming freshmen students consistently highlighted that students chose their program because they were "always interested in how things work", followed by "I am interested in designing things". It is evident that DIT students were primarily attracted to engineering by intrinsic features of engineering and their desire to understand and design. This motivation persisted through their studies and exhibited strongly in their responses to the final year student survey, in which they strongly identified (both qualitatively and quantitatively), *design* as a key competence of an engineer, a key differentiator between engineer and engineering technologist, and a key career activity for the engineer.

Although engineers consistently used *design* as a generic description of what they will do as professional engineers, their responses indicated they may not have a strongly developed understanding of the role of a design engineer. Nevertheless, *design* was used as a general descriptor of what the new graduate expected do upon entering the workforce. There was a clear disconnection between the students' identity as designers and their perception of their capacity to solve real world problems.

Absence of Identity While faculty members, engineering students and engineering technology students could all distinguish the role and function of engineers and technologists, there was weakly shared identity that was specific to students in engineering technology. They saw themselves as engineers but with different roles (see Land 2012). This finding is not surprising, given that up to half of these DIT students will eventually progress to an engineering program. In a sense, being an engineering technologist is not a goal for many of these students. But this absence of a strong identity can create difficulties in attracting and retaining students, because prospective students have little against which they can match their interests and aspirations. In the US context Land (2012) has made the point that "The lack of distinction (between engineers and technologists) has led to a number of persistent problems. Among them has been an inability of engineering technology programs to define themselves to potential students and their parents" (33).

Although faculty members can identity the role and function of technologists, they have not been able to convey a strong sense to students of the difference between them and engineering students. Indeed the faculty perceives difficulties the students have in understanding their role. This may raise an issue regarding the professional education of these students and the extent to which they are getting a broad education that will help them understand their specific role.

Conclusion

In this chapter we have explored the identities of engineering and engineering technology students in a large Institute of Technology in Ireland. A key finding is that faculty and students do differentiate between the two groups; the two are seen to have different roles and functions. The concept of engineers as designers emerged as a key characteristic distinguishing engineering from technology students. Both groups see engineers as career-oriented designers and both groups see technologists as more practical implementers. Technologists have a greater job orientation and a greater drive to apply knowledge in order to solve real world problems.

Despite these findings, the identity of 'technologists' is weak. According to faculty members, the engineering technology students see themselves as engineers. Our survey revealed that many of these technology students can't, or perhaps won't, distinguish themselves from engineers; they may not be *designers* but that does not mean they are not *engineers*. While design is a key issue, this does not seem to prevent technologists from seeing themselves as ‘engineers’. That choice seems related to how they understand and define the activity of ‘engineering’ itself. For them engineering is comprised of many different roles.

The commonly shared sense of identity is stronger among the engineering students. DIT students’ image of what an engineer *does* seems to be stronger than of what a technologist *does*. Overall, engineering students seem clearer about what they think the profession holds than technologists are. They probably developed a stronger professional understanding in college (after all, they have been here 1 year longer than the technologists and thus have had more time to construct a shared conception and/or adopt one handed to them by teachers and professional advisors). However, they also brought a stronger understanding with them when they arrived. These engineering majors had greater exposure to the profession than the technology students. They were more likely to have an engineer in the family and to have had positive experiences with an engineer in the past. As such, the engineers probably entered with a stronger sense of occupational identity than the technologists did.

As indicated earlier, we used Ireland’s accreditation standards as a guide in drafting some survey questions – to see if differences implied in the standards were clear to DIT students. These standards suggest technology is more applied and engineering is more theoretical and design-oriented. The students describe some key factors that professional bodies and their teachers see as distinguishing ‘technologists’ from ‘engineers’. Although they picked up on some differences, they did not distinguish more subtle delineations. Responses to “I can compare different technical solutions and make recommendations” and “I can use a range of engineering tools and methodologies” did not receive significantly different response rates, for instance. (Irish accrediting standards tag the first to technologist and the second to engineers.) In the net, however, we found evidence that occupational enculturation is part of the experience in DIT’s schools of engineering.

Our research suggests some challenges for DIT in addressing issues of professional identity in its engineering programs. Firstly, many engineering students come to DIT expecting a practical education. The perception that “DIT courses are more practical and applied” was significantly more important to engineers than to technologists. These engineering students have often chosen DIT over a university because of the appeal of its hands-on pedagogical approach. At the end of their educations, their identity as designers has been enhanced but they feel less prepared than technologists to solve real world engineering problems. This has implications for DIT in attracting and retaining students, because it suggests there could be a mismatch between the expectations of students who want a practical education and the more theoretical and analytical knowledge they ultimately feel they have received.

Society and school play important roles in shaping the professional identity of engineering students, but the same cannot be said for engineering technologists. It is not nearly as clear to students what technologists do and how technologists’ work differs from what engineers do.

Faculty members believe that students have a weak identity as technologists and do not distinguish themselves from engineers. Although faculty members articulate distinctions, a distinct professional identity, for technologists, has not been generated. This could be because the role is seen as somehow secondary to professional engineers. This could be unique to DIT, because the ladder system here allows students to easily move from technology into engineering. But it is somewhat worrisome that no clear identity is being offered to prospective technology students against which they could match their interests and aspirations.

The above presents a challenge for this multi-level institution as it seeks to grapple with the complexities of engineering identity and seeks to convey to prospective students the similarities and distinctions in the roles of engineers and technologists. The shared sense of role and professional identity of the *engineer* seem to be understood and communicated to students but the role of *technologist*, while understood, is not communicated as part of a wider professional identity.

Acknowledgments The authors, and in particular Mike Murphy, would like to thank Steen Hyldgaard Christensen for all of his support and encouragement to complete the work for this chapter.

References

Anderson, K. J., Boyett, C., Shaw, S., McGlamery, T., Nathans-Kelly, T. M., & Nicometo, C. (2010). Understanding engineering work and identity: A cross-case analysis of engineers within six firms. *Engineering Studies, 2*(3), 153–174.

Anderson, L. W., Krathwohl, D. R., & Bloom, B. S. (Eds.). (2001). *A taxonomy for learning, teaching, and assessing: A revision of Bloom's taxonomy of educational objectives*. Boston: Allyn & Bacon (Pearson Education Group).

Bloom, B. S., & Krathwohl, D. R. (1956). *Taxonomy of educational objectives: The classification of educational goals, by a committee of college and university examiners. Handbook I: Cognitive domain*. New York: Longmans, Green.

Beder, S. (1988). *The new engineer*. Sydney: Macmillan.

Chickering, A. W., & Reisser, L. (1993). *Education and identity* (2nd ed.). San Francisco: Jossey-Bass.

Conlon, E. (2006). Recruiting first year engineering students. *SEFI annual conference,* Uppsala, June 2006.

Downey, G., Lucena, J., & Carl, M. (2007). Engineering ethics and identity: Emerging initiatives in comparative perspective. *Science & Engineering Ethics, 13*(4), 463–487.

Hogg, M. A., & Terry, D. (2000). Social identity and self-categorisation processes in organisational contexts. *Academy of Management Review, 25*(1), 121–140.

Iacovos, N., & Conlon, E. (2013). The integration of sustainable development competencies in Irish engineering education: Findings of a curriculum content investigation of four engineering programmes. *Paper to Engineering Education for Sustainable Development Conference*. Cambridge, Sept 2013.

IEEE Spectrum and IEEE-USA. (2003). *Professional image study*. Dec 2003

Jamison, A. (2013). *The Making of Green Engineers*. San Rafael: Morgan & Claypool.

Land, R. E. (2012). Engineering technologists are engineers. *Journal of Engineering Technology, Spring 2012*, 32–39.

Leidner, R. (2006). Identity and work. In M. Korczynski, R. Hodson, & P. K. Edwards (Eds.), *Social theory at work*. Oxford: Oxford University Press.

Mann, L., Howard, P., Nouwens, F., & Martin, F. (eds.) (2009). Influences on the development of students' professional identity as an engineer. *Proceedings of the research in engineering education symposium 2009*, Palm Cove.

Meiksins, P., & Smith, C. (1996). *Engineering labour*. London/New York: Verso.

Meyers, K. L., Ohland, M. W., Pawley, A. L., & Christopherson, C. D. (2010). The importance of formative experiences for engineering student identity. *International Journal of Engineering Education, 26*(6), 1550–1560.

Mulcahy Report. (1967). *Steering committee on technical education*. Report to the Minister for Education. Dublin.

Pawley Alice, L. (2009). Universalised narratives: Patterns in how faculty members define "engineering". *Journal of Engineering Education, 98*(4), 309–319.

Pierrakos, O., Beam,T. K., Constantz, J., & Anderson, R. (2010, June 28). *On the development of a professional identity: Engineering persisters vs engineering switchers*. Draft manuscript submitted for the "engineering learning" workshop at ICLS 2010, Chicago.

Sheppard, S. D., Macatanguay, K., Colby, A., & Sullivan, W. M. (2009). *Educating engineers. Designing for the future of the field*. A report of the Carnegie Foundation for the advancement of teaching. San Francisco: Jossey Bass.

Smith, C. (1987). *Technical workers, class, labour and trade unionism*. London: Macmillan.

Stets, J. E., & Burke, P. J. (2000). Identity theory and social identity theory. *Social Psychology Quarterly, 63*(3), 224–237.

Strange, C. C., & Banning, J. H. (2001). *Educating by design: Creating campus learning environments that work*. San Francisco: Jossey-Bass.

Stevens, R., O'Connor, K., Garrison, L., Jocuns, A., & Amos, D. A. (2008). Becoming an engineer, toward a three dimensional view of engineering learning. *Journal of Engineering Education, 97*(3), 355–368.

Thompson, P., & McHugh, D. (2002). *Work organisations* (3rd ed.). Basingstoke: Palgrave.

Watson, T. (1997). *Sociology, work and industry* (3rd ed.). London: Routledge.

Wisnioski, M. H. (2012). *Engineers for change*. Cambridge, MA: MIT Press.

Mike Murphy Diploma from Dublin Institute of Technology, B.Sc. (Eng) in Electrical Engineering, Trinity College Dublin, M.Sc. and Ph.D. in Electrical Engineering, Stevens Institute of Technology. Fellow of Engineers Ireland. Member of the IEEE. He worked in Bell Labs and later at Bell Communications Research before returning to academia in 2002. Director of Dublin Institute of Technology and Dean of the College of Engineering & Built Environment. Member of SEFI Administrative Council, and Member of European Engineering Deans Council and current President. Area of special interest is engineering and technology education.

Shannon Chance B.Arch. & M.Arch from Virginia Tech, Blacksburg, Virginia, USA. Ph.D. in Higher Education focused on Policy, Planning and Leadership from The College of William and Mary, Williamsburg, Virginia, USA. 2012–2013 Fulbright Research Scholar to Ireland in Engineering and Design Education, hosted by Dublin Institute of Technology. Associate Professor of Architecture at Hampton University in Hampton, Virginia, USA. Adjunct Professor at The College of William and Mary and Dublin Institute of Technology. Research areas include: design process, change management, and cognitive and identity development. Teaching areas include: educational planning, architecture, urban theory and design, and environmental sustainability. Registered architect, LEED Accredited Professional, and member of the National Architectural Accrediting Board.

Eddie Conlon Assistant Head of Department of Engineering Science and General Studies – soon to be integrated in a new School of Multidisciplinary Technology – at Dublin Institute of Technology. He holds an M.A. in Sociology from University College Dublin. He is generally interested in the sociology of work but in recent years has published work on engineering ethics and the integration of sustainability in engineering education. He is the coordinator of a general entry programme for engineering technologists in the College of Engineering & Built Environment at DIT.

Chapter 4
Engineering as Profession: Some Methodological Problems in Its Study

Michael Davis

Gather round me, boys, and you will hear
The story of a brave engineer:
Casey Jones was that roller's name –
On a 68 wheeler, he won his fame.

– folk song

Abstract Engineering is a function, discipline, occupation, and profession to which the term "engineering" is only a rough guide. Some activities not called "engineering" – applied physics and naval architecture, for example – are plainly engineering in the sense relevant to this volume. Other activities called "engineering", such as driving a railway locomotive or overseeing the operation of a ship's boiler, are just as plainly not engineering in the relevant sense (despite the participation of "engineers" such as Casey Jones or a sailor rated "marine engineer"). (These examples are all from English, my own language but not one known for its logic. It is therefore worth noting that other languages seem to have similar difficulties – or, at least, so I have heard from their native speakers – Italians, Japanese, Greeks, and so on. I'll give one example here: The Dutch give the title "Ingenieur" to anyone who receives a bachelor's degree from a technological university, even if the degree is in political science or philosophy. Anyone with that title is free to use it, much as anyone in the United States with a Ph.D. is free to call herself "doctor". The Netherlands do not license or register engineers, yet everyone there seems to understand the difference between "engineers" who are just philosophers and "engineers" who are engineers strictly speaking.) The status of other activities is more controversial. Is "software engineering", "social engineering", "genetic engineering", "re-engineering", or "financial engineering" engineering in the relevant sense? What about architecture (strictly so called), computer science, industrial design, or synthetic chemistry? What separates those technological activities from engineering (in the sense relevant here)? The answer to such questions will (or, at least,

M. Davis (✉)
Humanities Department, Center for the Study of Ethics in the Professions,
Illinois Institute of Technology, Chicago, IL 60616, USA
e-mail: davism@iit.edu

S.H. Christensen et al. (eds.), *Engineering Identities, Epistemologies and Values*, Philosophy of Engineering and Technology 21,
DOI 10.1007/978-3-319-16172-3_4

should) determine what gets studied as "engineering" and therefore what conclusions we, those who study engineering, draw concerning our subject. What I propose to do here is summarize the answers I have elsewhere given these questions and then dispose of prominent objections to those answers. (See especially Davis M, Thinking like an engineer: essays in the ethics of a profession. Oxford University Press, New York, 1998; Davis M, Profession, code, and ethics. Ashgate, Aldershot, 2002; Davis M (Philosophia 37(2):211–225, 2009a); Davis M (The Monist 92(3):325–339, 2009b); Davis M, Distinguishing architects from engineers: a pilot study in differences between engineers and other technologists. In: van de Poel I, Goldberg D (eds) Philosophy and engineering: an emerging agenda. Springer, Dordrecht, 2010.) Most of the objections, it turns out, arise from disagreement about how to study engineering (method) rather than about ordinary facts concerning engineering (how certain people are trained, what they do, and so on).

Keywords Function • Discipline • Occupation • Profession • Ethics • Engineering

Function

Once, while unsuccessfully seeking a position at a certain large technological university, I briefly met with its president. To make conversation, I remarked how unusual it was for a university to have, as his did, both a School of Engineering and a School of Applied Science and Technology. How, I asked, did he decide which programs went to which school? His answer was: "The School of Applied Science and Technology consists of all those programs which look like engineering to me but not to the Dean of Engineering." I thought that answer showed considerable theoretical insight. I hope the insight will be clear by the end of this section. In any case, I propose to use the terms "technology" (and "technologist") in the same spirit – as a catchall that includes not only what I think is engineering (or an engineer) but also what others think is engineering but I do not.

By "technology", I simply mean any useful artifact embedded in a social network that designs, builds, distributes, maintains, uses, and disposes of such things. So, for example, while a hammer lost in space is only an artifact, a hammer at work in a factory is technology (part of a technological system). A technologist is anyone with a *significant* role in technology. A young child who lifts a hammer is not a technologist, but a carpenter doing the same is.

Like other technologists, engineers design, "build" (or, at least, manage the building), or otherwise contribute to the life (and death) of certain technologies. Indeed, designing, building, or so on is (some might say) "the function" of engineers, what engineers, and only engineers, exist to do. It is what defines engineering.

This way of defining engineering is, I think, a mistake because it has at least two undesirable consequences. First, equating designing, building, or the like with engineering makes distinguishing engineers from other technologists impossible – by a definition that clearly goes against usage. Architects, computer scientists, industrial chemists, and other technologists also design, build, and so on. Indeed, so may any inventor, however untrained, undisciplined, and isolated. Even beaver and corral are engineers in this sense. Any definition of engineering that counts such animals as engineers is, I think, plainly in deep trouble.

The second, and equally important, undesirable consequence of equating engineering with designing, building, or the like is that is gives a misleading picture of what engineers in fact do. Some engineers simply inspect; some write regulations; some evaluate patents; some attempt to reconstruct equipment failures; some sell complex equipment; some teach engineering; and so on. Whether all of these activities are properly engineering, indeed, whether any of them is, is a question that should not be settled by a mere definition. My point now is not only that inspecting, writing regulations, evaluating patents, and so on are what engineers frequently do (a mere statistical fact) but as well that such activities are what some engineers are supposed to do as engineers (a fact about how they are expected to function). So, for example, employers sometimes advertise for engineers rather than other technologists to do one or another of these things. I agree that design (or, rather, engineering design) is central to understanding engineering, but I do not see how designing can be the (defining) function of engineering or even one major element of an engineer's function – because, as I see it, there is no function that engineers, and only engineers, seem to have (except, of course, engineering itself, which is what we are trying to define).

If not defined by its function, what can define engineering? I have elsewhere given two answers to that question, one negative and one positive. (See, especially, Davis 1998.) The negative answer is that, if "define" means giving an abstract definition (for example, by genus and species), there are only practical definitions, useful for a particular purpose. There can be no philosophical definition, that is, one that captures the "essence", "nature", or "Platonic form" of engineering – because engineering no more has an essence, nature, or Platonic form than you or I do. Like you and me, engineering is a mere individual, a work of history rather than of logic or a priori reason. Therefore, all attempts at philosophical definition will either be too abstract to be informative, or circular (that is, define "engineering" using "engineering" or an equally troublesome term such as "technical"), or open to serious counter-examples (because they exclude from engineering activities clearly belonging or because they include activities clearly not belonging), or suffer a combination of these errors.

The positive answer helps to explain this negative one: Like other professions, engineering is self-defining (in something other than the classical sense of definition). There is a core, more or less fixed by history at any given time, which decides what is engineering and what is not. This historical core is not a concept but an organization of living practitioners who – by discipline, occupation, and profession – are undoubtedly engineers. They constitute the profession, that is, they admit or reject candidates for membership, using criteria such as similarity in education, method of work, and product. Often these criteria work like algorithms. So, for example, the ordinary physician or philosopher clearly is not an engineer (that is, competent to engineer), while the typical graduate of an ABET-accredited engineering program with a few years experience successfully working as an engineer just as clearly is.[1] But perhaps as often these criteria cannot be applied without exercise of judgment. Does someone with a degree in chemistry who, say, has successfully managed a large refinery for 5 years, count as an engineer because what she has been doing is, in effect, "chemical engineering"? (Davis 1998, Chap. 3; Davis 2010)

We can now understand what was going on at that unnamed technological university. The president, not himself an engineer, was applying a functional definition of engineering, one that could not distinguish between engineering and closely related technologies. The Dean of Engineering then applied engineering standards (especially, ABET's list of engineering disciplines), which did precisely that. These standards recognized naval architecture and applied physics as engineering but excluded "engineering technology" and "packaging science", as well as software engineering, social engineering, genetic engineering, re-engineering, financial engineering, architecture, computer science, industrial design, synthetic chemistry, and so on.

But, it will be objected, surely the theoretical question of what is and what is not engineering cannot be settled in such a practical way. There are good reasons for, say, excluding architecture and synthetic chemistry from engineering while including naval architecture or applied physics.

I agree. But those reasons are themselves a consequence of history, that is, a consequence of decisions that, over several centuries, made the discipline of engineering what it is today. The discipline might have been different, indeed, so different that it would not count as engineering at all.

Discipline

By "discipline", I mean any set of standardized ways of carrying on a specific activity, developed over time and taught in some structured way. Breathing is not a discipline but the breathing required for meditation is. Building is not a discipline but building according to the standards of the Guild of Masons was. Inventing is not a discipline, but engineering is.

The history of engineering is in large part the history of its discipline. The way I tell that story, the discipline began to take shape after the French created the *corps du génie* in 1676. Had the French given a different name to that organization (say, *corps de l'artifice* or *corps du mécanisme*), we might well have a different word for engineering (say, "artifice" or "machining"). Before 1676, the term "engineering" (or its equivalent) referred to a function (primarily, the management of sieges, whether defense or assault, and whatever skills were necessary for that function).

[1]ABET, Inc. (formerly the Accreditation Board of Engineering and Technology) is the non-governmental organization that accredits engineering programs in the United States. It also accredits some other technology programs, including computer science and applied science.

An engineer was simply someone who managed sieges (catapults, artillery, trenching, sapping, and so on). Within a few decades after 1676, the term engineering (or, rather, *le génie*) referred to the French way of doing such things. By then, engineering was a discipline.

To say that engineering (in the sense relevant here) did not exist before 1676 is not to say there were no technological achievements before then that might now count as engineering. There were, of course, for example: the invention of the ax, sling, and spear; the building of the Passage Tomb at Newgrange (3,200 BC), the Egyptian pyramids (2,575–2,150 BC), and the Beijing-Hangzhou Grand Canal (581–618 AD). To say that engineering did not exist before 1676 is to say instead both that no one called "engineer" did any of these things and that those who did do them did not work as engineers typically do but according to another discipline – or no discipline at all. The history of engineering is only a small part of the history of technology.

During the 1700s, the French slowly developed a curriculum – a sequence of formal courses – to teach the new discipline to those who were to be engineers in (roughly) our sense, *officieurs du génie* (not enlisted men with shovel or saw, an older sense of "engineer"). There was much curricular experimentation, some of which – from our perspective – may seem ridiculous, such as (for a time) including riding, dancing, and fencing in the curriculum. But, by the late 1700s, the curriculum was recognizably what it is today: calculus, physics, chemistry, mechanical drawing, statics, dynamics, and so on. There was a common core lasting 3 years; then, in the last year, the engineers specialized, choosing artillery, military engineering (fortification and sieges), mining, bridges and roads, cartography, or shipbuilding. Though the engineering curriculum has changed much since then (for example, adding electricity and computing), today's engineering curriculum resembles that of 1800 more than it resembles any other discipline's curriculum then or now.

Generally, it is this curriculum, or rather the distinctive discipline resulting from it, that distinguishes engineers at any time from the non-engineers around them, whether they have "engineer" in their job title or in the name of their discipline. So, for example, Benjamin Wright (2013), the Chief Engineer of the Erie Canal (2013) (1817–1828), was a self-taught surveyor with only a primary school education. Though he certainly functioned as an engineer (so much so that civil engineers like to claim him as the "father of American civil engineering"), he is in fact proof that one could still be a great builder without being an engineer. The nineteenth century had many other such builders, such as the gardener, Joseph Paxton who, though without any formal education, designed and oversaw construction of the Crystal Palace to house Britain's Great Exposition (1851). Indeed, it was only late in the nineteenth century that engineers (strictly so called) came to dominate large building projects. Today, the Crystal Palace could not be built without engineers involved at every stage after the initial sketch.

Have I not (it might be objected) put too much emphasis on the curriculum as the means of distinguishing the engineering discipline from other technological disciplines? The first year or two of the engineering curriculum today differs little from the corresponding curriculum in math, physics, or chemistry. In the last 2 years, it does differ from these, but the curricula of the major fields of engineering differ considerably too. Does that not make it hard to see engineering as a single discipline – without falling back on generalities that make it hard to distinguish engineering from the physical sciences? Indeed, the problem of line-drawing may be getting worse. One field of engineering, electrical and computer (ECE), seems to be abandoning courses that have helped to define the engineering curriculum since the eighteenth century, especially statics, dynamics, and thermodynamics. Surely, such changes do not individually, or even collectively, mean that ECE has ceased to be an engineering discipline. But, if they do not, then what does make engineering a single discipline – if it is.

This objection points to two unusual features of the way I have understood engineering. The first is that I have described the engineering profession as making decisions concerning the "similarity" (or difference) between the candidates for admission to the profession and those disciplines already in. Similarity is always a matter of degree. Matters of degree are often matters of judgment. Matters of judgment are subject to reasoned disagreement even among those competent to decide. There is no simple "fact of the matter". I therefore have no reason to be concerned if some people, even some whose judgment on matters of engineering I respect, have doubts about the engineering status of some discipline when I do not. What I have said about engineering stands as long as there is a historical core about which there is no dispute. That core can then make decisions about the others – decisions anyone, even a philosopher, can approve or criticize (just as judges make legal decisions which anyone may approve or criticize). But, just as with judges, so with the engineering professions: their judgments concerning membership matter in a way mine do not.

The other unusual feature of the way I have understood engineering is that inclusion of a discipline in engineering is (in part) a matter of history. That means, among other things, that what has already been included matters to what will be included. Consider ECE again. It might be that, if ECE were invented today, it would – like software engineering – not be recognized as engineering (strictly speaking). On the other hand, because of past decisions, ECE, already an engineering discipline, is likely to remain so in part at least because it is itself part of the comparison group. The departure of what was formerly an engineering discipline is unlikely to occur until the difference between that discipline and the rest of engineering has become so great that working as a single discipline seems too inconvenient. The inconvenience will consist in part of differences in curriculum (what engineers are supposed to know) but in part too in what happens after members of the discipline enter practice. Right now, the various disciplines of engineering do not seem to have trouble working together as engineers. Indeed, engineers not infrequently migrate from one field of engineering to another during their career. If, as a result of changes in curriculum, a certain engineering discipline can no longer work with other engineering disciplines without standing out as alien, then either the changes in curriculum will be abandoned or the former engineering discipline will eventually be accounted something other than engineering (strictly so called) – as happened, for example, with "scientific management" (which began within mechanical engineering and ended up as operations management and research, a business discipline).

So, I agree that recent changes in the ECE curriculum do not "mean" that ECE is no longer an engineering discipline. Of course, "mean" suggests that there is a sharp line between engineering and everything else, one making judgment unnecessary. Either recent reforms in the ECE curriculum have obviously passed that line or they have not. As I have defined engineering, though, not only is there no sharp line between engineering and everything else (except what engineers happen to draw) but also that there is a process of deciding a discipline's status as engineering that may take years to reach a conclusion. The objection thus seems to miss the point of my historical definition. Ultimately, it is history that decides – with abstract reasons (similarities and differences) constituting only some of the relevant considerations.

Occupation and Profession

Though the curriculum of engineering is recognizable by the late 1700s, engineering did not become an occupation until many decades later. This will seem a strange claim to anyone who does not appreciate how much is built into the term "occupation". By "occupation", I mean any fulltime activity defined (in part at least) by a certain body of knowledge, skill, and judgment (a discipline) by which one can (and a significant number of people do) earn a living. Not all disciplines are occupations. So, for example, fencing, though certainly a discipline, is (in the US at least) not a way to earn a living (though teaching fencing may be).

Engineering could not become an occupation until it ceased to be an exclusively military activity and became something more or less independent. Until then, engineers were a certain kind of military officer. They did not have a "calling" of their own. Engineering (strictly speaking) did not separate from military engineering much before the 1830s when railroads became the first important civilian employer of engineers. It was about then that the earlier distinctions between kinds of military engineering, including "civil engineering" (roads and bridges), became a distinction between military engineers (of all kinds) and civil engineering (in the modern sense), that is, the building of great works for civilian purposes, and mechanical engineering (that is, the building of boilers, pumps, and other machines).

But even after civilian engineering separated from military engineering, engineering still could not be an occupation. Engineers were still gentlemen. And, until well after 1830, a gentleman could not earn a living. To earn one's living meant "going into trade" or becoming "a hired man" (or, worse, a servant). For a gentleman to go into trade or become a hired man was to cease to be a gentleman. Gentlemen were supposed to have enough inherited wealth to live decently (or, at least, were supposed to act as if they did). Any money a gentleman received for what he did when following a "calling" was not earned (the way wages, pay, or salary is earned) but given as an honor (much like the modern "tip" but without its demeaning suggestion of subordination). What to us would clearly be payment for services rendered was a "pecuniary acknowledgement" (as physicians called it). Even today, professionals tend to refer to the price of their services as "my fee" – a word recalling "the knight's fee", that is, the land given to a knight so that he could afford weapons, armor, horse, and the time to fight for his lord. Gentlemen did not work to live but, if they worked at all, lived for their work, whether reimbursed or not. Engineering could not become an occupation until that conception of "gentleman" lost its force (or until engineers became tradesmen or manual laborers).

The term "gentleman" did not die – as, for example, its opposites, "villain" and "churl", did (more or less). Instead, gentlemanliness was reconceived as one or more of its former implications, especially, good manners, good character, and good education (college or its equivalent). In the rough markets of the late nineteenth and early twentieth century, being a gentleman in this sense (polite, decent, and well-educated) was not necessarily an advantage. Eventually certain occupations, those that tended to attract gentlemen, began to organize to help gentlemen earn their living as gentlemen (in something like the new sense of "gentlemen"). Each of these occupations was, or at least was intended to be, "a profession, not a mere trade or money-making calling".

The term "profession" has several senses today. In one, it is just a synonym for "occupation". A professional in this sense is the opposite of an amateur. In another sense, a profession is an honest occupation (one it is safe to profess, that is, to declare openly). In a third, a profession is a "learned art" (one requiring a knowledge of Latin and, hence, a university education). The opposite of a professional in this sense is a "mere artisan" or "mere mechanic". All three of these senses are quite old. During the late nineteenth and early twentieth century, "profession" came to have a new sense, one that provides an interpretation of the slogan, "a profession, not a mere trade or money-making calling". I have argued elsewhere that the following definition best catches this new sense: A profession is a number of individuals in the same occupation voluntarily organized to earn their living by openly serving a moral ideal in a morally-permissible way beyond what law, market, morality, and public opinion would otherwise require (Davis 2009a).

While each profession (in this sense) is a historical individual, profession as such is an ordinary concept (or conception), one developed by considering what the individuals apparently collected under the term are. Formal statements of the concept, that is, attempted definitions of it, might change over time both because the concept itself is changing or because our understanding of the concept has changed (or for both reasons). So, for example, the definition of "water" is different now from what it was, say, 300 years ago. That is in part because the concept no longer includes all clear, colorless, odorless, and tasteless liquids, but also in part because we have learned that water is H_2O (that is, that most of what was once called water consists of this chemical compound while some liquids once counted as water, such as aqua vitae, do not). Those who seek the meaning of profession in the origin of the term misunderstand how language works. Though the origin of the term can be suggestive, it can never be more than that. The concept a term names stands at the other end of its history.

Professions have been mocked as "gentlemen's clubs". Those so mocking them generally do not explain what is wrong with gentlemen's clubs. They should. After all, there is much to be said for a gentlemen's club if the alternative is, say, a criminal gang or illiterate clique. My guess is that what is supposed wrong is the criteria of membership. If, as with an ordinary gentlemen's club, membership in a profession were determined by sex, race, family, religion, and the like, then there would be something objectionable about professions. A gentlemen's club in which the membership is determined merely by sex, race, family, or the like marks of companionability would still be a gentlemen's club. Indeed, it might even be a good one. The purpose of a gentlemen's club is, after all, to please its members in a certain way (providing a home away from home, good company, and so on). A gentlemen's club makes no pretence of doing anything more exalted. Gentlemen's clubs differ in this respect from other similar voluntary associations, such as the Kiwanis or Lions Club, which have a higher purpose (charity). Professions also differ from a gentlemen's club in this respect. To be a profession, a voluntary association must – as the definition given above says – seek to serve a moral ideal (in a morally permissible way beyond what law, market, morality, and public opinion would otherwise require).

A moral ideal is a state of affairs every rational person (at his rational best) recognizes as a significant public good, that is, as something desirable enough that he wants everyone else to aid in achieving it, whether by positive support or merely by not interfering, even if their doing so would mean having to do the same. Among moral ideals are: justice, public health, knowledge, and beauty. The moral ideal of engineering is (roughly) improving the material condition of society. To serve that ideal as engineers, engineers must be competent in their discipline, honest in its practice, and so on. The sex, race, family, religion, class, or the like of an engineer is (more or less) irrelevant. Indeed, taking those factors into account in the selection of engineers is likely to exclude some candidates who would be good engineers or include some candidates who would not be (depending on which criteria are used and whether they are used positively or negatively). Hence, insofar as engineering seeks to serve its moral ideal, it should not select its members in the way a gentlemen's club properly selects its members. Selecting members by sex, race, family, and so on would tend to impede serving engineering's moral ideal.

Profession and Codes of Engineering Ethics

Like other professions, engineering seeks to serve its moral ideal by setting (morally permissible) standards that require more of engineers than law, market, morality, and public opinion otherwise would. These are the "higher standards" that are supposed to distinguish a profession from a mere trade or money-making calling. They are "higher" in the sense that they require (morally permissible) conduct that law, market, morality, and public opinion do not require (or at least, do not require until the profession has established the standards in question). These standards are "special" insofar as they apply to the profession in particular, not to all moral agents as such or even to all professions as such.

A profession's special standards are correctly identified as the profession's "ethics" and incorrectly identified with the profession's "code of ethics". I have argued elsewhere that professional ethics is best understood as those morally permissible standards of conduct that every member of a group (the profession in question) wants (at her rational best) every other member of that group to follow even if their doing so would mean having to do the same. See, especially, Davis (2002). Given this definition of professional ethics, it is, I think, obvious that the ethics of engineers includes a good deal more than what is called "the code of engineering ethics". Among the standards that are ethics in this sense are technical standards of safety, quality, and documentation. Or, to put the point another way, the entire discipline of engineering – apart from those few standards in dispute at any time – constitute the ethics of engineering. What engineers call "a code of ethics" is simply the most general statement of the discipline.

To say of some statement (or command) that it is an (actual) "standard of conduct" is to make two implicit claims. The first is that the statement generally guides conduct, that is, that its instructions are followed, that those it governs generally use it to evaluate their own conduct or that of others in the relevant group, and that members of the group generally use it to criticize publicly their own conduct or that of others in that group. If the standard does not generally guide conduct, it is an ideal (or model) standard, but not an actual standard – that is, not "really" a standard at all. An actual standard resembles a scientific law insofar as it allows us to predict (with reasonable success) what those it supposedly governs will do.

The other claim implicit in saying that some statement is a standard of conduct is that, though it generally guides conduct, the standard does not always. Statements that always "guide" conduct are not standards but scientific laws (strictly speaking). So, pointing to a few violations of a code of ethics does not refute the claim that it is an actual standard of conduct. A few violations may be explained away as, for example, the result of differences of opinion (rather than as indifference to ethics), as the result of factual mistakes, or simply as anomalies. To refute the claim that a code of ethics is a living practice requires showing that there are so many violations that the code tells us little, if anything, about what those whom the code supposedly governs will do.

I am therefore inclined to dismiss those critics of ethics codes who move from a few obvious violations of a code to the conclusion that the code in question is "mere window dressing". Certainly, codes are (or, at least, may be) "window dressing", that is, something put on display to potential customers in order to attract them into the store that lies behind the window. There is nothing wrong with window dressing as long as the store actually provides what it displays in the window. The problem is with *mere* window dressing, that is, with displays that mislead concerning the stock inside. On the evidence I have, codes of ethics in general, and codes of engineering ethics in particular, are not mere window dressing. I have myself interviewed several dozen engineers and found them to be serious about engineering ethics. I have also been assigning students in Engineering Ethics a paper requiring them to interview an engineer of their own choosing. Generally, they have found those they interviewed not only serious about engineering ethics but knowledgeable enough to give reasonably good answers to an engineering ethics case the interviewer posed to them. We definitely need empirical work on the question of how much engineers actually follow their ethics, including their technical standards, but absent such a study showing the opposite, I think the evidence points to the conclusion that engineering ethics is a living practice.

Indeed, it could hardly be otherwise – or, at least, otherwise for long. The public, including sophisticated businesses and governments, employ engineers for certain jobs when they could employ other technologists – and, in the past, did. Apparently, they do so because they suppose engineers to have certain ways of doing certain tasks different from their technological competitors. They suppose that because engineers have proved that they routinely do a better job than their technological competitors at those tasks (constructing large bridges, boilers, chemical plants, computer chips, and so on). Like a trademark, the term "engineer" is valuable only so long as individual engineers generally confirm the expectation that that term invites. Once engineering's special standards became mere window dressing, not much time would pass before only a fool would employ an engineer.

I have not claimed, please note, that most engineers have ever read their code of ethics, much less that they regularly consult it. The interviews that led me to the conclusion that engineers generally act as their codes of ethics require have taught me that most engineers cannot even recall seeing a code of engineering ethics. The engineering code seems to be "hardwired" into engineers. Of course, "hardwired" is a metaphor for a process we do not understand very well. Yet, we can be pretty sure that the process is not the self-selection by which students choose engineering. Those of us who teach engineering students in their first-year as well as in advanced courses can see that many of the attitudes we take for granted in fourth-year engineering students are not present in first-years. The hardwiring seems to occur during the 4 years of engineering school. Since few engineering courses (at least until recently) explicitly discussed engineering ethics, my best guess is that most engineers learn ethics through instruction in technical standards (which goes on almost everywhere in the engineering curriculum). The students learn engineering ethics much as native speakers learn their own language, that is, while doing something else.

Like many other professions, engineering seems confused about the moral status of its code of ethics (but not, I think, its technical standards). There are at least four reasons for that confusion. First, there is the question of how many codes there are. On the one hand, there seem to be dozens because so many engineering associations have their own code. The American Society of Mechanical Engineers has one; the American Institute of Chemical Engineers has another; ABET has another; and so on. Yet, these codes differ in language more than substance and even many differences that seem substantive at first disappear upon inquiry. (For example, engineers whose code of ethics does not yet include a provision on sustainable development seem to interpret the environmental or public welfare provision to include sustainable development.) I have therefore come to think of the many formal codes as much like the many dictionaries of (American) English. Though they differ, they are reporting the same underlying reality. One code simply omits what another includes because of a different purpose, style, or the like. One includes an interpretation that might be helpful in a certain context or fails to take account of recent change (because of editorial standards or date of publication). And so on. This variety in formal statement is consistent with (more or less total) agreement on the "unwritten code".

The second reason engineers have to be confused about the moral status of their code of ethics concerns the source of a code's moral authority. There are in fact at least two possible sources.[2] Some codes of ethics are supposed to be morally binding because those governed have taken an oath, made a promise or commitment, or otherwise given the code an "external sanction". (The IEEE's code of ethics is a good example of this sort: IEEE members "commit" themselves to it when they join the IEEE.) The other source of a code's moral authority is "internal" to the practice, much as the moral obligation to follow the rules of a morally permissible game arises from one's voluntary participation in the game. (A good sign that we have such a code before us is that it applies to "engineers" as such, rather than members of some formal association.) The idea is that, when a person voluntarily claims the benefits of a code of ethics – for example, the special trust others place in those whom the code binds – by claiming to be a member of the relevant group ("I am an engineer"), that person thereby takes on a moral obligation, an obligation of fairness, to do what the code says. Because a code of ethics applies only to voluntary participants in a special practice, not to everyone, a code, if it is generally followed, can create trust beyond what ordinary moral conduct can. It can create a special moral environment. So, for example, if engineers generally "issue public statements only in an objective and truthful manner [including] all relevant and pertinent information" (as the NSPE Code of Ethics, like most others, requires), public statements of engineers will generally (and justifiably) be trusted in a way those of politicians, lobbyists, and even ordinary private citizens would not be. Engineers will therefore have a moral obligation to do as required to preserve that trust. They will have a special moral obligation to provide all relevant and pertinent information even when others do not have such an obligation.

The third reason engineers have to be confused about the moral status of their code of ethics is controversy concerning whether – to be more than "mere window

[2] I ignore a third possibility here, that the code has moral authority because the code's content consists of rules derived (either by deduction or determination) from general moral rules (a kind of natural law approach rather than the two variations of social contract offered here). I ignore that possibility here because no modern code claims moral authority in this way. That was, however, not always so. The AMA code of 1847 presented itself as a work of "deontology" (Davis 2003). I also ignore other possibilities that will immediately come to a philosopher's mind, such as a grant from (or contract with) society, because they also do not seem to have anything do with present confusion among engineers about the moral status of their profession's code of ethics.

dressing" – the code must be enforced in the way laws are enforced, that is, by formal penalties (such as reprimand, fines, suspension, or expulsion). The legal (or "compliance") model of ethics often leads to calls for mandatory licensing of engineers, enactment of the code as "professional regulation", and an official body with the power to bar an engineer from practice for serious violation of the code of ethics. While there may well be good reason for legal enforcement of some aspects of the code of ethics, understanding ethics as primarily about law-like enforcement, that is, formal means of holding engineers accountable (such as expulsion from a professional association), simply confuses ethics with law. Law, custom, and other external guides to conduct do not claim to be standards everyone in the group (even at their rational best) wants everyone else to follow. Law, custom, and the like must, then, depend heavily on external enforcement. Ethics, on the other hand, need not. Insofar as individual engineers can see how everyone following the standards in question serves their interest, they have reason to do their share to maintain the trademark's value, that is, they have reason to act as engineers should. If they are dishonest, or simply indifferent to long-term consequences, they may (even at their rational best) find that reason unconvincing. They will therefore be incapable in principle of joining the profession (whatever their education and experience). In practice, they are likely to be driven from engineering by peer-pressure, employer avoidance, civil damages, or even criminal punishment. Most engineers, however, may be counted on to do their fair share (insofar as they understand it) because they are relatively rational and morally decent and understand that doing anything else would, all else equal, be morally wrong.

The fourth reason engineers have to be confused about the moral status of their profession's code of ethics is that different codes formally apply to different engineers. Some codes apply only to members of an association, some apply only to a class of engineers not defined by organization, and some apply to "engineers" generally. The IEEE's Code of Ethics is a good example of the first; the (Asian) Declaration on Engineering Ethics, of the second; and the code of ethics of the National Society of Professional Engineers (NSPE) (2007), of the third. The first sentence of the IEEE code says that IEEE members "do hereby commit ourselves to the highest ethical and professional conduct and agree" to the ten rules constituting the body of the code (IEEE 2013).[3] The suggestion is that, but for IEEE membership, the engineers in question would not have those obligations. The Declaration (adopted by the national engineering societies of China, Korea, and Japan in 2004) speaks instead of "Asian engineers". Interestingly, the only significant difference between the standards of the Declaration and the IEEE or NSPE code seems to be the last: "Asian engineers shall … Promote mutual understanding and solidarity among Asian engineers and contribute to the amicable relationships among Asian countries." (Asian Code 2004) The NSPE Code (2007), in contrast, speaks only of

[3]The IEEE is the organization formerly known as "The Institute for Electrical and Electronics Engineers".

"engineers". There is no distinction between ordinary engineers and (licensed) Professional Engineers, American engineers and others, or NSPE members and non-members. The suggestion is that the obligations arise from being an engineer, that is, from membership in the profession of engineering, not from membership in any technical, scientific, or professional association.[4] Only codes of ethics that apply to the profession as a whole are properly codes of professional ethics; the others are organizational codes (such as the IEEE's) or sub-professional codes (such as the Asian Declaration).

Conclusion

I have, I hope, now explained the importance of the distinction between function, discipline, occupation, and profession for the study of engineering ethics. While doing that, I tried to dispose of several objections commonly raised to this way of understanding engineering. Some of the objections seem to make the error of trying to refute a general claim with a few counter-examples, forgetting that general claims (which claim to be true "for the most part") cannot be refuted with a counter-example or two in the way that universal claims can be. The other objections seem to rely on empirical claims that, if true at all, remain to be proved. The error of these objections is putting the burden of proof on the wrong party.

References

Benjamin Wright (2013). *Wikipedia*. http://en.wikipedia.org/wiki/Benjamin_Wright. Accessed 14 Jan 2013.

Chinese Academy of Engineering. (2004). *Declaration on engineering ethics*. http://ethics.iit.edu/ecodes/node/5076. Accessed 14 Jan 2013.

Davis, M. (1998). *Thinking like an engineer: Essays in the ethics of a profession*. New York: Oxford University Press.

Davis, M. (2002). *Profession, code, and ethics*. Aldershot: Ashgate.

Davis, M. (2003). What can we learn by looking for the first code of professional ethics? *Theoretical Medicine and Bioethics, 24*, 433–454.

Davis, M. (2009a). Is engineering a profession everywhere? *Philosophia, 37*(2), 211–225.

Davis, M. (2009b). Defining engineering – From Chicago to Shantou. *The Monist, 92*(3), 325–339.

Davis, M. (2010). Distinguishing architects from engineers: A pilot study in differences between engineers and other technologists. In I. van de Poel & D. Goldberg (Eds.), *Philosophy and engineering: An emerging agenda*. Dordrecht: Springer.

Erie Canal Construction (2013). *Wikipedia*. http://en.wikipedia.org/wiki/Erie_Canal#Construction. Accessed 14 Jan 2013.

[4] Some professional societies, such as the American Medical Association (AMA), have actually gone back and forth between the first and third kind of code. For details, see Davis (2003).

Institute of Electronics and Electrical Engineers (IEEE). (2013). *IEEE code of ethics*. http://www.ieee.org/about/corporate/governance/p7-8.html. Accessed 14 Jan 2013.

National Society of Professional Engineers (NSPE). (2007). *Code of Ethics for Engineers*. http://www.nspe.org/Ethics/CodeofEthics/index.html. Accessed 14 Jan 2013.

Michael Davis Ph.D. in Philosophy, University of Michigan. Senior Fellow at the Center for the Study of Ethics in the Professions and Professor of Philosophy, Illinois Institute of Technology, Chicago. Before coming to IIT in 1986, he taught at Case-Western Reserve, Illinois State, and the University of Illinois at Chicago. Since 1991, he has held – among other grants – four from the National Science Foundation to integrate ethics into technical courses. Davis has published more than 200 articles (and chapters) and authored seven books, including: *Thinking Like an Engineer* (Oxford, 1998); *Ethics and the University* (Routledge, 1999); and *Profession, Code, and Ethics* (Ashgate, 2002). He has also edited or co-edited other books: *Ethics and the Legal Professions* (Prometheus, 1986); *AIDS: Crisis in Professional Ethics* (Temple, 1994); *Conflict of Interest in the Professions* (Oxford, 2001); *Engineering Ethics* (Ashgate, 2005); and *Ethics and the Legal Profession, 2nd ed.* (2009).

Chapter 5
Engineering Ethics and Engineering Identities: Crossing National Borders

Gary Lee Downey, Juan Lucena, and Carl Mitcham

Abstract This article describes and accounts for variable interests in engineering ethics in the United States, France, Germany, and Japan by locating recent initiatives in relation to the evolving identities of engineers. A key issue in ethics education for engineers concerns relationships between the identities of engineers and the contents and responsibilities of engineering work. These relationships have varied significantly over time and from country to country around the world. One methodological strategy for sorting out similarities and differences in engineers' identities is to examine who counts as an engineer, or what makes an engineer. The significant interest in engineering ethics in the United States has been linked to difficulties in adding professional identities to corporate employment. While engineering ethics has attracted little interest in France and formal education in the subject might very well be seen as insulting, German engineering societies have, since the conclusion of World War II, demanded from engineers a strong commitment to social responsibility through technology evaluation and assessment. In Japan, recent flourishing of interest in engineering ethics appears to be linked to concerns that corporations no longer function properly as Japanese "households." In each case, deliberations over engineering ethics emerge as part of the process through which engineers work to keep their fields in alignment with their changing images of societal advancement.

Keywords Engineering education • Ethics • History • United States • France • Germany • Japan

---

G.L. Downey (✉)
STS Department 0247, Virginia Tech, Blacksburg, VA 24061, USA
e-mail: downeyg@vt.edu

J. Lucena
LAIS Division, Colorado School of Mines, Stratton Hall, Golden, CO 80401, USA
e-mail: jlucena@mines.edu

C. Mitcham
Colorado School of Mines, Golden, Colorado, USA

Renmin University, Beijing, China
e-mail: cmitcham@mines.edu

S.H. Christensen et al. (eds.), *Engineering Identities, Epistemologies and Values*, Philosophy of Engineering and Technology 21,
DOI 10.1007/978-3-319-16172-3_5

Introduction

Professional ethics has become a well-established dimension of engineering education and practice in a number of countries – and may even be described as undergoing a process of globalization. Since the mid-1980s the U.S. Accreditation Board for Engineering and Technology (ABET) has increasingly required engineering programs to include the teaching of professional ethics – a requirement given new specificity in 2000, with the stipulation that 1 of 11 demonstrable outcomes should be "an understanding of professional and ethical responsibility." That same year the Japanese Accreditation Board for Engineering Education (JAABE) likewise began to require accredited programs to have an "understanding of … engineers' social responsibilities (engineering ethics)." Other expressions of commitment to engineering ethics can be found in the *Charte d'Éthique de l'Ingénieur* [Charter of Ethics of the Engineer] published by the Conseil National des Ingénieurs et Scientifiques de France (CNISF or National Council of Engineers and Scientists of France) in 2001, and the *Ethische Grundsätze des Ingenieurberufs* [Fundamentals of Engineering Ethics] issued by the Verein Deutscher Ingenieure (VDI or Association of German Engineers) in 2002.

Despite the obvious similarities of these initiatives, however, such emergent interests in engineering ethics are the products of distinct historical trajectories and have significantly different implications locally. For example, in issuing its new criteria, ABET, which traces its history back to the 1930s, was concluding a decade-long modification of its accreditation system. By contrast, JABEE, which was only created in 1999, was introducing the concept and practice of accreditation for the first time. In the French case, the *Charter for Engineering Ethics* was a new document issued by an engineering alumni organization, with no direct implications for any educational curriculum. The German *Fundamentals of Engineering Ethics* likewise has only indirect educational import, yet derives from a history that goes back to a post-World War II revival of professional engineering.

Such initial contrasts suggest that any real understanding of the global dimensions of engineering ethics requires further considerations of ethics in relation to differing trajectories of engineers' identities, including self-understanding (Downey and Lucena 2004). Ethical responsibility is a dimension of position and identity. Although similarities exist among lawyers, physicians, and engineers across national or cultural boundaries, there are often insufficiently recognized differences. For researchers and teachers interested in engineering ethics and students learning about professional practice in engineering, efforts to appreciate such differences can enhance their own self-understandings as well. In addition, the extent to which engineering ethics in different countries are now influencing one another, for example by the dissemination of ABET-like criteria around the world, depends in part on how national differences originally developed and how well such developments are appreciated.

United States: Engineers and Private Industry

As historian Edwin Layton (1971) has shown, engineers in the United States have long struggled with the fact that, unlike lawyers or physicians, they have been professionally divided into civil engineers, mechanical engineers, electrical engineers, and a host of other discipline and class specific groups. For instance, the American Society of Civil Engineers (ASCE, founded in 1852), adopted high membership standards and became an elite organization sometimes at odds with business interests. By contrast, the American Institute of Mining Engineers (AIME, founded in 1871) was more egalitarian and regularly allied itself with business interests in the mining industry. In response to ASCE professionalism and AIME commercialism, there appeared the American Society of Mechanical Engineers (ASME, 1880) and American Institute of Electrical Engineers (AIEE, 1884), attempting different blends of autonomous professionalism and economic pragmatism. The subsequent formations of the Society of Automotive Engineers (SAE, 1904), American Institute of Chemical Engineers (AIChE, 1908), and Institute of Radio Engineers (IRE, 1912) only intensified the fragmentation typical of professional engineering in the United States.

In response to these centripetal movements, there emerged a series of centrifugal efforts to unite the professional engineering community. In 1880 it was the creation of the Association of Engineering Societies for a national engineering congress. In 1911 it was formation of the Joint Conference Committee. Similar efforts can be found threaded through the Committee on Engineering Cooperation (1915), the American Association of Engineers (1915), the Engineering Council (1917), the Federated American Engineering Societies (1920), and the Engineers Council for Professional Development (ECPD, 1932) – the last of which has had a life longer than any other, eventually being transformed in 1980 into ABET. The very multiplicity of these efforts nevertheless indicates their weakness and attests to the fact that engineers in the U.S. are mostly not self-employed professionals, in contrast with physicians and lawyers, but employees of larger firms that benefit from engineering fragmentation. The individual engineer is typically not an autonomous or consulting engineer but one whose professional identity is defined in terms of the economic interests of private industry.

Parallel with such institutional efforts to unite the engineering profession were others to conceptualize the unique ideal that engineers pursue for the common good. The classic 19th century definition, that of the British engineer Thomas Tredgold (1788–1829), described engineering as "directing the great sources of power in Nature for the use and convenience of man" (Institution of Civil Engineers 2012/1828, p. 4). But in comparison with the ideals that inspire the practice of medicine and law – i.e., health and justice – "use and convenience" would seem to be lower-level goods subject to determination more by a client than by a professional. The dominant external interpretation of "use and convenience" in the U.S. has been the low cost and mass production of goods and services – an interpretation that the professional community has been challenged with since late nineteenth century (Downey 2007).

One influential but failed effort to articulate an ideal that would justify more professional independence for engineers from business interests focused on efficiency as promoted by the technocracy movement (Akin 1977). From the perspective of the technocratic ideology, what business actually wanted was not so much low cost but high profits. Low-cost-based design and manufacturing shortcuts coupled with manipulative advertising offended the engineering ideal of technical efficiency. In 1928, at the height of this dream of expanded engineering influence, on the basis of his public service achievements in post-World War I relief and the 1927 Mississippi River flood, Herbert Hoover was elected the first professional engineer president of the United States.

But the efficiency ideal was problematic on two counts. First, its elevation of technical expertise to public decision-making leadership tends to be at odds with locally-dominant images of democracy. The major European totalitarian philosophies of the mid-twentieth century, communism and fascism, often justified themselves by appeals to efficiency. Second, as a ratio of outputs over inputs, efficiency was context dependent, subject to multiple interpretations depending on how the inputs and outputs themselves are defined. Engineers, as employees, remained subordinate to commercial interests that defined efficiency in terms of economic profits.

In another approach to the enhancement of professional unity and autonomy, U.S. engineering societies began in the early decades of the twentieth century to formulate codes of professional ethics. Initial attempts at code creation, for instance, prohibited the engineering criticism of other engineers in ways that would undermine unity, and as part of its unifying mission, the ECPD was tasked with drafting a code of ethics to bridge those of different member organizations. The first ECPD code of 1947 actually constituted a watershed in U.S. engineering ethics development by explicitly introducing responsibility for public safety, health, and welfare as a basic consideration. Then in a 1974 revision, the first of seven fundamental canons became: "Engineers shall hold paramount the safety, health and welfare of the public in the performance of their professional duties."

Over the next two decades, high-profile cases such as two major DC-10 crashes (Paris in 1974 and Chicago in 1979) and a series of fatal accidents with the Ford Pinto automobile (manufactured from 1971 to 1980) – both associated with problematic engineering designs – conspired to raise public concerns about engineering safety. Beginning in the 1970s, the U.S. National Science Foundation promoted collaborative research between engineers and philosophers to better analyze engineering ethics issues – such as those associated with whistle blowing, autonomy, and the "paramountcy clause" – and to develop appropriate materials for teaching engineering ethics. When the Challenger shuttle launch disaster of 1986 further revealed limitations on engineering independence, ABET was thus able to draw on existing work to promote more strongly requirements for ethics education as a component of all accredited engineering programs. Yet more recent attempts to integrate issues of sustainability and social justice into engineering ethics discussions have proven less successful, likely because employers tend to judge them in conflict with a dominant normative commitment to maximize sales and profits.

The engineering profession in the United States has been a world leader in promoting engineering ethics code development and associated educational activities. But this leadership has grown out of a long, and continuing, struggle and desire for professional unity and autonomy.

Japan: Engineers and Households

In Japan the promotion of professional engineering ethics instruction began in the late 1990s, stimulated in part by some high-technology accidents such as a sodium leak at the Monju fast-breeder reactor (1995) and a disastrous criticality accident at the Tokaimura reactor (1999). In this respect, there are similarities to the situation in the United States, where DC-10 crashes and the Challenger accident contributed to the rise of ethics education to promote professional autonomy. But for the Japanese engineering community, it was not so much autonomy that was at issue as reforming a disordered household and meeting the perceived challenges of internationalization.

To understand the distinctive features of engineering ethics in Japan, it is important to appreciate the role harmony plays for Japanese engineers in both domestic and international professional relations (Luegenbiehl and Fudano 2005). A key concept here is that of the 家 (*ie*, pronounced ee-aa, commonly translated into English as "household.") *Ie* has multiple meanings and can refer both to a physical home or family estate and to a family genealogy. It can also involve economic and socio-religious implications. As described by anthropologist Dorinne Kondo, in Japan a distinctive vision of the *ie* marks it as a basic social phenomenon and a site of obligation and responsibility. "The *ie* is not simply a kinship unit based on blood relationship, but a corporate group based on social and economic ties." Personal identity does not exist prior to or independent of the household but is defined by one's position within it "Subordinating one's individual desires to that of the household enterprises takes on the character of moral virtue," Kondo observed. "Pursuing one's own plans and disregarding the duties toward the household smacks as selfish immaturity" (Kondo 1990, p. 131; see also Traweek 1993, p. 401; Lanham 1979, p. 5). The household serves as a center for attachment, or *uchi*.

Students begin competing to demonstrate their appropriateness for corporate household attachments long before entering higher education, in kindergarten or even pre-school. The country is widely known for what the Japanese call "examination hell," the extended preparation for higher education entrance examinations that determine life career paths (Vogel 1971). But examinees are not just revealing individual achievement so much as demonstrating a mature other-directedness developed through the disciplined acceptance of hardship. In this sense preparation for the exam is about "polishing the heart" [*kokoro*]. As Kondo puts it,

> In Japanese society generally, hardship is considered one pathway to mature selfhood…. [E]ndurance and perseverance are among the most frequently cited virtues in Japanese society.... Learning to stick to a task, no matter how difficult or unpleasant, thus strengthens the *kokoro*. (Kondo 1990, p. 109)

Those who achieve the highest entrance examination scores are able to enter engineering programs at prestigious national universities. But in contrast with engineering students in the United States, once matriculated they need to do little more to warrant good employment. Typically, Japanese students regard their university years as a well-deserved vacation from hard work. Although those in engineering work more than others, even for them university life constitutes something of a time-out from household duties. Having departed from the family household of origin, they are transitioning to the corporate household that will serve as the basis of identity and obligation for the balance of their working lives.

This distinctive approach to reckoning identity and responsibility through household-like social groups has a long history, one nodal point of which was establishment of the Japanese nation state under what is known in the West as the "Meiji Restoration" (1866–1869) (Chizuko 1996). Undertaken in response to the challenge of the West – as manifested, for instance, by U.S. Commodore Matthew Perry's forced opening of Japan to world trade in the 1850s – the new imperial government explicitly restructured Japan as a "family state" (Shibata 2004, p. 76). Survival could best be assured through the fulfillment of obligations to a family state that has made possible an unusual openness to adaptations from the West such as industrialization, provided such imports are subsequently given a Japanese form.

One key example of an import followed by a process of Japanization was technical education. In 1868, Yozo Yamao, who had been studying abroad in Glasgow, returned to become vice minister of education with the goal of establishing an engineering school. The Imperial College of Engineering was founded in 1873 with Scotsman Henry Dyer imported to serve as head. The government then systematically replaced British professors with Japanese graduates until finally, in 1886, the College was merged into what became the University of Tokyo.

The University of Tokyo engineering program in turn has been a major source for managers and directors of the most powerful technology-based corporations (Morok and Nakamura 2003). One of the first replacement faculty members, Fujioka Ichisuke, helped found Toshiba. Hitachi had 11 directors prior to 1941, all but one being engineering graduates from the University of Tokyo. Other graduates founded Toyota and Nissan (Odagiri 1998, pp. 143–146). Although the post-World War II occupation authorities dismantled the militaristic hierarchies of the Japanese *ie*, their practices also stimulated new forms of household formation, including those that included university-corporation partnerships for the development of science and technology. As Prime Minister Suzuki Kantaro already proclaimed shortly after Emperor Hiroito announced Japan's surrender in 1945: "It is essential that the people should cultivate a new life spirit of self-reliance, creativity and diligence in order to begin the building of a new Japan, and in particular should strive for the progress of science and technology, which were our greatest deficiency in this war" (Morris-Suzuki 1994, p. 161).

During the postwar period, according to Kondo, the "company as family" became the basis of the Japanese employment system, which has been characterized by welfare paternalism, seniority promotions, lifetime employment, and worker identification with the firm. But the great post-war success of the new corporate *ie*, which extended corporate households beyond the boundaries of the nation, also raised certain challenges. Extended household loyalties weakened and public scandals occurred as individual self-interest took precedence over harmonious service to a common goal. Given the household model, traveling down such a path risks, to invoke Kondo's terms, "disregarding the duties toward the household," failing to demonstrate "moral virtue," and even "selfish immaturity" (Kondo 1990, p. 131; see also Luigenbiehl 2004, p. 9).

Actions by professional engineering societies attempted to respond to such challenges. When working engineers attend continuing education classes in engineering ethics, they receive a booklet documenting their accomplishments. But these professionals are not being trained to become whistle blowers who risk job and career in the name of individual honesty and autonomous judgment (Wokutch and Shepard 1999). Although more than one commentator has described engineering professionals as learning "to judge what they should do or not to do according to the engineers' ethics" (Ohashi 2000; Kawashima et al. 2004, p. 101), there are subtle ways in which professional attachment is being emphasized as much as individualism. In the course of promoting ethical decision making, professional engineering societies are also offering themselves as *uchi*, new centers of belonging responsible for defining the identify of engineers in order to help them struggle with change. Ethics is also a means "to secure international acceptance of engineers' qualifications" (Kawashima et al. 2004, p. 101).

The JABEE criteria for engineering education that include explicit mention of engineering ethics further support this interpretation. In first place among the eight new standards for assessing engineering education programs is teaching "the ability and intellectual foundation for considering issues from a global and multilateral viewpoint." Second place goes to an ethics-related standard of learning to appreciate "the effects and impact of technology on society and nature, and of engineers' social responsibilities." In this image of the Japanese engineering curriculum, what is primary is the development neither of abilities in mathematics and science nor of skills in engineering analysis – the first outcomes listed in the ABET accreditation scheme – but learning to consider issues from a point of view that rises above self-interest, overcomes selfish immaturity, and locates one's concerns and interests in relation to those of others engaged in the general pursuit of harmony.

Further evidence for the importance of *ie* in the new engineering professionalism can be found in the 1999 action by the Japanese Society of Civil Engineers to replace its "Beliefs and Principles of Practice for Civil Engineers" with a new "Code of Ethics for Civil Engineers." The "Beliefs and Principles" had not been updated since 1938 and had been of relatively little consequence since its formulation. Although the new "Code of Ethics" admonishes engineers to "adhere to the ethical principles of self-disciplined moral obligation when applying advanced technology" it also repeatedly stresses responsibilities to society at large. The first provision, for example, states that the civil engineer shall "[a]pply his/her technical skills to create, improve, and maintain 'beautiful national land,' 'safe and comfortable livelihood,' and 'prosperous society,' thus contributing to society through his/her knowledge and virtue with an emphasis upon his/her dignity and honor." A sense emerges of the professional engineering society as a household through which obligations can legitimately be formulated and fulfilled for the common good.

The national movement to promote professionalization and ethics among engineers may thus be read as an innovative move to establish the viability of a new household, the engineering profession, that functions both as an aid to corporate households and retains a primary obligation to the national household – one that may also serve as a pathway for engineers to work around those corporate households that have failed to fulfill their obligations at the national level (Wokutch and Shepard 1999). As Hideo Ohashi eloquently put it,

> We need a revolution of our consciousness, from ignoring to respecting professionals … The recovery of competitiveness should not be the final target. We dream of a society whose keywords are safe, reliable, healthy, peaceful, and heart-warming. (Ohashi 2000)

France: Engineers and Social Order

As noted, the French Charter for Engineering Ethics was not meant to become part of standard engineering curricula. The CNISF, which did not create even the first version of this charter until 1997, coordinates the activities of alumni associations for engineering schools and has no oversight responsibilities for practicing engineers. Indeed, there is little evidence that most engineers in France have ever heard of CNISF (Didier 1999, 2000). At the same time, it is notable that this Charter explicitly links engineering with the concept of progress, describing engineers as the source of innovation and the engine of progress: "L'ingénieur est source d'innovation et moteur de progrès" – a view that is undoubtedly held by many engineers in the United States, although this is not a statement that would ever be thought appropriate to an ethics code.

The key to understanding the larger disinterest in ethics in engineering education in France rests with the longtime elite status of French engineers (Alder 1997). Unlike in the United States, engineers in France do not have to struggle for social respect. As the French journalist Jean-Louis Barsoux (1989) explains: "In France, engineering education does not play second fiddle to medicine, law, or architecture – it is the recognized way to the top, both socially and professionally." Barsoux is referring to a special category of so-called "state" engineers, i.e., those who work as high-level civil servants in the national government. Although state engineers have been in the minority at least since 1900, their status has cast a favorable glow over all French engineers.

The ethics of French state engineers is both established and demonstrated by their participation in a rigorous exam system. Students who aspire to become engineers have first to complete a *baccalauréat*, or high school diploma, with appropriate emphasis in mathematics and science. They then undertake two years of math-intensive study in *classes préparatoires*, often held in the same buildings in which they completed the baccalaureat. At the conclusion of this process, prospective students compete for positions in the elite schools, the so-called *grandes écoles*, by sitting for the *concours*, a combined written and oral exam whose scores are published in local newspapers and determine who will be granted admission to which schools. In this respect, there are similarities with the Japanese system.

But in what sense is one's morality demonstrated by successful completion of this process? Important clues lie in the fact that the process of gaining entry into one of the elite schools is not called "admission" but "promotion," and that eventual graduates will forever identify themselves as cohorts based not on the year of graduation but on the year of matriculation. Furthermore, the rankings continue throughout their studies, at the conclusion of which the highest-ranked graduates remain on pathways leading ultimately to senior positions in government ministries. By entering an engineering school, prospective state engineers join a system in which they serve as both leaders and embodiments of French society.

In contrast with the challenge of progress prominent in the United States – which aims for free market individualism maximized in the low-cost mass production of goods and services – since the Enlightenment the dominant view among French state engineers has been that the goal is rational social order achieved through sound mathematical principles. Such rational unification takes place best in government, protected from the diverse economic perspectives and interests of private industry. Examples of this commitment to rational planning are legion. As historian Cecil Smith writes, "Ever since the birth of the Corps des Ponts et Chaussées in the eighteenth century, French state engineers have promoted the complementary notions of rational public administration in the general interest and planning on a national scale." For instance, in the 1820s, when Corps des Ponts director Louis Becquey gained approval for a national system of canals in France, private companies applied to construct the projects, following the practice in Great Britain. But Becquey successfully "defended the interests of state engineers by arguing that the plans 'are in the public interest, for without [state engineers'] supervision, private companies would indulge in the meanest economizing'"(Smith 1990, p. 659). At the end of the century, a Corps des Ponts chief engineer successfully resisted the encroachment of private interests into plans for the electrification of France as "ignorant greed [which] threatens to squander a national resource" (Smith 1990, p. 685).

During the early twentieth century, a group of graduates from the most elite of the technical schools, the École Polytechnique (aka "*L'X*"), established the think tank *X-Crise* to promote an alternative philosophy to capitalism, communism, and fascism. They called it "planism." Among them was Jean Coutrot, an engineer-intellectual and founder of the Centre d'Études des Problèmes Humains [Center for the Study of Human Problems]. According to Coutrot, the leadership of engineers was rooted in engineering analysis: "It is to the engineers, today, that it falls to construct better societies because it is them and not the legalists or politicians who hold onto the necessary methods" (Clarke 2001, p. 81). As historian J. Clarke explains, for Coutrot and other engineers who were concerned about the dehumanizing effects of mass production, communist collectivism, and fascist centrism, "the central problem of their time was the question of how to organize a society that was both rational and human" (Clarke 2001, p. 84). In some respects, what French engineers achieved in this instance was the rationalist ideal of the technocracy movement that was growing in the United States during the same period.

After World War II, state engineers secured complete jurisdiction over electricity, train transportation, and atomic energy, all in the name of rational national planning in the general or public interest. As Smith explains, "they acted as planners, economists, urbanists – 'inter-ministerial generalists,' drafting legislation and then the decrees to implement it" (Smith 1990, p. 692). The influence of state engineers spread through a greatly enlarged "para-public" sector that included electric power, gas, coal, banks, airlines, telecommunications, Renault, and the French national rail system SNCF. "As true as it is that public engineers acted as an elite all too confident in the power of 'superior light' [lumières supérieures] to determine the 'general interest,'" Smith concludes, "it is no less true that for 250 years they sustained an ethos of public service rarely found elsewhere" (Smith 1990, p. 693). This is an ethos acquired at the *grandes écoles*.

Since their eighteenth century founding, engineering educators in the most elite *grandes écoles* – that is, the École des Ponts et Chaussées (1747), École des Mines (1783), and École Polytechnique (1794) – have placed the highest value on mathematical knowledge. As historian Wolfhard Weber explains, Gaspard Monge, the "father of the École Polytechnique," explicitly saw mathematical theory as the key for steering the present by enabling clear descriptions of the future. "Monge himself insisted that descriptive geometry was an answer to the French nation's requirements." This new science made it possible to represent three-dimensional objects in two dimensions, which was crucial for designers, and could fix the exact location of objects and the relations of their parts. By these means it "brought together a series of factors fundamental … for progress" (Weber 1986, pp. 21–22). The names of subsequent mathematician-engineers who taught at the top schools and served in the civil service constitute a virtual Who's Who of the engineering sciences: Joseph Fourier (1768–1830), André-Marie Ampère (1775–1836), Siméon-Denis Poisson (1781–1840), and Sadi Carnot (1796–1823), to name only a few.

For French engineers, demonstrating the ability, commitment, and discipline to become proficient in the mathematical foundations of engineering is to demonstrate that one has the moral character and reliability to warrant the trust of the Republic. Students who have been promoted into the national system of rational deliberation and action geared toward increasing social order have already demonstrated everything necessary to warrant a position of national leadership. They have mastered all the principles and values that constitute engineering ethics in France; indeed, one might find considerable support for the claim that rationalist engineering constitutes the dominant ethic of France. For students who have already demonstrated their character through their competence, to then have to enroll in a course in engineering ethics would seem ludicrous, if not insulting. It should be no surprise, then, that the annual military parade on Bastille Day, which publicly celebrates the accomplishments of the Republic, is led by second-year students from the École Polytechnique.

Why then did the collective organization of alumni associations feel pressure to formulate and disseminate a code or charter? This move may perhaps be understood as one of many efforts in and around French engineering education to adapt to the increasing value accorded the private sector as a measure of national worth after the end of the Cold War. A U.S.-led shift in the dominant image of international relations replaced a grand conflict between two philosophies of political economy with a model of economic competitiveness that pictures countries competing for economic dominance. This shift has forced other countries to adapt to a North American model of progress oriented toward the free market production of low-cost goods for mass consumption. In response, the *grandes écoles* reluctantly initiated international student exchange programs and new career pathways oriented more toward private industry. In particular, expecting engineers to participate increasingly in international workplaces beyond Europe, schools have also begun expanding the non-technical dimensions of engineering education.

It is in this context that "ethical reflection on the engineering profession" has gained a modest foothold. In 1995, the Commission des Titres d'Ingénieurs [Engineering Titles Commission], established in 1934 to protect the formal title "graduate engineer," updated its non-technical requirements to include "foreign languages, economic, social and human sciences and a concrete approach to communication problems as well as providing openings to ethical reflection on the engineering profession" (Centre d'Études sur la Formation des Ingénieurs 2000). This modest effort nevertheless did not generate significant concrete activity at the elite schools.

Germany: Engineers and *Bildung*

The *VDI* Fundamentals of Engineering Ethics stresses in unique ways that although engineers must "know the relevant laws and regulations of their countries" they should honor them only "insofar as they do not contradict universal ethical principles." Moreover, in cases of value conflicts, engineers are admonished to choose "the values of humanity over the dynamics of nature," "human rights over technological implementation and exploitation," and "public welfare over private interests." How did universal ethical principles become such a major commitment – one much stronger than the U.S. commitment to protecting public safety, health, and welfare – and what does it mean for the German engineers themselves?

The immediate interest among German engineers in the impacts and effects of technologies on humanity can be traced to the post-WWII period. Having been co-opted by the National Socialists during the 1930s, the *VDI* was revived in 1947 with an international engineering education conference on "*Technik als ethische und kulturelle Aufgabe*" [Technology as ethical and cultural task]. The problem for the members of the VDI was precisely that they had accepted the ideal of what engineer-historian Thomas Hughes (1980) calls "culture-determined technology," in which they failed to challenge Nazi cultural leadership. A major post-war task was thus to break free from such a determination, a project that began with this conference and continued through an active collaboration with anti-Nazi German philosophers in a series of four additional meetings between 1950 and 1955 on the general theme of technology and humanity. Indeed, a strong collaboration with philosophers is itself a distinctive feature of the lives and work of engineers across Germany.

As its contribution to this effort, the 1950 conference drafted an Engineer's Confessions that employed a distinctly religious rhetoric to offer a vision of engineering as a spiritual vocation (*Verein Deutscher Ingenieure 1950*). According to the Confessions, the engineer "should place professional work at the service of humanity … [and] should work with respect for the dignity of human life and so as to fulfill his service to his fellow men without regard for distinctions of origin, social rank, and worldview." To include an explicit commitment to humanity as a whole constituted a self-criticism by German engineers, who previously had understood themselves as advancing civilization by serving Germany. At the same time, a significant continuity was the idea of technology as a vocation, the understanding of which points toward the distinctive German notion of *Bildung*, formal education oriented toward spiritual growth and perfection.

In the mid-1800s, German culture and education became a major vehicle for the expression of German aspirations for unification. Indeed, already in 1807, philosopher Johann Gottlieb Fichte had argued in his *Reden an die deutsche Nation* [Lectures to the German Nation] the significance of *Bildung* as a means to unify and develop Germany. Germany could become great and contribute to human development through a *Bildung* that was, however, conceived as grounded in and an extension of Greek and Latin cultural life.

Throughout most of the nineteenth century, the professions of law, medicine, philosophy, and theology monopolized *Bildung* in this strong sense because of the preparatory curriculum they demanded in the elite secondary schools, or Gymnasia. Only those students who had mastered classical studies in Greek and Latin philology were thought able to manifest the *Geist* or spirit that was the perfection of human nature (Masschelein and Ricken 2003). The significance of this *Bildung* actually derived in part from its contrast with technical training and work. *Techniker*, or technologists, who actually functioned in ways similar to what in other countries were called "engineers," underwent an educational program separate from that of the gymnasium, with the gymnasium degree or *Abitur* being the only path into the university.

Early attempts to enhance the cultural prestige of technical learning and work included creation of the Association for the Promotion of Technical Activity in Prussia (1821) by Prussian Finance Minister Christian Peter Beuth. Understanding *Bildung* and cognizant of negative effects of industrialization on English workers and landscapes, Beuth sought to promote a distinctively German industrialism that imbued technology with art and emphasized aesthetics as an evaluative criterion (Brose 1992). According to Beuth, industrialization would be acceptable in Germany as a site for the emancipation of *Geist* by means of a new form of *Bildung*. He thus stipulated, unsuccessfully, that art and aesthetics be included in the curricula of nascent technical schools serving the lower classes of society.

An educational movement that proved more immediately successful established *Technische Hochschulen*. These were technical post-secondary schools or institutes that included among their responsibilities fundamental research on *Technik*, a concept that included both technical products and the technological processes for their production (Manegold 1978). First established during the middle part of the century, the new institutes gained greater visibility and status after the 1870s, during the unification of Germany under the Prussian-led Second Reich. Advocates for the technical institutes also established a new form of quasi-academic secondary education in *Oberrealschulen*, whose "realism" included teaching modern rather than classical languages.

In 1885 a commission of the *VDI* (which had been founded in 1856) concluded a review of the structure of German education and its implications for engineers by demanding that the courses students followed into and through the technical institutes have the same legal standing as those through the *Gymnasium* to university. "The engineer in the eyes of many," according to the commission,

> was – and partly still is – an advanced artisan, neither requiring nor deserving the higher *Bildung* offered by the *Gymnasium*. We declare that German engineers have the same needs with respect to their general *Bildung* and wish to be subject to the same standards as the other higher professions. (Gispen 1990, p. 146)

William II approved this request in 1892 by giving *Oberrealschulen* graduates the right of admission to the engineering corps, in 1899 accepting them as eligible for employment in the civil service, and in 1900 granting them equal status to graduates of the classical *Gymnasium*.

In the early twentieth century, members of this new professional engineering community defended the thesis that the emancipation of the human spirit included not just classical culture but also *Technik*. In *Lebendige Kräfte* (1904), for instance, the German engineer Max von Eyth even argued contra Hegelian philosophers and Prussian lawyers that technology rather than reason should be seen as the vehicle for the unfolding of Geist or mind/spirit (von Eyth 1903). As historian Jeffrey Herf summarized Eyth's view,

> there was more Geist in a beautiful locomotive or electric motor than in the most elegant phrases of Cicero or Virgil. Technology, like poetry, dominates matter rather than serves it…. [T]echnology was actually more cultural than culture itself. (Herf 1986, p. 159)

Feeling empowered by an increasing national commitment to industry, engineers openly challenged the value of the universities and "praised their own achievements as 'national' ones and engineers as 'pioneers of German value and culture'" (Herf 1986, p. 156).

Elite German engineering intellectuals thus engaged in a kind of cultural politics that historian Karl-Heinz Ludwig (1974) has described as the "anticapitalism of technicians." This philosophy held that "technology emanated from the deepest impulses of German Kultur"; that contemporary crises in German society, especially after World War I, "were not due to the machine but to its misuse by private capitalist interests"; that "the welfare of the national community could be protected only by a strong state"; and that "engineers had a central role to play in providing the expertise necessary for Germany in an age of technological warfare" (Ludwig 1974). This engineering point of view found the development of National Socialism compatible with its goals, because the new political movement claimed to be oriented toward emancipating a German essence that promised to overcome the misdirections of a self-interested aristocracy and a disordered free-market capitalism by relying on a charismatic individual.

During the Third Reich, engineers tolerated and even supported antisemitism on the grounds that Jews were not essentially German and served as purveyors of free-market capitalism. Through a deliberate political neutrality oriented only to technical work, they stumbled into the role of collaborators who sanctioned through inaction, and sometimes obedience, a willful and active misuse of *Technik* to destroy humanity rather than develop it. When a reconstituted *VDI* was struggling to understand what had happened and how engineers should reposition themselves as positive contributors to society, they therefore had to extend their vision beyond any hypothetical German essence to include humanity as a whole. *The Engineer's Confessions* stipulated that "The engineer should not bow down to those who disregard human rights and misuse the essence of technology; he should be a loyal co-worker for human morality and culture." Engineers now had to re-conceptualize *Technik* to acknowledge that technology could have serious negative consequences that would not constitute societal advancement of any kind.

In the 1970s this new sense of social responsibility was expressed in efforts by German engineers to influence the emerging discipline of technology assessment. During this period German engineers sought to embody their broad ethical responsibilities in assessing technologies according to eight metrics of value in three categories, including functionality, economy, and material standard of living; safety, health, and environmental quality; and development of individual personality and quality of social life. The very use of the term *Technikbewertung* [Technology evaluation] as a translation of "technology assessment" tended to stress going beyond the kind of limited cost-benefit analysis that became the norm in the United States. Moreover, individual engineers were not left alone to evaluate technologies on the basis of personal conscience but were presented with guidelines that had been authorized by the engineering community as a whole (Huning and Mitcham 1993).

Why then a commitment to updating and simplifying these guidelines in 2002? Like the Japanese and the French, Germans were working to adapt to a world increasingly dominated by images of economic competitiveness, with an emphasis on low-cost production for mass use. On the one hand, German engineers were struggling to construct new practices in which technology evaluation was not only an ideal but also reduced costs (Legg 1990). On the other, a reaffirmation of a responsibility to engage in technology assessment offered evidence that *Technik* was still about emancipating *Geist*. Simplifying and reaffirming universal ethical principles was a way to achieve both ends.

Conclusion: How Engineering Ethics Follows Different Trajectories

As these comparative cases suggest, the progressive concern for engineering ethics in different countries may well be one manifestation among engineers of what is today called "globalization." Because of their situation in the largest economy in the world, within which competition on the basis of low-cost production for mass use has a long history, leadership in engineering ethics development in the United States undoubtedly influenced advocates for engineers and engineering in other countries. But border crossing also produces transformations. The fact that engineering ethics has been pursued in the United States to promote professional unity and autonomy does not mean that others would pursue it in similar ways in other countries.

In Japan, the early twenty-first century interest in engineering ethics among professional societies and the promotion of ethics education by a new Japanese engineering accreditation organization offers a case of consciously-imported influence, in part to achieve international recognition of domestic engineering programs. But engineering ethics in Japan can also be interpreted in terms of its relationships to the uniquely Japanese social institution of the *ie* and efforts to develop the engineering profession as a household center of belonging alongside existing corporate households. The professional engineer in Japan appears to be emerging as someone with a new, untarnished pathway to fulfilling obligations to the national household.

In France, formal education in engineering ethics has attracted little interest. Explicit courses in engineering ethics are easily seen as unnecessary if not insulting to those elite engineers whose dedicated study led to a higher education committed to civil service in pursuit of rationalist national progress. Indeed, in such a context, for non-elite schools to adopt education in engineering ethics might even be interpreted as an open admission and acceptance of subordinate status – although embraces of global competitive pressures, as well as new pan-European efforts could well lead in this direction.

In Germany, a post-World War II reassessment of the relation between engineering and the traditional ideals of humanistic *Bildung* has led to a new commitment of engineers to the good of humanity as a whole. A longtime commitment to social responsibility through the production of high quality technology further led to the adoption of technology assessment as a major feature of engineering ethics. For German engineers, engineering ethics and technology assessment constitute a spiritual contribution to globalization.

Recognition of how engineering ethics follows diverse local trajectories with distinctive implications across particular countries has implications for how to think about engineering ethics within any country. Who openly advocates instruction in engineering ethics? Who passively ignores such initiatives? Who openly resists? Asking questions such as these may serve to indicate something about both the positioning of ethics in engineering identities and the complexities of struggles among those who are content with their current identities and those who might be seeking change. In this sense, following debates over engineering ethics can provide a means of mapping and understanding some of the contemporary flows of globalization as engineers interpret and engage them.

Acknowledgments We have drafted this essay with classroom readers in mind. It is an edited and adapted version of Downey et al. (2007). We draw text from that article with permission. Gary Downey's contribution is based upon work supported by the National Science Foundation under grant #DUE-1022898. Any opinions, findings, and conclusions or recommendations expressed in this material are those of the author(s) and do not necessarily reflect the views of the National Science Foundation. We thank the editors of this volume for inviting this contribution and for their efforts in seeing it to publication.

References

Akin, W. E. (1977). *Technocracy and the American dream: The technocrat movement, 1900–1941*. Berkeley: University of California Press.

Alder, K. (1997). *Engineering the revolution: Arms and enlightenment in France, 1763–1815*. Princeton: Princeton University Press.

Barsoux, J.-L. (1989). Leaders for every occasion. IEE Review, 26 Jan.

Brose, E. D. (1992). *The politics of technological change in Prussia*. Princeton: Princeton University Press.

Chizuko, U. (1996). Modern patriarchy and the formation of the Japanese nation state. In D. Denoon, G. McCormack, & T. Morris-Suzuki (Eds.), *Multicultural Japan* (pp. 213–223). Cambridge, UK: Cambridge University Press.

Clarke, J. (2001). Engineering a new order in the 1930s: The case of Jean Coutrot. *French Historical Studies, 24*(1), 63–86.

Comité d'Études sur la Formation des Ingénieurs. (2000). *Engineering education in France*. http://cri.ensmp.fr/cefi/plaquet.html. Accessed 17 Sept 2013.

Conseil National des Ingénieurs et des Scientifiques de France. (2001). *Charte d'éthique de l'Ingénieur*. Paris: CNISF.

Didier, C. (1999). Engineering ethics in France: A historical perspective. *Technology in Society, 21*(4), 471–486.

Didier, C. (2000). Engineering ethics at the Catholic University of Lille (France): Research and teaching in a European context. *European Journal of Engineering Education, 25*(4), 325–335.

Downey, G. L. (2007). Low cost, mass use: American engineers and the metrics of progress. *History and Technology, 22*(3), 289–308.

Downey, G. L., & Lucena, J. (2004). Knowledge and professional identity in engineering: Code-switching and the metrics of progress. *History and Technology, 20*(4), 393–420.

Downey, G. L., Lucena, J. C., & Mitcham, C. (2007). Engineering ethics in comparative perspective. *Science and Engineering Ethics, 13*(4), 463–487.

Gispen, K. (1990). *New profession, old order: Engineers and German society*. Cambridge, UK: Cambridge University Press.

Herf, J. (1986). *Reactionary modernism: Technology, culture, and politics in Weimar and the Third Reich*. Cambridge, UK: Cambridge University Press.

Hughes, T. P. (1980). National socialist ideology and German engineers. In H. Friedlander & S. Milton (Eds.), *The Holocaust: Ideology, bureaucracy, and genocide* (pp. 165–181). Millwood: Kraus International Publishers.

Huning, A., & Mitcham, C. (1993). The historical and philosophical development of engineering ethics in Germany. *Technology in Society, 15*(4), 427–439.

Institution of Civil Engineers. (2012/1828). *Royal charter, by-laws, regulations and rules*. http://www.ice.org.uk/getattachment/2006ff1e-d4e4-49a7-955e-21a82234c903/ICE-Royal-Charter,-By-laws,-Regulations-and-Rules.aspx. Accessed 19 Sept 2013.

Kawashima, K., Ikeda, S., Hirotani, A., & Katayama, K. (2004). Roles of the Japan society of civil engineering on the continuing education for engineering ethics. *Proceedings of the 9th world conference on continuing engineering education*, Tokyo, 15–20 May, pp. 101–106.

Kondo, D. K. (1990). *Crafting selves: Power, gender, and discourses of identity in a Japanese workplace*. Chicago: University of Chicago Press.

Lanham, B. B. (1979). Ethics and moral precepts taught in schools of Japan and the United States. *Ethos, 7*(1), 1–18.

Layton, E. J. (1971). *The revolt of the engineers: Social responsibility and the American engineering profession*. Cleveland: Case Western Reserve University Press.

Legg, G. (1990). Measuring the cost of quality: German engineering at the crossroads. *EDN, 35*, 59–62.

Ludwig, K.-H. (1974). *Technik und Ingenieure im Dritten Reich*. Düsseldorf: Droste Verlag.

Luegenbiehl, H. C. (2004). Ethical autonomy and engineering in a cross-cultural context. *Techne: Journal of the Society for Philosophy and Technology, 8*, 1–19.

Luegenbiehl, H. C., & Fudano, J. (2005). Japanese perspectives. In *Encyclopedia of science, technology, and ethics* (Vol. 2, pp. 1071–1074). Detroit: Macmillan.

Manegold, K.-H. (1978). Technology academized: Education and training of the engineer in the 19th century. In W. Krohn, E. T. Layton, & P. Weingart (Eds.), *The dynamics of science and technology: Sociology of the sciences* (pp. 137–158). Dordrecht: D. Reidel Publishing Company.

Masschelein, J., & Ricken, N. (2003). Do we (still) need the concept of *Bildung*? *Educational Philosophy Today, 35*(2), 139–154.

Morok, R., & Nakamura, M. (2003). *The history of corporate ownership in Japan*. European Corporate Governance Institute working paper series in Finance. http://www.ecgi.org/wp/f2003.pdf. Accessed 17 Sept 2013.

Morris-Suzuki, T. (1994). *The technological transformation of Japan: From the seventeenth to the twenty-first century*. Melbourne: Oxford University Press.

Odagiri, H. (1998). Education as a source of network, signal, or nepotism: Managers and engineers during Japan's industrial development. In W. M. Fruin (Ed.), *Networks, markets, and the Pacific Rim* (pp. 141–153). New York: Oxford University Press.

Ohashi, H. (2000). From organization-based toward individuals-based engineers – Current reforms in Japan. *ASME 2000*. Orlando: American Society for Mechanical Engineers.

Shibata, M. (2004). Controlling national identity and reshaping the role of education: The vision of state formation in Meiji Japan and the German Kaiserreich. *History of Education, 33*(1), 75–85.

Smith, C. O., Jr. (1990). The longest run: Public engineers and planning in France. *The American Historical Review, 95*(3), 657–692.

Traweek, S. (1993). Cultural differences in high-energy physics: Contrasts between Japan and the United States. In S. Harding (Ed.), *The racial economy of science* (pp. 398–407). Bloomington: Indiana University Press.

Verein Deutscher Ingenieure. (1950). *Engineer's confessions*. Düsseldorf: VDI.

Verein Deutscher Ingenieure. (2002). *Ethische Grundsätze des Ingenieurberufs*. Dusseldorf: VDI.

Vogel, E. F. (1971). *Japan's new middle class: The salary man and his family in a Tokyo Suburb*. Berkeley: University of California Press.

von Eyth, M. (1903). *Lebendige Kräfte: Sieben Vorträge aus dem Gebiete der Technik*. Berlin: Julius Springer.

Weber, W. (1986). German '*Technologie*' versus French '*Polytechnique*' in Germany, 1780–1830. In M. Kranzberg (Ed.), *Technological education – Technological style* (pp. 20–25). Cambridge, UK: Cambridge University Press.

Wokutch, R. E., & Shepard, J. M. (1999). The maturing of the Japanese economy: Corporate social responsibility implications. *Business Ethics Quarterly, 9*(3), 527–540.

Gary Lee Downey Alumni Distinguished Professor, STS, Virginia Tech. B.S. in Mechanical Engineering, B.A. in Social Relations, Lehigh University. M.A. and Ph.D. in Cultural Anthropology, University of Chicago. Author of *The Machine in Me: An Anthropologist Sits Among Computer Engineers* (Routledge 1998), co-author of Engineers for Korea (Morgan & Claypool Press 2014), co-editor of *Cyborgs and Citadels* (SAR Press 1997) and *What Is Global Engineering Education For?* (Morgan and Claypool 2011). Founder of *Engineering Cultures* course (https://www.youtube.com/user/downeygary). Founding Organizer of International Network for Engineering Studies (INES). Editor of *Engineering Studies* journal (Routledge), the *Engineering Studies* book series (MIT Press) and the *Global Engineering* book series (Morgan and Claypool). President, Society for Social Studies of Science (2013–2015). Main interest in engineering knowledge and personhood. (www.downey.sts.vt.edu).

Juan Lucena B.S. in Mechanical and Aeronautical Engineering from Rensselaer Polytechnic Institute. Ph.D. in Science & Technology Studies from Virginia Tech. Juan is a Professor at the Colorado School of Mines where he teaches Engineering & Sustainable Community Development and Engineering & Social Justice and directs the Humanitarian Engineering program. His books include *Defending the Nation: U.S. Policymaking to Create Scientists and Engineers from Sputnik to the 'War Against Terrorism'* (University Press of America, 2005), *Engineering and Sustainable Community Development* (Morgan and Claypool, 2010, with Jon Leydens and Jen Schneider) and *Engineering Education for Social Justice: Critical Explorations and Opportunities* (Springer, 2013).

Carl Mitcham B.A. and M.A. in Philosophy from University of Colorado, Boulder. Ph.D. in Philosophy from Fordham University. Professor of Philosophy of Science and Technology, Renmin University of China, Beijing; Liberal Arts and International Studies, Colorado School of Mines, Golden, Colorado. Scholarly contributions have been directed toward the philosophy and ethics of science, technology, engineering, and medicine and to science, technology, and society (STS) studies. Teaching areas: ethics, STS, and science and technology policy.

Chapter 6
Identifying Engineering: The Need for Better Numbers on Human and Related Resources and Policy

Tony Marjoram

Abstract Engineering and technology surrounds us and identifies the world, in terms of social, cultural and economic development and change. Yet engineering, partly because of this ubiquity, remains relatively poorly understood and identified. This chapter begins with a background discussion why engineering is as it is, with reference to engineering ethics, values and changing modes of knowledge generation, transfer and application. It continues with discussion of the need for better understanding, definition and measurement – without adequate data there is limited identification, and with limited identification there is limited interest, understanding and priority. The need for better epistemology and models of engineering, science, technology and innovation is then explored. The chapter concludes with a discussion of the need to develop of engineering studies to support this, as part of S&T studies.

Keywords Engineering • Identity • Identification • Epistemology • Development • Statistics • Indicators • Metrics • Innovation • Human resources • R&D • Models • Knowledge production • Engineering studies

Introduction

Engineering, in terms of knowledge and design, production and use of tools and infrastructure, has changed and continues to change the world and the way we live, eat, drink, work, travel, communicate, care for, love and fight each other. Engineering defines and identifies us, and is the vital driver of social, cultural and economic development. But what exactly is engineering, what is an engineer and how is engineering knowledge produced, disseminated, applied, learnt, measured and

T. Marjoram (✉)
UNESCO Centre for Problem Based Learning in Engineering Science and Sustainability, Department of Development and Planning, Faculty of Engineering and Science, Aalborg University, Aalborg, Denmark

S.H. Christensen et al. (eds.), *Engineering Identities, Epistemologies and Values*, Philosophy of Engineering and Technology 21,
DOI 10.1007/978-3-319-16172-3_6

managed? These are both philosophical and highly practical questions. Engineering is often unmentioned in the study of, models of, and policy making relating to 'science', 'science and technology', 'S&T', or more recently 'science, technology and innovation', STI, and associated indicators. Even though engineering is generally accepted as a part of science, and it is mainly engineers who create, apply and innovate technology. Engineering is also generally overlooked in economic development and economics theory, apart from the interest in innovation and evolutionary economics. Engineers, engineering, science and technology are routinely confused in the media. Why is this so? Does this situation need addressing and, if so, what can be done to address it?

These questions are highly practical because engineering is at the centre of meeting the challenges we face in terms of sustainable development, climate change mitigation and adaptation, improving the quality of life of the 1.3 billion people who live in poverty, and the other Millennium Development Goals. At the same time, engineering faces its own challenges in terms of declining public and policy understanding and interest, which is reflected in declining student interest and enrolment, increasing shortages of engineers reported around the world and consequent brain drain and impact on development, especially in developing countries. This also reflects the very success of engineering in the production of ever more reliable technology. Engineering has over 30 main fields and over 100 sub-branches, various levels and modes of professional practice, and is increasingly a black box, with less 'user serviceable parts' and understanding (Whitley 1972).

Engineering, together with science and technology, is in serious need of better understanding. This chapter begins with a background discussion of why engineering is as it is, with reference to engineering ethics, values and changing modes of knowledge generation, transfer and application. It continues with discussion of the need for better understanding, definition and measurement – without adequate data there is limited identification, and with limited identification there is limited interest, understanding and priority. The need for better epistemology and models of engineering, science, technology and innovation is then explored. The chapter concludes with a discussion of the need to develop of engineering studies to support this, as part of S&T studies. This is an urgent priority, if the challenges of sustainable development, climate change, poverty reduction and the understanding, application and development of engineering are to be recognised and addressed.

Engineering and Development

Throughout history, engineering has been closely linked to human, social, cultural and economic development. The history and pre-history of humanity – the way we live, interact with nature and each other, and the world we see today, is the history of engineering. Much of the direction, pace, shape and framing of human, social and economic development and change relates to engineering and the design, use and innovation of tools and technology. Engineering is one of the oldest professions in the development of knowledge, technology and infrastructure. Engineers built the Pyramids, Angkor Wat, Borobudur, Machu Pichu, Great Zimbabwe and the medieval cathedrals. Beginning over 150,000 years ago, the Stone, Bronze, Iron, Steam and Information Ages and associated civilisations were underpinned by engineering and innovation. And the change from one Age to another was not because our ancestors ran out of stones, cooper, tin, iron or steam – it was due to engineering, technology and innovation. Engineering drove the first Industrial Revolution, and the five major waves of technological innovation over the last 200 years- from iron and water power to steam power, steel and electrification, oil and automobiles and mass production, to electronics and computers. Engineering is driving the sixth wave of new knowledge in ICT, biotechnology, nanotechnology and new materials, and will drive the seventh wave of sustainable development, climate change mitigation and adaptation. All waves of technological innovation and development are also accompanied by new modes of knowledge generation, dissemination and application, which require new approaches to learning and education.

Amid these broader waves of revolutionary technological, industrial, social and economic development, engineering has also played a central role in the incremental development of infrastructure in transportation, communications, buildings, water supply, sanitation, energy generation, distribution and use. These technological, industrial, social and economic revolutions originated in Europe and spread around the world, initially in the periods of exploration and colonisation, later in trade and development – indeed, the concept of "development" and "developed" countries has been closely identified with the development of technology, industry and infrastructure. Although many "developed" countries now have large tertiary service sectors as well as secondary industry, and primary resource sectors, much of the service sector is also built and depends upon engineering and technology, as does the primary sector. The concept of development remains largely linked to the development of industry and infrastructure and standard of living, although continues to be measured by such indicators as Gross Domestic Product, GDP per capita and, more recently, by the Human Development Index, HDI. Indeed, given the role of engineering in development and, conversely, the link between underdevelopment and the lack of engineering and technology (Stewart 1977), it could be argued that engineering is as important an indicator of development as GDP, growth and other economic data.

Models of development originated from the Western model of development in modernisation theory, dependency theory and World System Theory, although development continues to be defined economically, rather than sociologically, politically or structurally, and, along with GDP/capita, economic growth remains a dominant indicator. Industrialisation continues as a key policy objectives, together with basic needs and, more recently, human and sustainable development. Many models of development depend on engineering and technology transfer, not only in industry and infrastructure, but also to address basic needs and the UN Millennium Development Goals, particularly poverty reduction and sustainable development, together with climate change mitigation and adaptation. This includes technology transfer at lower as well higher levels, into, between and within developing countries

(S-S cooperation). The importance of technological adaptation and development within developing countries also needs to be emphasised to promote technological appropriateness and "learning by doing". The focus in most universities in developing countries is on education, with limited resources for research and development on local issues, problems and challenges. University staff have often completed research degrees in developed countries, and promotion is usually based on Western university models – for example on research and papers published, particularly in international journals on international issues. There is a particular need to promote research and development, university cooperation and access with local communities on local issues.

These relate particularly to addressing basic needs, poverty reduction and sustainable development. Engineering, technology and engineering education are vital in addressing basic human needs, poverty and sustainability. Although this is often more honoured in the breach, than in the observance, in the comments of world leaders on knowledge societies and economies, and in the declarations of international conferences and world summits. Despite this, however, engineering is routinely overlooked in the context of development policy and planning – it is hardly mentioned in relation to the Millennium Development Goals or in many Poverty Reduction Strategy Papers (PRSPs) – documents that aid donors and international finance organizations require from low income countries in order for them to receive debt relief and financial assistance.

Given the importance of engineering in social, cultural, economic and humanitarian change and development, ethics, norms and values play an important role in the development of engineering, and the identity of engineering. Engineers, like medical practitioners, are one of the professions to be bound by ethical codes of practice. A model code of ethics and professional values was formulated from 1990 by the World Federation of Engineering Organisations and adopted in 2001 (WFEO 2001). This was based on and developed by various WFEO members in relation to the need for engineers to demonstrate integrity, practice competently, exercise leadership and promote sustainability (Engineers Australia 2010). Present and future professional ethics and values are also reflected in graduate attributes and professional competencies of the Washington Accord and its members around the world. These attributes relate to engineering knowledge, problem analysis and investigation, design and development, use of tools, engineering and society, environment, ethics, teamwork, communications, project management and life-long learning. As can be seen – half the attributes that define and identify engineers relate to non-technical factors.

Needs and Numbers: Engineering Metrics and Indicators

How many engineers does a country need? How many engineers does a country need to produce to keep up with this need? What types of engineer does a country need to produce, and at what levels? Is there a shortage of engineers? Are developed countries, such as the United States and in Europe, failing to produce enough engineers, compared to rapidly developing countries, such as China and India? Do other developing and least developed countries have enough engineers, are they producing enough or losing too many to brain drain to be able to promote development, reduce poverty and tackle major issues regarding climate change mitigation and adaptation? What are the consequences of these questions for development around the world, and what are the implications for education policy, for engineering education at tertiary level, and for science education at secondary and primary school? What do engineers, policy makers, planners, aid donors and international agencies and organizations need to do, and what can they do about it?

Such questions are being asked increasing urgently in many countries, for various background reasons. These are, in fact, rather complex questions, for which there are no simple or straightforward answers. This is partly, and perhaps surprisingly, due to a shortage of metrics, statistical data and indicators at national and international level to enable analysis and comparison of response, in both developed and developing countries. Many of the questions, for example, are based on such widely broadcast estimates in the media that the US only graduates 70,000 engineers a year, compared to India at 350,000 and China at 600,000. This is one reason why there is also much reference to quantitative, qualitative and anecdotal evidence from universities, industry and professional engineering organizations regarding the supposed shortage of engineers, now and into the future. This concern was reflected in the production of the report, "Rising above the gathering storm: energizing and employing America for a brighter economic future", by the National Academy of Engineering in the US in 2007, and ensuing debate (National Academy of Engineering 2007).

These questions are complicated by the fact that there are different fields, types and levels of engineers (engineers, technicians and technologists, academic, professional and consulting engineers, with diplomas, degrees and doctorates), and also by differing needs for engineers in different sectors, fields, types and levels at different places (countries and regions within them), as technologies and industries develop and decline. There are also different needs for engineers over time – for example, the increasing use of CAD software has made civil and structural engineers more productive, requiring less support staff. Engineering and technology are the incremental and disruptive drivers of change in society and in engineering, and the understanding, policy making and planning of engineering requires a knowledge of such transverse and longitudinal changes over time in engineering.

These questions are further complicated by different definitions and understandings of what an engineer is. In Germany, for example, there are around 50 definitions of an engineer. In many countries the term "engineer" is also used commonly and in the media to refer to almost anybody that does anything technical – so that people referred to as technicians and technologists in some countries are defined as engineers in others. This is often reflected in official metrics and statistics, where, for example, degrees of different length (3, 4 or 5 years) may be similarly accredited, as may degrees in computer science and IT, which are not included as engineering degrees in some countries. This infers that the US, far from lagging behind

India and China, may actually be producing more engineers than India absolutely, and more than China on a per capita basis. On the other hand, this picture requires further clarification when the numbers of overseas students studying and researching at developed country universities are included in the human resources, research and publication data. Various reasons may also underlie a desire to manipulate reporting, including the interest in some countries to over-represent the number of engineers to promote national status, prestige and attract investment, whereas the interest in others may be to downplay the numbers to lobby for increased government support for engineering education and research, or maybe to justify immigration and the outsourcing of engineering services and related overseas investment.

In reality, there are indeed shortages of engineers in some fields, at various levels of qualification, age and experience, in some sectors, regions, countries and continents (UNESCO 2010). This includes newer, expanding fields of engineering and industry (where the demand is increasing faster than supply), and older established fields (where supply is decreasing relative to demand due to the retirement of post-war, baby-boom engineers). It also relates to the increasingly diversified labour market and declining interest, enrolment and retention of young people in engineering, especially women. While supply and demand may be to a degree self-correcting, it can take up to 10 years to train an engineer, and many governments around the world have been sufficiently concerned to promote public awareness, interest and recruitment in engineering. This includes, for example, addressing the "leaky pipeline" of women in engineering, and the need to attract other under-represented groups into engineering. One area of expansion and rapidly increasing importance is that of sustainable and green engineering – of increasing importance in the context of climate change mitigation and adaptation, one of the major areas of need and growth for engineering, and one of the greatest demands and challenges that engineering and the world has ever faced. One of the first challenges is to ensure enough appropriately qualified and experienced engineers to meet this demand – which will require the development of new courses, training materials and systems of accreditation. Young people are attracted to such courses, and this will help to raise overall awareness of the role and importance of engineering in sustainability, build capacity and a carbon-free future (Morton 2009).

Need for Better Metrics and Indicators of Engineering and Innovation

Engineering, science and technology (EST), and associated applications and innovation, are fundamental parts of a knowledge system that drives human, social and economic development. In this context, 'innovation' refers to technological innovation and the introduction of new technology, although the term has expanded to include such subjects as innovation in education and pedagogy – also of interest in this chapter. In order to properly understand, formulate and implement policy, manage and promote the public awareness of engineering and technology in development, we need good information, data and indicators on EST, especially in developing countries. Numbers are important in advocating evidence-based policy and practice – as emphasised in the context of gender indicators for EST – "no data, no visibility; no visibility, no priority" (Huyer and Westholm 2007). Metrics and indicators also relate to needs as well as numbers – how many engineers, scientists, technologists and technicians do countries actually need, in what fields and at what levels? And what engineers and scientists will be needed in 10, 20 and 50 years time? Despite the importance of such numbers, there has been little interest in engineering metrics and indicators – "No international engineering data of the very detailed kind have ever been published; the only (and usually still rather scarce) information available is for R&D expenditures and personnel in the higher education and private non-profit sectors" (Westholm 2010; UNESCO 2010).

Metrics and indicators on engineering education are part of a wider set of data and statistics on engineering, science, technology and innovation – what has been more commonly referred to as "science and technology", or, more recently, "science, technology and innovation" or STI. Many richer, OECD, countries produce their own information on EST, often in great detail (for example, the National Science Foundation in the United States). This information is then used to contribute to data collection at the international level. The main collectors of data at the international level are the OECD (a leader in the field, with a long-term interest in the field, for OECD member and non-member countries), the World Bank, Eurostat (for the EU), UNESCO (at the global level, mainly following the OECD data collection methodology for EST) and the ILO (for labour and employment data). OECD gathers data broadly on issues relating to economic development, and began collecting data on 'science and technology' in the 1960s. In this field, OECD data collection relates particularly to human resources (and more recently includes interest in the careers of doctoral holders), research and development, and innovation. UNESCO began collecting data on science and technology in the 1960s, in 'capitalist', 'socialist/communist' and 'developing' countries. UNESCO made a particular contribution in the 1978 "Recommendations Concerning the International Standardisation for Statistics on Science and Technology", covering R&D, science and technology education and training (STET) and scientific and technological services (STS). One difference between OECD, World Bank and UNESCO statistics is that OECD and World Bank data is collected and "cleaned" to address possible inaccuracies and irregularities, whereas UNESCO is obliged to use data directly, as submitted by Member States.

Statistical data collection is based on various classifications – the UN Systems of National Accounts (SNA), the International Standard Classification for Education (ISCED), the International Standard Classification for Occupations (ISCO) and the International Standard Classification of all Industrial Activities (ISIC), and is collected with a focus on subject – for example, numbers of graduates and human resources, amount of financial resources, investment and expenditure, and the number of papers published, copyrights and patents issued. It is important to note that metrics and indicators relate to information that is requested or required at national and international level (in line with the need for internationally harmonious and comparable data). Statistics of science and technology relate to numbers that are readily available and measurable, and are not necessarily metrics and indicators that give the best picture of engineering, science or technology. How good the information and data is, and how good a picture it presents of EST and engineering education, depends essentially on the metrics and indicators used, and the quality and quantity of data collected. While what is measured may not give the best indication or picture of engineering, the statistics actually measured in developed, OECD countries are generally accurate, less comprehensive and accurate in developing and especially least developed countries. There are also issues with the methodology of data collection – engineering and science may be defined and measured differently in different countries.

There can be particular differences in the data for human resources – engineers and scientists are often defined differently in different countries and, to compound matters, engineers and scientists are usually aggregated together in data and statistics – so that it is not possible to examine and compare data relating to engineering and engineers in general, or the various branches of engineering. The quality of the data for research and development depends on the definition of R&D. As defined, R&D is mainly carried out in richer, OECD countries, and is measured by such indicators as financial outlay (public and private), the number of patents issued and papers published. R&D may have less relevance for painting a picture of engineering, where fewer papers are published, and less relevance to developing and least developed countries. Similarly, the data for innovation depends on definition, and the measurement of innovation at the formal level relates particularly to intellectual property (IP) and to patents and copyrights issued, which again focuses on richer, OECD countries, whereas much innovation takes place in engineering (Metcalfe 2009), at the informal, non-measured level, in both developed and developing countries. As noted above, the conventional view emphasises knowledge production, application and dissemination in terms of what can be measured in terms of patents and papers, rather than reality, and again paints a poor picture of engineering, especially in developing countries.

To address this situation, a better definition and measurement of engineering is clearly required, with a better disaggregation of data with particular reference to engineers and engineering, branches and levels of engineering. Better definition and measurement of research and development are also required, especially regarding development (rather than research), and better definition and measurement of innovation, especially at the informal level in developed and developing countries. It is only better indicators of engineering and innovation that will provide a better understanding of the role of engineering in development, what sort and how many engineers will be needed in the future. The statistics relating to human resources, R&D and innovation relating to engineering will now be discussed in more detail.

Human Resources and Engineering

Human resources in engineering are part of the international data set produced by OECD in the "Frascati family" of documents on the "Measurement of Scientific and Technological Activities". The definition and collection of this data followed the first meeting of the OECD group of National Experts on Science and Technology

Indicators (NESTI), in Frascati, Italy, in 1963. The Frascati family of documents includes the OECD/Eurostat "Manual on the Measurement of Human Resources Devoted to S&T", otherwise known as the "Canberra Manual", first published in 1995. The particular interest of the Canberra Manual relates to "Human Resources in Science and Technology" (HRST – a new term coined in the Canberra Manual) and the "stocks and flows of S&T personnel". These relate to the numbers of HRST personnel in a country (sometimes referred to as Qualified Scientists and Engineers – QSEs), and the flow of people into and out of that 'stock'. These are based on ISCED standard degree-level qualifications in six categories: the natural sciences, engineering and technology, medical sciences, agricultural sciences, social sciences and humanities. ISCED also divides personnel based on occupation into researchers, technicians and support staff, based on qualification – Ph.D., basic degree, tertiary, post-secondary, secondary diplomas and other qualifications. In reality, data collected under Canberra Manual guidelines is difficult and inaccurate to interpret due to differences of definition and comparison of ISCED data sources, and the fact that many countries group "scientists and engineers" together in data collection. There is a particular need for the refinement of methodology and data collection by national offices and bureaux of statistics, for which advocacy and lobbying will be required by national and international engineering organisations. It is interesting to note that some movement in this direction of disaggregation has taken place, with recent interest in the careers of doctoral holders – which revealed that around 20 % of doctoral holders are in engineering, and that 75 % of them work in higher education (UNESCO 2010).

Research and Development

Research and development statistics are part of the international data set produced by OECD under the "Proposed Standard Practice for Surveys of Research and Experimental Development", otherwise known as the "Frascati Manual", first published in 1963, with the sixth edition in 2002. A subsequent Revised Field of Science and Technology in 2007 and Annex on Measuring R&D in Developing Countries, published in 2012, focused largely on methodology rather than actual measurement of R&D in developing countries – reflecting concern regarding R&D in developing countries. The Frascati Manual focuses on financial and human resources in R&D, in the six ISCED fields noted above (natural sciences, engineering and technology, medical sciences, agricultural sciences, social sciences and humanities), further divided by occupation and qualification as noted above (including "Research Scientists and Engineers" – RSEs), and into four sectors: business enterprise, government, higher education and private non-profit. Issues relating to the collection and use of data regarding R&D include defining what exactly constitutes research and development (which includes 'basic', 'fundamental' and 'applied' research, and various interpretations of 'development'), and how this should be measured, when this may not be a full-time activity (hence the need for head-counts when measuring human resources in R&D, and the use of 'full-time equivalents').

Innovation

Innovation relates to the introduction, dissemination and use of an idea, method, product or process that is new to the user or user group, although may not be absolutely new. Innovation initially related specifically to technological innovation, although the term is now used to refer to any new idea, method, product or process – such as innovation in education. Innovation is closely linked to engineering, as most innovation derives from engineering and engineers, and should therefore be a vital part of engineering education – which is not generally the case at present. Innovation statistics are part of the international data set produced by OECD under the "Measurement of Scientific and Technological Activities, Proposed Guidelines for Collecting and Interpreting Technological Innovation Data", also known as the "Oslo Manual", was first published in 1992, most recently in 2005. The Oslo Manual focuses on guidelines for the collection and use of data on innovation activities in industry, with particular reference to intellectual property – patents, copyrights and license, expenditures on R&D, machinery and equipment, staff training and marketing. The Oslo Manual includes reference to innovation process and economic development in OECD member and non-member countries, including non-technological innovation and the linkages between different types of innovation, and most recently includes an annex on the implementation of innovation surveys in developing countries (OECD 2010). Issues relating to the collection and use of data regarding innovation include the need to maintain a focus on technological innovation and engineering (when the definition of innovation has widened from technology products and processes to any new idea, method, product or process; recognising that much innovation derives from engineering), and the need to measure innovation in the formal and informal, corporate and non-corporate sectors.

Need for Better Numbers

The data and examples discussed above illustrate how the statistics and indicators for engineering are in serious need of research, refining and redefining. The data for engineering is collected under guidelines developed particularly by the OECD NESTI group, and is collected and analysed at such an overall level as to be of limited usefulness in answering many of the questions relating to engineering raised above. These guidelines have been developed with particular reference the situation in OECD countries relating to R&D (Frascati Manual), innovation (Oslo Manual) and human resources (Canberra Manual). Other information is also used – for example for enrolment and graduation from education data (International Standard Classification of Education, ISCED) and labour force surveys (ILO).

As noted above, the data combine "scientists and engineers", without disaggregating into science and engineering, or the various fields of science and engineering, and focus on R&D without specifying the division of research and development and the respective roles of science and engineering. Data on patents, scientific publications and innovation are presented without reference to the origin of patents and papers in science or engineering (and the fact that publishing papers is less of a career priority for engineering than scientists), what constitutes innovation, who does it and where it takes place. While the origins of international trade in high-tech products in various fields of engineering is more apparent, a clearer indication and attribution of this, and the origins and destinations of exports and imports, would be most useful in analysing issues related to the technological balance of payments.

These indicators are of limited use in analyzing the need for, types and numbers of engineers required at national and international levels. Statistics and indicators need to be refined and in some cases redefined to allow better disaggregation between science and engineering and the various fields of engineering and engineering employment (e.g. industry, teaching, research). This will facilitate a better understanding of the role of engineers in R&D, patenting, publishing and innovation, the contribution of engineers and engineering to international trade and the role of engineering in development. It will also help dramatically in providing better data for policy makers and planners.

Considering the importance of engineering, science and technology in the knowledge society and economy, it is surprising that better data is not available on one of the most important drivers of social and economic development. If the data that exists can be disaggregated by gender, surely a better disaggregation into science and engineering, and various fields of science and engineering, should also be possible. Policy makers need to ask national bureaux of statistics for such data and indicators (following the example of gender disaggregated data – that became available in response to requests from policy makers).

Engineering Indicators and Identity

Significant structural as well as cyclical changes are taking place in and in relation to engineering. Government R&D funding in many developed countries is declining in real terms, R&D facilities are downsizing, outsourcing and off-shoring to cheaper locations, as reflected in the declining publication of research papers. These changes may be a disincentive to promoting awareness, understanding and the recruitment of young people into engineering, and may be a factor leading to the shortage of engineers in developed countries. As restructuring takes place, lower level skills are displaced (with an attendant need for re-skilling), although this may be counterbalanced by the greater numbers of young people in higher education in many developed countries. There are still significant differences, however, between higher and lower income countries. UNESCO data shows that developed, industrialized countries have between 20 and 50 scientists and engineers per 10,000 population, compared to around 5 scientists and engineers on average for developing countries, down to one or less scientist or engineer for some poorer African countries. The low numbers of scientists and engineers also reflects the low investment in R&D, the low numbers of research papers published, low level of innovation and patents, and the high level of brain drain from some countries. The engineering, science and technology capacity of many African countries has declined since independence and the post-ward period of decolonization. Given the importance of EST in development, this background will have serious consequences for the future of developing countries.

The actual and impending shortage of engineers has been identified and emphasized by Governments and professional engineering bodies around the world as a national and international priority. In the UK, for example, Engineering UK (formerly the Engineering and Technology Board, ETB), reported in 2010 that the UK has a shortage of engineers, is not producing enough engineers, and estimates that 600,000 new engineers will be needed this decade to help build and maintain new and growing industries (Engineering UK 2010). Engineering UK pointed to a particular shortage of technicians and manufacturing engineers (the UK is the sixth largest manufacturing economy in the world) and an impending decline in the medium and longer term. Concern was expressed regarding the pending shortage of engineers in all areas as many engineers approach retirement (30 % of engineering lecturers and academics have retired in recent years) and decline in birth rate, despite increasing numbers of young people in tertiary education overall. A high demand was identified for mechanical, civil, medical and biochemical engineers in the infrastructure, industry and health sectors, and an impending demand for engineers in (re)emerging industries associated with nuclear power, renewable energy and, particularly, climate change mitigation and adaptation. The importance of attracting the interest of parents, careers advisors as well as young people themselves to raise the status of engineering was also emphasised, and the role of practical, project-based work at school and related out of school activities.

Other governments and engineering bodies have identified the shortage of engineers as a national priority. These include South Africa – where a target was set in 2008 of producing up to 2,500 engineers per year, as part of the (then) government's Joint Initiative on Priority Skills Acquisition, launched in 2006 to help address the skills shortages considered to be a key constraint to economic growth (JIPSA 2006). Roads, electricity, water and housing were to be developed to increase economic capacity – for which a dramatically increased supply of engineers was required. Engineering graduations increased from 1,200 graduations per year in 2000 to 1,500 in 2008 (Lawless 2005). In Morocco, to counter the shortage of engineers a plan was introduced to train 10,000 engineers per year in 2007. In Malaysia, the President of the Institute of Engineers Malaysia, Prof Datuk Chuah Hean Teik, emphasised in 2009 that Malaysia would need to increase the number of engineers from 60,000 engineers to 200,000 engineers by 2020 (Datuk Chuah Hean Teik 2009).

In considering the needs and numbers issue in engineering, it is useful to benchmark in comparison with other professions, such as doctors, lawyers and teachers. While the demand for lawyers is more variable and elastic around the world, and the demand for teachers is more directly determined by school age populations, the need for doctors, physicians and associated senior health service professionals provides the closest and most interesting model for engineers and engineering. The numbers of doctors per capita is also similar to that for engineers – on average 13 doctors per 10,000 people in richer countries, rising to over 20, in poorer countries falling to below 5 per 10,000. Doctors and physicians are also more in the public eye, however, as improvements in health are linked to the supply, and shortage, of health-care professionals. Models have been developed to estimate the need and demand for and supply of doctors, and the identification of potential shortages. It would be useful to explore such models and the forecasting of need, demand for and supply of doctors and physicians, with the view of developing and applying similar models to engineering. An assessment of national technology and infrastructure is already facilitated in many countries with the development of infrastructure report cards, with obvious links to engineering needs and numbers (UNESCO Report 2010).

Indicators and an Epistemology of Engineering

The changing nature and increasing complexity of engineering, science and technology, the rapid diffusion of new information and communication technologies and increasing globalisation of R&D and technological applications have created a new context and demands for metrics and indicators of EST. New challenges regarding education, employment, human resources, R&D, knowledge generation, application and non-patent, lower-tech, sector-based innovation need to be researched and addressed. Existing metrics, statistical collection and analysis, despite progress, is inadequate for analysing the linkages and dynamics of science, engineering, technology and innovation in an increasingly globalised world. There is an increasing need to research and develop a new generation of metrics and indicators to measure EST and associated outputs of knowledge-societies, with particular reference to the data required for human, social and economic development in developed and developing country contexts. This will facilitate the management, planning and policy-making to promote engineering, science and technology for development.

The general objective of such activity would be to research and review the current status, usage and effectiveness of indicators of engineering, science and technology, to research and review current needs in terms of the changing nature of EST and changing needs for associated indicators. This will provide a forum to exchange information and experience regarding the changes that have taken and are taking place in engineering, science and technology indicators and likely future changes and needs.

Specific objectives of activity to develop new metrics and indicators would be to:

- Research and review current approaches to the collection and analysis of EST indicators
- Research and review changes in EST, need for and development of new indicators
- Research and review the need for new indicators of EST in the development context
- Define potential future needs and orientations for EST indicators
- Promote networking and ongoing discussion on the development of EST indicators.

In the developing country context, such activity should also research and examine the particular need for the development of existing and new metrics and indicators of EST to cover applications and innovation, especially that linked to development and addressing basic needs, poverty reduction and grass-roots environmental sustainability, climate change mitigation and adaptation.

Change in Knowledge Production, Application and Dissemination

In terms of R&D and innovation, significant changes in knowledge research, production, application and dissemination have taken place over the last decade. Further changes in knowledge production and R&D systems will continue with the development of ICTs and globalisation. In engineering, science and technology, a major area of R&D, interest and support has moved from a "traditional" system based on a unidisciplinary, hierarchical academic model with public sector funding, to a transdisciplinary, non-hierarchical and heterogeneous system of mixed public and private sector inputs and funding. This is from what is termed Mode 1 knowledge production to Mode 2 (Gibbons et al. 1994; Nowotny et al. 2001; Etzkowitz and Leydesdorff 2000). These changes have been accompanied by changes in research goals, activities and methodology. This has particular implications for R&D, research policy, planning and management, and for wider fields of economic and education policy and planning in engineering, innovation and university-industry cooperation. These changes have particularly important implications for engineering education and innovation in developing countries.

Developing countries have particular needs for R&D to promote social and economic development and poverty reduction, yet face serious constraints of limited funds and weak institutional capacity for R&D (and limited support from aid donors and development agencies for "research"). Developing countries need to strengthen research policy, planning and management to promote R&D in response to the challenges and opportunities presented by ICTs and globalisation. This requires research, institutional development and capacity building, the development and sharing of information, international networking and cooperation. To facilitate this, analysis of knowledge and innovation systems regarding R&D and research policy, planning and management is required, in conjunction with studies and analysis of research and research needs in developing countries. This includes research and research needs in engineering, science and technology, education, the social sciences, information-communication and other areas.

Activity in this area is required to address issues relating to research and research needs, information-sharing and international cooperation, with a focus on information, education and training in engineering research, research management and innovation in developing countries. Activity would be facilitated in a range of linked intersectoral activities, such as the development of information, workshops and training activities. Activity should include related initiatives, including the development of information, learning and teaching materials and the organisation and support of discussion at international, regional and national level to promote institutional development and capacity building in research management.

The main overall objective of such activity would be institutional development and capacity-building in research and research management in developing and economically transitional countries. This would be achieved by the development, production and distribution of information (for awareness-raising and advocacy) and learning/teaching materials for use in education and training relating to research and research management, and the organisation of associated regional and national discussions on research management. Such discussions would focus on information exchange and training activities. A related objective of such activity would be promoting the importance of research on engineering, science and technology, information production and advocacy to aid donors and development agencies, who frequently overlook these topics in the development and poverty reduction contexts. Promoting and popularising EST and linking this to economic development and poverty reduction is of the utmost importance in overcoming the popular notion that EST is disconnected from development. Research and the discussion of the (non-linear) linkage between EST and economic development is of particular relevance in the research management context, and it is useful to note that the World Bank and various regional Development Banks, international, regional and national organisations and agencies have been developing an interest in this area.

Models of Engineering Education and Innovation

Engineering changed through the nineteenth and into the twentieth Centuries, and with it engineering education. This reflected the rise of the 'engineering sciences' and the increasingly close connection between engineering, science and mathematics (the engineering sciences rose in Europe, and reflects linguistic and cultural differences in the understanding and status of 'engineer' and 'engineering' – similar, for example to 'SET' in the United States, which reflects the cultural esteem of engineering in that country, compared to the UK, where the 'E' is often silent). By the end of the nineteenth Century, most of what was becoming industrialised countries had established their own engineering education systems, based on the liberal, student-centred model introduced by Wilhelm von Humboldt at the University of Berlin – combining theory and practice, focused on scientific research. The "Humboldt model" went on to influence the development of universities in France and elsewhere, although the emphasis on practice as well as theory was often later overlooked. In the twentieth Century, the professionalization of engineering continued with the development of learned societies and the accreditation of engineers through qualification and continued professional development, with universities and professional societies facilitating education, research and the flow of information through journals, technical meetings and conferences. These processes continue with the development of international accords, standards and accreditation for engineering education, and the mutual recognition of engineering qualifications and professional competencies. These include the Washington Accord (established in 1989), Sydney Accord (2001), Dublin Accord (2002), APEC Engineer (1999), Engineers Mobility Forum (2001) and the Engineering Technologist Mobility Forum (2003), and the 1999 Bologna Declaration relating to quality assurance and accreditation of bachelor and master programmes in Europe.

The Humboldt model, of the need for theory and practice in university education, was transferred, innovated and developed with an increasing focus on theory, less on student-centred practice. This development of the 'post-Humboldtian' model is one of the factors that lead to the present day decline of interest and enrolment in engineering at university level. The mathematical base became regarded by many young people as too abstract, out of touch, hard work and boring. This lead to a questioning of the post-Humboldtian model, and, ironically, increasing interest in problem- and activity-based learning – part of the original theory/practice model of Wilhelm von Humboldt. The post-Humboldtian model, with the emphasis on theory, also underpins the "linear model of innovation" – the first and major conceptual model of the relation between research-lead science, technology and economic development. The linear model is based on the post-Humboldtian notion that pure, disinterested, basic scientific research, followed by applied research and development, leads to knowledge applications, production, innovation and diffusion.

Early models of engineering, science, technology and innovation include softer and harder versions of technological determinism – the view that technology drives social and cultural development. This was accompanied by various perspectives on the history of science and technology (and engineering). More sophisticated models followed, based around science, technology and society (STS) and S&T studies, and increasing interest in the social construction and shaping of technology. This interest reflected the growing number of university departments focusing on undergraduate and postgrad courses and research in STS and S&T studies in the US, UK and Europe from the 1960s (stimulated particularly by the publication of Kuhn's "The Structure of Scientific Revolutions" in 1962 and the concept of scientific paradigms and paradigm change). This also stimulated an interest in science determinism and the linear model of innovation. Other models of science, technology and innovation (STI) followed, based on systems theory. More recent STI models based on systems metaphors and analogies include the "ecosystem" model. While models, metaphors and analogies have facilitated and enriched understanding in science and engineering (Hesse 1963/1966; Black 1962), the ecosystem model adds little epistemological insight and has confused casual observers in understanding the construction (and social construction) of reality (Berger and Luckman 1966).

Models of engineering, science, technology and innovation began with the linear model of innovation, although there is some question whether this model ever actually existed, and was more a social construction to fill such a role (Edgerton 2004; Godin 2006). The linear model of innovation and technological change posits that innovation derives from basic research in a linear fashion through applied research and engineering to technology and innovation, emphasizing basic research in "technology push" or "market pull" versions. Later, more complex and richer models of innovation derive from Actor-Network Theory (Latour 1987), and the social shaping of technology (Mackenzie and Wajcman 1985). Contemporary models of innovation include open innovation and user innovation – relating, respectively, to openness to internal and external ideas, products, processes and services, and to user-developed ideas, products, processes and services.

The linear model became the common world view on innovation, due largely to its beguiling simplicity for the public and policy makers, and the support of the science funding lobby. The precise origins of the model are unclear, although many accredit the emphasis in 1945 of Vannevar Bush, an electrical engineer and head of the U.S. Office of Scientific Research and Development during World War II, on the role of basic science in technological development and wartime success, underpinned by statistics based on and reinforcing the conception of the linear model. The linear model became the paradigm for "science and technology policy" and postwar economic development, as embodied in the Marshall Plan and later the work on "science and technology" indicators by OECD and UNESCO, despite various critiques. Chief among these were that the linear model overlooks the role of engineering and engineering education in innovation. Science and technology statistics and indicators overlook engineering, for example, in not differentiating or disaggregating data on science and engineering in relation to numbers of graduates, employment and research – where many engineers may be recorded as doing 'science', and many scientists may actually be doing engineering (the space programme and 'rocket science' was essentially engineering). The linear model therefore gives a misleading and inaccurate picture of science, engineering and technology, and largely overlooks the role of engineering in development, and in science and technology policy.

An accurate and up-to-date model more representative of actual and changing modes of knowledge production, application and innovation is urgently required. Science and engineering are part of a system, combining research, application and innovation, encompassing government, universities and industry, and an accurate model should be based on a systems conceptualisation of science, engineering, technology and innovation. The limitations of such an approach should also be recognised – in many developing countries, for example, there is less of a system at national level, where a model of knowledge transfer, application and innovation is more accurate and appropriate. Science and engineering education needs to be based on such a systems conception, as does policy understanding of the role of engineering in development. There is a particular need to address post-Humboldtian notions underlying the 'fundamentals' approach to engineering education as well as the linear model of innovation.

There is a need to emphasise the particular contribution of engineering, over science, to innovation and development, to underline the weakness of the linear model. There is a need to develop engineering studies and associated policy to facilitate this, and to support research to better facilitate understanding of innovation and technology transfer, at all levels, especially in developing countries. In the developing country context, there is a particular need to put engineering on the development agenda by focusing specifically on the important role that engineering and engineering education plays in addressing the UN Millennium Development Goals, especially poverty reduction and sustainable development. Engineering education needs to respond to the seventh wave of sustainable 'green' engineering and technology, climate change mitigation and adaptation, with a focus on environmental and eco-engineering and associated design, manufacturing and infrastructure.

Promoting public and policy understanding and interest in engineering will result from a better appreciation of the contribution of engineering to development, sustainability and poverty reduction. This will be facilitated by information, case studies, advocacy and the inclusion of engineering studies in educational curricula at all levels. At the university level, for example, there needs to be better course content and project activity relating to the relevance of engineering in addressing contemporary concerns and better linkage of engineering with social and ethical issues, sustainability and improving the quality of life around the world. The efficacy of such an approach is demonstrated by the success of such activities as the Daimler-UNESCO Mondialogo Engineering Award and growth of Engineers Without Borders groups around the world – which are attractive to students concerned about such issues. Such initiatives help engineering enrolment, public and policy awareness of the importance of engineering in social, economic, international and humanitarian development. Engineering has changed the world, but is professionally conservative and slow to change. To attract young people, and to help them in facing the challenges of the future, engineering education needs to put fun back with the fundamentals. Curricula and pedagogy needs to be transformed from a formulaic approach, that turns students off, with the use of information and experience in active, project and problem-based learning, combining just-in-time theory and hands-on applications that turns them on.

Engineering Studies, Policy and Planning

The background to the transformation of engineering education relates particularly to government and academic interest in science and science policy and planning, which has neglected engineering. Despite the importance of engineering in development, engineering is generally overlooked in economic development, also in science studies, the social studies of science, "S&T", science policy and planning, where engineering is seldom mentioned. Engineering studies, policy and planning needs to be developed to facilitate the transformation of engineering and engineering education. Interest in science and science policy and planning developed particularly after 1945. Reflecting this interest, courses and then departments focusing on science and technology studies, policy and planning were established in the 1960s at universities around the world. Business schools also developed an interest in science, technology and innovation. Most of this interest focused on science or 'science and technology' policy, with little reference to engineering. The study of engineering, and engineering policy, remained a neglected area of interest and emphasis. This is reflected in the limited public, media and policy awareness, perception and understanding of engineering today. The main reasons that science and technology policy has a focus on science rather than engineering relate to classical economics, public and research policy, and the popular perception and 'linear model' of science and innovation.

In classical economics, technology is regarded as residual, subordinate to the three factors of production – land, labour and capital. Science policy developed from public and research policy, and the principle that decisions regarding the allocation of research funds should be made by researchers rather than politicians – thus favouring science rather than engineering. In the 'linear model' of innovation, basic science research is imagined to lead, through applied science and engineering, to technological application, innovation and diffusion. This model, promoted by Vannevar Bush in the postwar period, and has endeared and endured with scientists and policy-makers on grounds of simplicity and funding success, although many science and technology policy specialists now regard the 'linear model' as inaccurate and misleading. This is partly due to the recognition that many innovations derive from engineering, rather than basic science. Interest in the role of science, engineering and technology in international development also evolved towards the end of the colonial period in the 1960s, with the development of universities in developing countries, again with a focus on science, rather than engineering, replicating universities in developed countries.

Given this background, and the rapid change in knowledge production, dissemination and application, there is a particular need to develop a more holistic view of science, engineering and technology, better integrating engineering into the narrow, linear model focusing on the basic sciences. To achieve this, there is a need to emphasize the way engineering, science and technology contributes to social and economic development, and the vital role engineering will plays in promoting sustainability, climate change mitigation and adaptation. There is also a need to better integrate engineering into science and technology policy and planning, and of better integrating engineering, science and technology into development policy and planning, to provide a more useful and accurate reflection and model of reality. This also applies in the development context – engineering, science and technology drive development, are vital in promoting sustainability and poverty reduction, and need to be placed at the heart of policies addressing these issues, at the national and international levels. As noted above, engineering is vital in addressing basic human needs in water supply, sanitation, housing, energy, food production and processing, transportation, communication, income generation and employment creation. Development policy and planning would benefit from a broader approach and 'evidence-based' analysis of the way engineering and technology drives development and reduces poverty. These considerations also relate to the need to transform engineering education to facilitate innovation and development, and the important role of PBL and humanitarian engineering in encouraging the interest, enrolment and retention of young people in engineering in developed and developing countries.

References

Berger, P. L., & Luckmann, T. (1966). *Social construction of reality: A treatise in the sociology of knowledge*. Garden City: Anchor Books.

Black, M. (1962). *Models and metaphors: Studies in language and philosophy*. Ithaca: Cornell University Press.

Datuk Chuah Hean Teik. (2009). *Presentation at the Institute of Engineers, Malaysia (IEM) 50th anniversary*. Reported in: *Engineering: Issues, challenges and opportunities for development*, redaction Tony Marjoram, UNESCO Report, 2010.

Edgerton, D. (2004). 'The linear model' did not exist: Reflections on the history and historiography of science and research in industry in the twentieth century. In K. Grandin, N. Wormbs, & S. Widmalm (Eds.), *The science–industry Nexus: History, policy, implications*. New York: Watson.

Engineering, U. K. (2010). *Annual report 2002–2010*. London: Engineering UK.

Engineers Australia. (2010). *Code of ethics*. http://www.engineersaustralia.org.au

Etzkowitz, H., & Leydesdorff, L. (2000). The dynamics of innovation: From national systems and "Mode 2" to a Triple Helix of university-industry-government relations. *Research Policy, 29*(2), 109–123.

Gibbons, M., Limoges, C., Nowotny, H., Schwartzman, S., Scott, P., & Trow, M. (1994). *The new production of knowledge – The dynamics of science and research in contemporary societies*. London: Sage.

Godin, B. (2006). The linear model of innovation: The historical construction of an analytical framework. *Science, Technology & Human Values, 31*(6), 639–667.

Hesse, M. (1963/1966). *Models and analogies in science* (revised edition). London: Sheed and Ward.

Huyer, S., & Westholm, G. (2007). *Gender indicators in science, engineering and technology*. Paris: UNESCO Publishing.

JIPSA. (2006). *Joint Initiative on Priority Skills Acquisition, part of the Accelerated and Shared Growth Initiative for South Africa (ASGISA)*, Government of South Africa.

Latour, B. (1987). *Science in action – How to follow scientists and engineers through society*. Milton Keynes: Open University Press.

Lawless, A. (2005). *Numbers & needs: Addressing imbalances in the civil engineering profession*. Halfway House: South African Institution of Civil Engineers (SAICE).

Mackenzie, D., & Wajcman, J. (Eds.). (1985). *The social shaping of technology*. Milton Keynes: Open University Press.

Metcalfe, S. (2009). *Keynote presentation at the OECD-UNESCO International Workshop, "Innovation for development: Converting knowledge To value"*. Paris: OECD, 28–30 Jan 2009.

Morton, O. (2009). Wanted: Green engineers. *The Economist*. The World in 2010, Dec 2009.

National Academy of Engineering. (2007). *Rising above the gathering storm. Energizing and employing America for a brighter economic future*. Washington, DC: The National Academies Press.

Nowotny, H., Scott, P., & Gibbons, M. (2001). *Re-thinking science: Knowledge in an age of uncertainty*. London: Polity.
OECD. (1963/2002). *Proposed standard practice for surveys of research and experimental development – "Frascati Manual"*. Paris: OECD.
OECD. (1992/2005). *Measurement of scientific and technological activities, proposed guidelines for collecting and interpreting technological innovation data – "Oslo Manual"*, Paris: OECD.
OECD. (2010). *Annex to OECD Oslo manual*. Paris: OECD
OECD/Eurostat. (1995). *Manual on the measurement of human resources devoted to S&T – "Canberra Manual"*. Paris: OECD.
Stewart, F. (1977). *Technology and underdevelopment*. London: Macmillan.
UNESCO. (2010). *Engineering: Issues, challenges and opportunities for development*. Paris: Redaction Tony Marjoram, UNESCO.
Westholm, G. (2010). Engineering indicators: Measurement and metrics. In UNESCO (Ed.), *Engineering: Issues, challenges and opportunities for development* (pp. 70–123). Paris: Redaction Tony Marjoram, UNESCO.
WFEO (World Federation of Engineering Organisations). (2001). *Code of ethics*. Paris: WFEO.
Whitley, R. (1972). Black boxism and the sociology of science. *Sociological Review Monograph, 18*, 61–91.

Tony Marjoram Guest Professor at Aalborg University, Honorary Fellow at Melbourne University and *Manchester Institute of Innovation Research*. Responsible for the Engineering Programme in the Division of Basic and Engineering Sciences at UNESCO from 2001 to 2011. Ph.D. from the University of Melbourne on Technology and Development in the Pacific Islands, M.Sc. on Science and Technology Policy from Manchester University, B.Sc. Hons in Mechanical Engineering from UMIST. Conceived, edited and produced the UNESCO Report *Engineering: Issues, Challenges and Opportunities for Development* in 2010 – the first UNESCO Report on engineering. Worked for UNESCO for 18 years, first at the Regional Office for Southeast Asia and the Pacific in Jakarta. Prior to UNESCO was a Senior Research Fellow at the International Development Technologies Center of the University of Melbourne, Senior Development Fellow at the University of the South Pacific, and Research Fellow at Manchester University. Published over 50 papers, books, articles and reports.

Part II
The Epistemological Basis of Engineering

Introduction

Anders Buch and Stig Andur Pedersen

In the tradition of Cartesian philosophy, epistemology, ontology and ethics deal with separate domains of reality and these branches of philosophy are often pursued independently of one another. Ontology reflects on and specifies the metaphysics of existence and reality, epistemology is concerned with the nature and scope of our knowledge of reality, and ethics is preoccupied with what is right and wrong in our conduct. However, the study of engineering practices challenges this departmentalization of philosophical reflections, and it seriously challenges the Cartesian ambition to find a firm epistemological basis for our knowledge claims about reality. Studying Engineering work reveals a plethora of complex processes involving many different kinds of assumptions, knowledge, techniques, values, and procedures.

As Stig Andur Pedersen demonstrates in Chap. 10, engineers are quite often involved in difficult decision processes involving serious uncertainties where relevant knowledge in many cases is lacking. On the other hand, as William Grimson and Mike Murphy show in Chap. 9, engineers can also be overwhelmed by data and information. Contrary to how it is in natural science, as in nuclear physics, for example, engineers are not able to restrict their work to a clearly delimited domain that can be studied by applying a restricted class of well-defined methods and procedures (Hendricks et al. 2000). In many cases engineering decision processes do not only concern genuine technical problems, but are entangled with social, economic, ethical, and political issues, as suggested by Pieter Vermaas in Chap. 8. So,

A. Buch
Department of Philosophy and Learning, Aalborg University Copenhagen,
A.C. Meyers Vænge 15, Copenhagen, SV 2450, Denmark
e-mail: buch@learning.aau.dk

S.A. Pedersen
Section for Philosophy and Science Studies, Roskilde University,
Box 260, Roskilde 4000, Denmark
e-mail: sap@ruc.dk

when discussing epistemological problems related to engineering work and research it is necessary to bear in mind that engineering problem solving often takes place in situations where (1) serious uncertainties are involved, (2) relevant knowledge is lacking, or is undiscernable due to the overload of data, and (3) the problem under consideration has essential social, economic, ethical, or political consequences. Engineering work is thus often conducted in complex and heterarchical settings (Stark 2009) that give no straightforward directions for performance. An epistemology of engineering has to address these kind of complications.

Engineering decision-making and problem solving take place in all kinds of practical situations in society. It is the practical situation that defines the scope of an acceptable solution. If a possible solution does not fit into customary practice, it will not be accepted unless it is modified in a way that fits practice, or until customary practice itself is changed. So, engineering decision-making and problem solving are heavily practice dependent and reliant on the normativities inherent in situations. That means, for instance, that the intention of the engineer does not necessarily coincide with intentions that dominate activities in practical life outside the engineering profession. Such different intentions must converge in order for a solution to be accepted as valid. Consequently, when discussing validity of engineering reasoning it is important to make clear that criteria of validity in many cases cannot be defined in a purely technical sense, but must pay regard to the various views and normativities that dominate practical life. In engineering, questions of right and wrong – questions of normativity traditionally dealt with in the discipline of moral philosophy and ethics – de facto affects the directionality and pursuit of engineering knowledge.

Newberry's wide-ranging Chap. 11 is working at the boundary of epistemology and axiology, thus transgressing the split between Part II and III in this volume. His chapter may be said to highlight the complexity or ambivalence of such important goals for engineers as "efficiency" and "optimization." On the one hand, efficiency is a presumed good of high importance: what engineer, or what human being, seeks inefficiency? On the other hand, what kind of efficiency do we seek? As Newberry writes, quoting Billy Koen, "In a society of cannibals, the engineer will try to design the most efficient kettle". Meditation on efficiency and optimization as unqualified goods leads to the conclusion that they are not: their goodness depends in part on the ends or goals for which they, in turn, are sought. If we do not examine these ends, we might enjoy "micro-efficiency", but this certainly does not ensure a general result in which we can be confident or of which we can be proud. To the contrary, the efficient manufacture of inefficient products is no one's ultimate goal.

It is of course the main focus of engineering work and research to efficiently solve practical – and often very complex – problems as they appear in our modern society, but complexities in society lead to further complexities and create new kinds of problems that require new kinds of engineering. Modern technical problems are so complicated that new advanced scientific methods and theories are needed in order to cope with them. That means that engineering decision-making and problem solving are, and must be, science based. So, engineers must not only have a clear understanding of the practice they are engaged in, their analyses and activities require advanced scientific methods and theories. In a sense, one distinctive feature of a modern engineer is that he or she must be capable of identifying a practical problem, putting it into an abstract scientific framework, and then coming up with a scientific solution that lives up to the practical needs. This leads to other complications of engineering epistemology. Engineering work comprises both abstract theoretical and normativized concrete practical knowledge, and both kinds of knowledge are relevant and must be interlinked. However as Stig Andur Pedersen shows in Chap. 10, scientific abstraction and engineering concretization pull in opposite directions, for which reason the interlinking of these two forms of knowledge is a serious challenge.

As engineering work situations are part of complex social practices there are often many other groups of professionals involved. In many cases engineers must consult political decision-makers, economists, scientists, and craftsmen. In order to work out a design or solve a problem the engineer must be able to communicate with other players within a heterogeneous project group and navigate within complex organizational structures. A design proposal or work plan is thus often a compromise made within heterogeneous groups, organizational textures, and interrelated social formations. It is an important engineering skill to be able to understand and compose compromises that eventually will lead to sustainable solutions. Understanding these complex patterns of professional conducts, performances, interactions, and negotiations places other demands on studying engineering activities. Drawing on practice theoretical perspectives, Buch argues in Chap. 7 that understanding engineering work would benefit from replacing the Cartesian epistemology of the individual knower with a relational and ontological perspective on engineering knowledge that focuses on the situated doings of engineers – thus developing a different ontology of knowing.

It is important to notice that many engineering designs and constructions address complex problems with huge social, economic, and political consequences. They may lead to great benefits but at the same time they may be very risky and might lead to serious catastrophes. As a consequence of this, engineering work will as a rule involve ethical issues. In fact, it radicalizes ethical problems due to the fact that engineering often proposes novel solutions that have far-reaching consequences and implications. That is evident if one looks at the consequences of the weapons industry, the use of pesticides and other modern polluting chemicals, or development of nuclear energy production. But also, less dramatic engineering designs – e.g. in software development, surveillance technologies – may affect human conduct in ways that call for ethical reflections. So, ethical issues are an integral part of an engineering problem situation. They are related both to the evaluation of useful consequences of possible solutions and to negative ones such as possible risks, issues of sustainable production, possible violation of human rights, surveillance, etc. How ethical and other value judgments are integrated in and cooperate with other forms of engineering knowledge is an important open epistemological issue.

Epistemology is concerned with the nature of knowledge. But there are many different forms of knowledge(s) and it is impossible to give a common and unambiguous characterization of all forms independently of how they are situated

(Harraway 1991) and enacted within social practices. To understand engineering knowledge we need to contextualize it. As we have seen, engineering work requires knowledge on many different levels and it is important that these very different forms do cooperate during engineering work. So, engineering epistemology concerns practical, context-dependent as well as abstract, supposedly context-free knowledge. It must consider questions of objectivity as well as subjective, situation -dependent, or value-dependent questions, and it must cope with uncertain and partial knowledge. Karen Barad has in fact suggested that instead of speaking of epistemology, ontology, and ethics as compartmentalized perspectives, we should preferably pay attention to the interrelation of these domains:

> The separation of epistemology from ontology is a reverberation of a metaphysics that assumes an inherent difference between human and nonhuman, subject and object, mind and body, matter and discourse. Onto-epistem-ology – the study of practices of knowing in being – is probably a better way to think about the kind of understandings that we need to come to terms with how specific intra-actions matter. Or, for that matter, what we need is something like an ethico-onto-epistem-ology – an appreciation of the intertwining of ethics, knowing, and being – since each intra-action matters, since the possibilities for what the world may become call out in the pause that precedes each breath before a moment comes into being and the world is remade again, because the becoming of the world is a deeply ethical matter. (Barad 2007, p. 185).

How to integrate all these different forms of knowledge in a rational and responsible way in concrete problem solving situations is thus the big issue for engineering epistemology. The chapters in this part all deal with this complex phenomenon.

Anders Buch in Chap. 7 reflects on studies of engineering practices and proposes a research agenda inspired by practice theory for advancing engineering studies further. In the practice theoretical perspective, epistemological questions are transformed into ontological ones. Following practice theory he proposes that it is necessary to study the situated lived lives and social practices of engineers in order to understand the ways engineers perceive, interact with, and reflect on their environment. He outlines the theoretical and methodological presumptions of the practice theoretical perspective and points to the advantages of adopting this perspective in engineering studies. The practice theoretical perspective is preoccupied with identifying and describing mechanisms of change and stability in social practices, in organizations, and in social reality in general. It can thus serve as a valuable framework for addressing issues and concerns in relation to engineering education reform initiatives and interventions and design efforts in engineering work practices.

In Chap. 8 Pieter Vermaas contemplates the changing role of the engineer in design processes. Designing has been viewed as a quintessential and defining characteristic of engineering practices, but Vermaas documents how this trademark has in fact undergone fundamental changes over the last five decades. Originally, the engineer was positioned in the role of an assistant supplying technical solutions in design processes. But roles have shifted and engineers have increasingly involved themselves in 'non-technical' elements in the design process, such as needs formulation and problem formulation in the design process, to in fact taking upon themselves to actively suggest and identify latent needs of users. This suggests that the engineers have come to play a more significant role in all phases of the processes, but it equally suggests that engineering can no longer be confined to the technological domain. Design processes have developed in ways that give still more authority to engineers, but at the price of pushing engineering to non-technical domains. The development Vermaas documents in design practices mirrors what Rosalind Williams (2002) in more general terms has diagnosed as the 'expansive disintegration' of engineering. Increasingly engineers comes to occupy new domains and functions in developing and producing solutions, but in transgressing the traditional confines the engineers risk blurring the unique and defined disciplinary boundaries of engineering.

William Grimson and Mike Murphy point out that engineers use whatever knowledge is relevant, whatever its origin, to address a particular challenge. It is an epistemological problem that there is more relevant knowledge than can be absorbed. The body of knowledge relevant for an engineer can be organized in pyramidal structure. The bottom layer represents fundamental knowledge such as mathematics and natural science. The middle layer comprises engineering domain knowledge such as engineering analysis and design methods. Finally, at the top level one finds knowledge associated with the competence required of chartered or professional engineers. This system of knowledge is not uniquely determined and could easily be expanded. It is in a sense a mélange. Consequently, it is not possible to find a unique epistemological foundational basis for engineering. But although the knowledge content of the structure changes, sometimes abruptly, the structure is still there. The middle layer represents the core of what engineering is, namely the ability to analyze, design, test, evaluate, etc. The top layer of knowledge characterizes the deep understanding of the underlying layers by which engineering work can be carried out using appropriate methods and when necessary devising new approaches.

The chapter by Stig Andur Pedersen discusses the tension between scientific idealization and engineering concretization. Modern science and engineering design are two closely related activities. Both mathematics and empirical science need advanced technical devices such as computers and laboratory equipment, and modern engineering design would be impossible without advanced scientific theories and methods. However, in spite of this close interdependence they are two different activities with their own specific logics. The main goal of science is to identify and study the most general laws of nature. This requires comprehensive abstraction and idealization, and, as a consequence of that, advanced mathematical and physical theories are only valid of highly abstract and isolated systems. But engineering design is concerned with concrete constructions in real contexts. Engineers are able to build models that describe and explain systems and mechanisms under abstract and idealized conditions. However, technological devices must live outside controlled laboratory conditions. In such open contexts our knowledge is uncertain and imperfect. Hence, the engineer must face the very complicated epistemological problem of building a useful foundation for decision-making in situations where certainty and completeness is impossible.

Finally, in Chap. 11, Byron Newberry reflects on different notions and meanings of efficiency and how they relate to engineering values. Newberry points out that although efficiency plays an important role as a norm and value in engineering work, it is far from clear what should be understood by the term. Also, he points out, it is a mistake to associate efficiency specifically to engineering practices since notions of the term are also closely related to thinking in other disciplines, e.g., evolutionary biology and economics. He explores the many meanings of the term, both technical and broader, and argues that efficiency is not a specific signature value of engineering per se. Only in a microscopic sense is efficiency closely linked to engineering practices – but this seems also to be true in relation to economic activities in general.

References

Barad, K. (2007). *Meeting the universe halfway. Quantum physics and the entanglement of matter and meaning*. Duke University Press.

Harraway, D. (1991). Situated knowledges: The science question in feminism and the privilege of partial perspective. *Feminist Studies*, *14*(3), 575–599.

Hendricks, V. F., Jakobsen, A., Pedersen, S. A. (2000). Identification of matrices in science and engineering. *Journal for General Philosophy of Science. 31*, 277–305.

Starck, D. (2009). *The sense of dissonance. Accounts of worth in economic life*. Princeton University Press.

Williams, R. (2002). *Retooling: A historian confronts technological change*. MIT Press.

Anders Buch M.A. in Philosophy, University of Copenhagen, and Ph.D. in Educational Studies, Roskilde University. He holds an associate professorship in Techno-Anthropology at Aalborg University Copenhagen at the Department for Learning and Philosophy and he is affiliated to the Centre for Design, Innovation and Sustainable Transitions (DIST). He has published articles and books on knowledge, learning, education, and the professional development of engineers. He is presently involved in the strategic research alliance PROCEED: "Program of Research on Opportunities and challenges in engineering education in Denmark."

Stig Andur Pedersen M.Sc. in Mathematics, Physics and Philosophy, all from University of Copenhagen. Professor of Philosophy of Science at Roskilde University. He has published articles and books on mathematical logic, epistemology, philosophy of mathematics, philosophy of natural sciences, philosophy of medicine, philosophy of technology and engineering science. He is involved in the research alliance PROCEED, "Program of research on Opportunities and Challenges in Engineering Education in Denmark".

Chapter 7
Studying Engineering Practice

Anders Buch

Abstract The study of engineering practices has been the focus of Engineering Studies over the last three decades. These studies have used ethnographic and grounded methods in order to investigate engineering practices as they unfold in natural settings – in workplaces and engineering education. However, engineering studies have not given much attention to conceptually clarifying what should be understood by 'engineering practices' and more precisely account for the composition and organization of the entities and phenomena that make up the practices. This chapter investigates and discusses how a 'practice perspective' can make a contribution to Engineering Studies by clarifying the theoretical and methodological presumptions behind this widely used – but only vaguely conceptualized – study of practices. The chapter highlights the inspirations of practice theory and delimits practice theory from other accounts of human activity in order to clarify what a practice perspective suggests. Further, it clarifies the concept of practice and highlights how practices are fundamental in understanding the fabric of social orderings. Having accounted for these theoretical perspectives of practice theory the chapter will draw out some methodological consequences and discuss the ramifications of a practice theoretical approach for Engineering Studies.

Keywords Engineering studies • Practice theory • Methodology • Context

Introduction

The reproduction, development and transformation of engineering work and culture have been the focus of a number of theoretical and empirical studies over the last 60 years or so (Barley 2005). In the 1950s and 1960s the predominant perspective was that of the engineering profession studied by sociological methods including studies of engineers serving authoritarian regimes. In the 1970s the perspective shifted to

A. Buch (✉)
Department of Philosophy and Learning, Aalborg University Copenhagen,
A.C. Meyers Vænge 15, Copenhagen SV 2450, Denmark
e-mail: buch@learning.aau.dk

S.H. Christensen et al. (eds.), *Engineering Identities, Epistemologies and Values*, Philosophy of Engineering and Technology 21,
DOI 10.1007/978-3-319-16172-3_7

Marxist inspired discussions of the engineering profession in relation to class structure, in parallel to studies of engineering education and skills from a perspective coming from 'industrial sociology'.

Over the last 30 years the studies have – to a large extent – used ethnographic and grounded methods in order to investigate the specifics of engineering work practices in situated perspectives. This trend has – in many respects – led to a richer and empirically sensitive perspective on engineering work and culture. Thus, detailed studies of engineering work practices provide new material for a richer understanding of engineering culture (e.g. Bucciarelli 1994; Vinck 2003; Henderson 1999; Kunda 2006; Downey 1998; Barley and Kunda 2004; Johri 2010). The situated studies of engineering work practices have reflected on the organizational and corporate embedding of engineering work and described the minute negotiations that take place on a day-to-day basis in the reproduction of engineering culture.

A smaller body of engineering studies has supplemented the situated diachronic perspectives on engineering work practices by introducing synchronic perspectives that illuminate the broader life-worlds of engineers – thus reflecting on the subjective dimensions of engineering practices as narrated through the life-stories of practicing engineers (e.g. Buch and Christensen 1998; Buch 2002; Mellström 1995). Likewise, a small number of other studies have given accounts of the process of *becoming* an engineer and the process of neophytes entering engineering culture at engineering schools and universities (e.g. Downey and Lucena 1997; Tonso 2007).

What unites all these ethnographies is the awareness of context in studying engineering – an awareness that the phenomenon of engineering should be studied as and through situated practices; i.e., that engineering should be seen as a bundle of activities immersed in, influenced or determined by, and composed of various entities and phenomena. Although not all of the mentioned ethnographies explicitly talk about engineering in terms of situated practices, it is clear that all of them view engineering as social and material activities that are situated in time and space, part of a 'wider scene' and characterized by relatively durable ways of doings and sayings, rules, conventions, specific tools, equipment, procedures, analytical preferences, etc., that we recognize as part of engineering culture. This approach comes close to a practice perspective by recognizing the heterogeneity and complexity of the sites where engineering culture is enacted, reproduced, or even transformed. But it is also clear that engineering studies have not given much attention to conceptually clarifying what should be understood by 'engineering practices' or more precisely accounting for the composition and organization of the entities and phenomena that make up the practices.

In what follows I will investigate and discuss this 'practice perspective' – or the 'practice lens' as it is often referred to (Feldman and Orlikowski 2011; Corradi et al. 2010) – and thus make the theoretical and methodological presumptions behind this widely used – but only vaguely conceptualized – approach to the study of engineering culture more explicit. However, it is not the ambition of this chapter to review the body of literature of engineering studies or to analyze the theoretical and methodological approaches of existing contributions to the field. Instead I will suggest and outline a proposal for a research agenda for engineering studies by drawing on insights from the emerged interdisciplinary research tradition of practice theory. In doing so, caution must be made not to reify or hypostasize practice theory – in fact it is more precise to talk about practice theories in the plural. Practice theory is not a unified theory and methodology. Practice theory is better described as a set of theoretical and methodological insights that are historically affiliated and bear conceptual similarities. I will start by highlighting the inspirations of practice theory and delimit practice theory from other accounts of human activity in order to clarify what a practice perspective suggests. I will then continue to clarify the concept of practice and highlight how practices are fundamental in understanding the fabric of social orderings. Having accounted for these theoretical perspectives of practice theory I will draw out some methodological consequences and discuss the ramifications of a practice theoretical approach for engineering studies.

Practices

Practice theoretical approaches have made their entry in the social sciences and humanities over the last 30 years. Still more scholars in different disciplines and with different research interests and backgrounds have focused on the day-to-day practices of actors in their studies. Philosophers like Theodore Schatzki (1996, 2002), Joseph Rouse (2007) and Andreas Reckwitz (2002a, b) have sketched out the fundamental ontological and epistemological presumptions of practice theories in relation to agency, the social, and society, and described how practice theories draw on philosophical insights from mainly the late Wittgenstein and the younger Heidegger, but also significantly the early Giddens, Bourdieu, Butler, and the late Foucault. In organizational studies, social scientists like Wanda Orlikowski (2000, 2002), Silvia Gherardi (2006), Davide Nicolini (2013) and others have theorized and analyzed the role of technology within organizational development and change, and learning theorist like Paul Hager et al. (2012), Jean Lave (1988, 2011), Jean Lave and Etienne Wenger (1991) and Etienne Wenger (1998) have demonstrated how learning processes are best understood as transformations of and within practices. In another intellectual tradition, namely activity-theory, Yrge Engeström (1999) and others have studied work practices and stressed the interplay with the material environment and the role of tools as essential features of human practices. The practice theoretical approaches have spread to other areas of research like consumption (Shove et al. 2012; Warde 2005) and sustainability studies (Shove and Spurling 2013; Cohen et al. 2013). In Science and Technology Studies (STS), practice theoretical approaches have appeared most notably in the works of Karin Knorr-Cetina (1985, 1999) and Joseph Rouse (1996, 2002), but practice theoretical approaches are held in common with many STS approaches, e.g. in the traditions of ethnomethodology, actor-network theory, and other posthumanist perspectives (e.g. Pickering 1995). Several announced scientific journals have devoted special issues to the discussion of the new practice approaches within the social sciences (e.g. *Organization* 2000, *The British Journal of Sociology* 2002 and *Human Affairs* 2007) – thus practice theoretical approaches have come to the fore and significantly influenced contemporary social science. Many scholars have observed this impact and describe the increasing attention to social practices as a 'practice turn' in social science (Schatzki et al. 2001) or a 'bandwagon' of practice based studies (Corradi et al. 2010).

But what have made the practice perspective so attractive to these social scientists? What are the general assumptions that draw researchers of different intellectual origin and tradition together in studying such diverse phenomena as 'consumption' and 'scientific knowledge production' by using the 'practice lens'? In a newly published introduction to practice theory Nicolini (2013) characterize the general assumptions within practice theoretical studies. He points to five assumptions (2013, 1ff.): Firstly, it is a characteristic of practice theories that they focus on the lived social life of actors. Thus the social activities and work processes of actors are studied and the routinized rule governed and institutionalized characters of performances are given special attention – not as explanatory devices, though, but as accomplishments of human activities. Thus the practice theoretical approach stresses the productive and reproductive aspects of human activities in understanding stability and transformation within social formations. Learned skills, rituals, procedures, etc., are central foci for investigations. Secondly, practice theoretical approaches try to do away with dichotomies and refuse to understand human activities in binary terms of agency/structure, subjective/objective, and body/mind. Practice theories stress that human action is embodied, and temporally and spatially situated in material environments. In order to understand human action it is thus mandatory to reflect on the specific physical and material settings within which the actions take place. Thirdly, practice theories do not conceptualize human agency in line with the classical conceptions of either the *homo economicus*, i.e. the autonomous rational individual with purposes, intentions, etc., or the *homo sociologicus*, i.e. the norm-abiding or rule-following 'cultural dope'. Instead practice theories conceptualize human agency in terms of the *homo practicus* – the human agent that 'carries', but also 'carries out' social practices (Reckwitz 2002a, p. 256). Practice theory is a branch of culturalist theories, but it deviates from structuralist and subjectivist (phenomenological and interpretative) cultural accounts by focusing on the enactment of practices. Fourthly, practice theories resolutely reject representational theories of knowledge, meaning, and language. 'Knowledge' is not a property of the mental states of individuals, but is better understood as 'knowing' that is produced and shared within concrete activities and practices. Likewise, discourse does not belong to a separate non-material structural realm, but is an integral part of unfolding human material practices. Finally, practice theories foreground that the dynamics of practices should be understood in terms of power relations, interests, negotiations, conflicts, etc. Thus the perspective recognizes the highly contingent features of human affairs and stresses the interruptions, contestations and unevenly distribution of resources and privileges in social life.

These characteristics of practice theoretical perspectives are, of cause, highly interwoven in theoretical accounts, as they are in concrete studies informed by the practice theoretical approach, and it is also the case that practice theorists have different concrete interpretations of the tenets outlined above. It would be presumptuous – and in fact misleading – to stipulate that *a* practice theoretical position can be identified. Instead, more authors (e.g. Nicolini 2013) – borrowing Wittgenstein's concept – have pointed to the fact the theories bear a 'family resemblance' to one another: No strict communalities can be found in all of the theories, but many similar features can be traced in many of them. It would thus be more precise to speak of practice theories – in the plural. But the aforementioned approaches do share a common awareness of the fundamental relational character of being-in-the-world. Individuals are not isolated observers or agents that occasionally interact with other individuals or the material environment, nor are relations construed as abstract structures that bind actors together. Actors construe relations as links between particular and specific entities – both human and non-human. Likewise, humans do not interpret each and any occurrence in order to experience the world they live in. Practice theory recognizes that things are mostly always-already-interpreted; we have become familiar with the world through the training, routines, socialization, ways of life, etc., that makes us human. What pre-occupies researchers within the tradition is thus, according to Martha Feldman and Wanda Orlikowski (2011, p. 1240), to investigate the 'why', 'how' and 'what' of practices. The 'why' question is primarily dealt with by philosophical reflections over the ontological and epistemological status of social life. Here practice theoretical approaches argue for the ontological primacy of practices. The 'how' question of practice theories is answered by the specific practice theories of, say, Pierre Bourdieu (1990), Lave and Wenger (1991), or Engeström (1999). They specify the dynamical mechanisms that explain how relations within and between practices are enacted, reproduced, and transformed. Finally, the 'what' question deals with the empirical findings of practice theories. It will take us too far to elaborate on the 'how' and 'why' questions. Practice theories comprise a broad range of theoretical approaches and span numerous fields of research. To illustrate what practice theories can offer Engineering Studies I will instead focus on the 'why' question and account for the ontological thesis of practice theory. Here I will primarily refer to the work of Theodore Schatzki.

The Primacy of Practices

As this book testifies, the role of context is central to the study of engineering (work) practices. Firstly, engineering studies recognizes that engineering is part of a 'wider scene' and that engineering is not just about technical specificities. There is seemingly a broad consensus in engineering studies that engineering and technology must be studied as complex phenomena through ethnographic methods that are sensitive to the complexities of the endeavor. But the consensus stops when scholars try to answer questions about the complexities. What exactly is this 'wider scene' and how do the complexities impact, shape, or determine engineering (work) practices? Context is often invoked to indicate that engineering is not a self-sufficient, self-determined, and self-explanatory phenomena, but, on the contrary, part of something more that 'surrounds' it. The use of context thus indicates that something – the text: engineering – is part of and entangled in something more that surrounds it. Secondly, context indicates that forces of determination are at play within this entanglement that somehow give structure and establish orderings. And thirdly, that the entanglement is made up of various entities that are interconnected (Schatzki, 60ff.). These characteristics vaguely and only formally make the use of context intelligible. We need to learn more about in what sense engineering is surrounded by 'something', how that 'something' is affecting engineering, and the character of the various entities that make up the context. Engineering studies are thus challenged to be more precise and explicit about specifying the ontological and epistemological presuppositions of contextual investigations.

I suggest that engineering studies can benefit from the development of practice theoretical accounts. Practice theory sees practices as fundamental units of analysis and investigates the specific activities and the organizing of the activities in detail. I suggest that this focus can help engineering studies to be more specific about invoking contextual analysis and thus be more explicit about outlining what elements and mechanisms are at play in engineering contexts. It is true that the concept of practice has been used in a variety of ways (cf. Turner 1994) and it thus might seem a poor qualifier for being more precise about the use of context. But through the work of Thedore Schatzki (1996, 2002, 2003) the concept has been specified to give more precision. I will follow Schatzki's account of practices and illustrate how the notions he introduces are relevant for understanding engineering (work). Schatzki broadly characterize practices as sets of doings and sayings (2002, p. 73). Practices thus comprise bodily actions as well as linguistic utterances, gestures, etc., and thus subsume what in other theoretical traditions are labeled as behavior and discourse. What unites these actions and linguistic utterances into sets of doings and sayings are the specific tasks and projects that impose orderings of the actions. What makes us characterize a reading of a thermometer or the reporting of temperature increase as part of engineering practices are by reference to the tasks (e.g. doing experiments) and the project (e.g. developing enzymes) of which they are a part. Practices are thus composed as hierarchically ordered wholes that have certain duration in time. The regularity of the doings, sayings, tasks, and projects does not have to be constant over time in order to qualify as practice. Practices can change and innovate over time and it is a matter of empirical investigation to trace these changes as they unfold. But for doings and sayings to qualify as part of a practice it is essential that regularities can be detected and disruptions are outbalanced by continuities.

Practices thus indicate that human activities are linked through certain *normative orderings*. One essential ordering element is the *practical understandings* of the actors. Actions are considered competent and qualified according to standards and procedures – mostly implicit and tacit by nature. The bio-chemical engineer who is engaged with the development of a new enzyme must know how to deal with experimental settings and among a lot of other things know how to read a thermometer. Furthermore she must be able to identify why and when it is appropriate to read the thermometer and how to respond to an increase in temperature in the experimental situation. She must be able to see things like an engineer (cf. Goodwin 1994) and frame problems and (research) questions accordingly. Practice theory emphasizes that these activities are founded in the practical skills and know-how that actors acquire through participation in practices and through drill. Practical understanding displays an ability of knowing 'how to go on' and having 'a feeling for the game', thus acting according to the prevailing standards of the practice. Bourdieu stresses (1990, Chap. 4) that the acquisition of the skills is very much a matter of bodily incorporation and Wittgenstein highlights the importance of drill and training in learning how to follow rules and partake in 'a form of life' (1958, §218ff.). From a practice theoretical perspective it is important to understand the *processes of becoming* an engineer and understand how the practice of engineering is reproduced through learning and training activities. That might be in engineering schools and universities but also very significantly in work practices. Practices thus only exists as continual (re)productions or accomplishments.

Another ordering element is of cause the *explicit rules*, regulations, instructions, standards, and procedures that are pertinent for specific practices. Engineering is a profession that is regulated by professional bodies, legislation, corporate rules, standardization of equipment, safety procedures, etc. The institutional role of engineering as a profession in society is regulated through myriads of restrictions and allowances that shape and order the labor processes through e.g. the division of labor among professionals, and the incentive structures in wage or contract labor. These explicit regulations are very much based on conventions and bear huge national differences. But they are essential in shaping the practices of engineering education and work. Gary Downey and Juan Lucena (2005) for example demonstrate how the ongoing internationalization of engineering work has ramifications for engineering education and thus the formative training of engineers into the profession.

According to Schatzki a third ordering element that links doings, sayings, tasks, and projects is the *teleoaffective structures* of practices. "A 'teleoaffective structure' is a range of normativized and hierarchically ordered ends, projects, and tasks, to varying degrees allied with normativized emotions and even moods (Schatzki 2002, p. 80)." These structures need not be explicitly conscious goals to, or ends in view for the actors, but should rather be seen as structural signifiers that give an overall sense to actions. Schatzki emphasize that these structures are recurring effects of actions and should not be conflated with structuralist accounts. The teleoaffective structures emerge when there is general agreement about what is acceptable or unacceptable to do in situations. The presence of teleoaffective structures does not exclude controversy or disagreement about specificities but provides an overall sense of purpose and direction for the activities. The structures both produce the practice and are produced by the practice. Louis Bucciarelli and Sarah Kuhn (1997, p. 212) describe the 'object worlds' that engineers live within in the following words:

> .....the goal of storytelling and scenario making is to achieve closure: arrive at a design that is fixed, repeatable, stable, unambiguous, and internally consistent. Object world thinking is thinking about the rigidly deterministic. [...] The engineer's ability to abstract from a concrete situation, to see an object as a collection of forces, or as a network of ideal current generators connected in series and in parallel, is key to problem solving and to managing complexity within object worlds. One of the crucial skills conveyed as part of disciplinary training is the ability to look at a design, or at a collection of objects, and to see them as an abstraction to which scientific principles can be applied.

These observations of the overall goals that inform engineering work both describe the overall teleology installed in engineering practices and clearly demonstrate the normativities, values, and virtues that actors subscribe to in engineering practices.[1]

A final ordering element relates to the *general understandings* that are available to and shared by actors within a practice, though these general understandings, as the word indicates, are not proprietary of specific practices, but are generally shared norms and values. However, they are also active in structuring specific practices. Engineers like all other members of a community endorse certain religious, ethical, ideological, or political norms. Many of these are codified in codes of conduct within companies or professional societies and associations (cf. Van de Poel and Royakkers 2011, Chap. 2), but they need not be explicitly stated to be conductive. These general understandings thus often span different practices and can make them overlap at specific junctures in history.

These ordering elements of practices are not meant to be jointly exclusive or exhaustive characteristics. On the contrary the elements are combined in the doings, sayings, tasks, and projects of the practice in complex and interwoven ways. Thus the specific constellation of these – and maybe other – elements compose the uniqueness of the practice. Furthermore, practices are always situated in specific orders or arrangements that comprise both practices and non-human/material objects. The arrangements and the social practices thus jointly constitute the overall site where things exist and events happen (Schatzki 2002, p. 63). Sites are a special kind of contexts – namely the kind where practices unfold in activities and events. To put this point another way, sites are the kind of contexts where actors' ends and human intentions matters. Sites are thus not only locations in objective time and space or even activity-place space, but they are also significantly teleological located. Sites are part of 'wider scenes' of events and activities. The bio-chemical engineers reading of the thermometer is an activity that is part of the event of the experiment. Likewise, the experiment is part of a project about the development of new enzymes, and this project, in turn, a part of a company's ambition to develop new products that can increase profits, etc. Sites are thus nested. Finally, for an event or activity to occur within a site is tantamount to that event or activity being a constituent part of that context. Activities and events are thus both contained in the site, but also an integral part of the sites makeup.

This site ontology forefronts and gives special attention to human activities and social practices by highlighting the teleological and intentional dimensions of activities. While recognizing that practices are intrinsically interwoven with material objects, and that objects in significant ways order, prefigure, and causally impact practices, the ontology is reminiscently humanist. It gives special attention and priority to human endeavors.

[1] I am quoting Bucciarelli's and Kuhn's description of the 'object worlds' of engineers not to make a general point about concept of 'object worlds' in relation to teleoaffective structures, but to illustrate the overall teleology and normativity that is installed within engineering practices.

Methodological Pluralism

What are the consequences of a practice perspective for engineering studies? It is obvious that the practice perspective introduces a new ontology of sites, orderings and practices and thereby envisions the research object – engineering – in new ways. Engineering is not to be studied as either encompassing structures or individual achievements, through the lens of technological determinism or subjective voluntarism, or as an act of intellectual or manual work. Instead the practice perspective suggests that engineering should be studied as an ongoing practice of day-by-day skillful and goal oriented social and material reenactments of procedures and (codified or tacit) rules. Thus the practice theoretical perspective not only suggests a new ontology, but – by implication – methodological approaches:

> In the end, I believe, one should adopt a neo-Quinian picture of social investigation in which (1) ontologies are part of the conceptual armature of social investigation and (2) arguments about ontological issues are part of the overall enterprise of social research, another part of which is the methodic gathering of data. (Schatzki 2003, p. 189)

Schatzki thus, by alluding to Quine's doctrine of confirmation holism (Quine 1961), suggest that the conceptualization of practice ontologies should be informed by methodological considerations and vice versa. Although Schatzki (2002) throughout the development of his ontological suggestions gives empirical illustrations, he does not develop a methodology. Others within the practice tradition have, however, elaborated detailed accounts (e.g. Gherardi 2012; Nicolini 2009, 2013). Nicolini seems to agree with Schatzki that methodologies and ontologies are closely interwoven and that the researcher in his/her investigations must develop a sensitivity and flexibility to adopt the right tools for the right job. In recognizing the complexity, heterogeneity, and uniqueness of practices and the varying research interests of researchers, different methods and approaches must be adopted. In the broad spectrum of practice theoretical approaches different research agendas have appeared. In the traditions of discourse analysis and conversation analysis special focus has been given to the role of language and communication in practices (e.g. Fairclough 1995; Richards 2001), theorist like Orlikowski (2000) and Activity-Theorist (e.g. Engeström 1999) have, respectively, paid much attention to the role of technological artefacts and the role of tools within practices, Lave and Wenger (1991) have stressed the role of identity and belonging to communities of practice, and ethnomethodological research (e.g. Garfinkel 1967) has given special attention to the minuteness of day-to-day activities as accomplishments of practices. Practice theorists have thus adopted different methods and approaches according to the specific research interests and the specific character of the practices investigated. Nicolini (2013, p. 213) suggests that the practice theorist adheres to a methodological pluralism in research:

> [...] I will embrace a [...] strategy that can be described as a form of programmatic eclecticism or, more simply, a toolkit approach. My main tenet is that to study practice empirically we are better served by a strategy based on deliberate switching between theoretical sensitivities.

Just as the study objects of practice theorists – the constituents of practices – are situated in and reenacting complex practices, so are the researchers themselves. Researchers select their study objects according to specific purposes, goals, interests, perspectives, and motives and their research practices are prefigured by material, technological, institutional, and economical restraints and affordances. The character of the theoretical perspective, the chosen nature of interaction and intervention in relation to the objects of study, and the chosen methods of data interpretation all underlines the performative, partial, and perspectival nature of research. The deliberate and reflexive consideration of the choice and use of methods and theories can be seen as strengthening validity and transparency in the research process. But the practice theorist must insist that there is no one privileged perspective or method that can represent the totality or complexities of practices. In understanding the character and dynamics of practices, research will always be on its way to find more apt and more sensitizing questions and concepts for investigation as well as reconsidering the overall usefulness of methods.

It must be realized that practices always exist and develop in relation to other and wider practices. Practices are nested and bear relational ties of causal, spatial, intentional, restrictive, and affording characters to the arrangements they are part of (Schatzki 2002, 38ff.). To understand these entanglements and relations adequately Schatzki proposes (2002, p. 41) that the accounts

> .....of social relations must satisfy at least two desiderata. First, it must construe relations as links among particular entities, as opposed to types of hypostasized abstractions. Second, it must cover the full range of connections among components of arrangements through which human lives hang together, not just links that join humans directly.

These desiderata point to the fact that practices are impacted by and have ramifications on events and happenings beyond the practice considered. Schatzki thus calls for methods of study that are able to understand the situated and contextual character of the practices. Nicolini suggests a research method that honors Schatzki's requirements. He suggests that the repertoire of practice theories is mobilized according to the specific character of the research field and the specific interests of the researcher. This calls for a reflexive, flexible, and innovative use and combination of tools available. He does, however recommend that the research follow a pattern of zooming in and zooming out on the practices under investigation. Starting by zooming in on the located practice – i.e. the doings and sayings of the participants in the practice, describing the temporal flow of the practice, accounting for the practitioners' general understandings and horizons – Nicolini argues (2009, p. 123) that the researcher can start organizing the ethnographic research process. The zooming in should then be followed by a process of zooming out in time and space in order to 'follow the practice' wherever it has ramifications. This process of zooming out is motivated by the same reasons George Marcus (1998) laid down for doing multi-sited ethnographies, namely the increasingly dispersed and network character of human lives. The zooming out is thus laying out the rhizomatic nature of practices and describing the texture of connections between practices. The process of zooming in and zooming out should be iterated until the researcher feels comfortable explaining why the practices are the way they are.

Studying Engineering Practices Through the Practice Lens

Feldman and Orlikowski (2011) point to two advantages in adopting the practice lens. Firstly, practice theory does not pretend to produce theoretical generalizations and give universal explanations. Instead practice theoretical studies are preoccupied with the situated dynamics of practices. "[But a]lthough each context of study is different, the dynamics and relations that have been identified and theorized can be useful in understanding other contexts. In this way, theoretical generalizations are powerful because they travel" (Feldman and Orlikowski 2011, p. 1249). Secondly, Feldman and Orlikowski stress that practice theoretical generalizations can be of practical use in identifying organizational levels of change and supporting or restricting specific microdynamics, e.g. by highlighting the reproductive effects of identified practices.

Let me try to exemplify the potentiality of the practice perspective vindicated by Feldman and Orlikowski by introducing two very different research projects of relevance to engineering studies. Neither of these projects is conducted under the aegis of 'practice theory' in any strict sense, but they serve to illustrate problematics that are both central to engineering studies and that can be framed in terms of practices. The first project is the ongoing PROCEED project (Program of Research on Opportunities and Challenges in Engineering Education in Denmark)[2] that strives to elicit the challenges facing engineering (education) today and analyze the response strategies taken towards these challenges. The other one is an experiment with the human practices in synthetic biology undertaken by Paul Rabinow and Gaymon Bennett (Rabinow and Bennett 2012). This project aimed to develop ethical practices among groups of bio-engineers that did research in synthetic biology.

Challenges Facing Engineering Education

The literature on the challenges facing engineering is vast.[3] Although most observers agree that the challenges are many there is no consensus about the nature of the challenges. Some observers stress that labor market demands call for engineers to be more business oriented and flexible in order to guarantee employability and competitive advantages on a personal, organizational and national level. Other observers call for engineers to recognize their professional responsibility and to conduct their engineering professionalism in ways that serves humanity and the environment. Still other observers stress that the disintegration and proliferation of technological knowledge in modern society calls for a new brand of hybrid engineers that can synthesize technical and social elements. Thus, according to the observers, engineering education has to change its curriculum and didactical principles to

[2] For more information visit: http://www.proceed.dk/?languageId=1

[3] The points made in this sections are further developed in Buch (2012).

accommodate the challenges as perceived by the respective observers. The challenges are thus construed in an ontology that stipulates them as objective and irredeemable. This construal installs a one-way causality that demands certain changes within engineering education in order to accommodate the objective challenges facing engineering. Reformers thus contemplate how curriculum and didactics should be changed in order to educate either more flexible, more responsible, or more hybrid engineers.

Now, reframing this problem in the light of the practice lens construes the problem in a different way. First of all, the ontological status of the challenges should be reconfigured and situated in relation to specificities of the observers' normativities and positions in society. The challenge perceptions should be understood on the basis of the interests, privileges, and power relations associated with the observers' positions. Furthermore, the manifestations of the challenge perceptions should be studied as material-discursive practices – and so should engineering education. Thus, secondly, the one-way causality between presumed societal, normative, and epistemic challenges to engineering on the one hand and engineering reform on the other must be questioned. Challenge perceptions (i.e., the initial framing of what engineering is and what is wrong with contemporary engineering education) and response strategies (i.e., indications of how engineering education should be reformed) are intimately linked and co-constitutive. It is not possible to establish a 'view from nowhere' to identify challenges and suggest reform initiatives. The challenges to engineering are always perceived from somewhere, e.g. the perspective of commercial enterprises, the engineering profession, or academia. Furthermore, these vistas are formations of enacted material-discursive practices that privilege certain virtues – such as e.g. profit, professional autonomy, or intellectual reflection.

The PROCEED project studies practices in engineering work by ethnographic methods that elicit the practical understandings, the rules, the teleoaffective structures, and the general understandings of the engineering practices. This is *not* done in order to establish a 'more realistic corrective' to engineering reform initiatives, although it does qualify imageries about what engineering work 'really' is nowadays. Instead, the intention of doing engineering ethnographies – seen from the practice theoretical perspective – is to identify dynamics and relations at play in engineering practices that can be theoretically generalized. One significant thesis of my research is that in order to understand engineering practices adequately the specific relations between the constitutive relationships of engineering educational practices and engineering work practices must be illuminated. My studies in engineering work practices indicate that the professional preferences, perspectives, and aspirations of engineers significantly points to formative processes, identity formations, and socialization processes initiated during engineering education. Accordingly, the practice theoretical methodology recommends to 'follow the practice' around – significantly, I would argue, from engineering educational practices to engineering work practices. Adopting a practice theoretical perspective in engineering studies thus calls for undertaking more longitudinal studies of transitions between engineering education and engineering work. It is vital to understand the ramifications and dialectical interplay between educational practices and work practices in engineering when educational reform initiatives are discussed. The discussion gets off on the wrong foot when challenges to engineering work practices are reified and engineering education is perceived as an independent variable in construing a 'match' between demands for engineering competencies and the production of engineering capabilities in education. Adopting the practice lens can rectify this deficient perspective and provide a richer and more dynamic way of framing the discussions on reforming engineering education.

Designing Engineering Practices

Feldman and Orlikowski's second point has to do with change and how practice theoretical studies can contribute to stimulate changes in practices by highlighting the micro-dynamics of the practices. It is obvious that the analytic identification of dysfunctionalities within practices can provide a good starting point for interventions. The question is whether the practice theoretical approach has potentiality beyond the mere analytic identification of micro-dynamic dysfunctionalities. It is not possible to settle this question here, but I will point to an interesting research project conducted by Rabinow and Bennett (2012, 2013) in synthetic biology. I leave it as an open question whether Rabinow and Bennett's approach describes a way forward for practice theoretical interventions.

Rabinow and Bennett (2012) report on an intervention 'experiment' they conducted at the Synthetic Biology Engineering Research Center (SynBERC) from 2006 to 2010. They were invited to participate in a NSF project and develop bioethical procedures and reflections for the new research traditions of synthetic biology. Instead of framing the task as a question of providing ethical criteria and codas for scientific conduct, their ambition was to make a design for human practices in research processes that could lead to human flourishing in the sense of the ancient Greek concept of eudaimonia. For reasons we do not have to go into here the project failed and the initiative at SynBERC was abandoned, but it is worth considering the general idea of an interventionist practice theoretical approach. Rabinow and Bennett suggests that this approach should be outlined as an 'anthropology of the contemporary'. Unlike Foucault's method of 'a history of the present' that problematizes present constellations and practices and demonstrates their inherent contingencies by using archeological and genealogical methods, an 'anthropology of the contemporary' proceeds through different rationales:

> [...] techniques for demonstrating contingency and for opening up possibilities, such as the history of the present allows, are not the principal aim and necessity. Rather, analytic modes are needed for giving form to under-determined and emergent relations, and for specifying the significance of these relations (Rabinow and Bennett 2013, p. 2).

Rabinow and Bennett suggest that research engineers and anthropologists join up in collaborative practices in order to reflect on possible blockages and opportunities in research. The common task at hand is thus to reframe the blockages and opportunities in new ways that opens for new solution spaces. When the anthropologist enters a practice and engages in collaborative reflections with the practitioner, new avenues of actions are made available for enactment. Thus Rabinow and Bennett suggest a new research agenda where anthropologists and other social scientists concerned with the study of human practices join up with researchers in the natural and technical sciences as co-researchers in order to incorporate ethical reflections in the unfolding research process. This practice theoretical proposal indicates a shift from downstream to upstream or midstream research where the role of social scientists are changed and the performativity of the research enhanced. I will refrain from discussing the viability of Rabinow and Bennett's proposal. But the example helps to illustrate the performative potentialities of practice theoretical approaches.

Conclusion

Engineering studies is a relatively new research field. Although there is a rich literature on engineering work and engineering education, it is only recently that efforts have been made to establish engineering studies as a research field in its own right with scientific journals, conferences, etc. In this chapter I have suggested that the practice theoretical research approach could serve as an impetus for engineering studies. Although it must be recognized that practice theory is not a monolithic theory or a unified methodology I have argued that it has potentials that can support and propel engineering studies. By stressing that the phenomenon of 'engineering' should be conceived as enactments of practices of skillful work, routines, rules, rituals, and procedures, and by paying attention to the normativities of these practices, the complexities and dynamics of engineering can be studied without resorting to reified conceptions. Likewise, practice theoretical efforts to avoid dichotomies can help understand engineering practice as an embodied activity that unfolds in materially situated contexts. I have proposed that Schatzki's outline of a site-ontology could serve as a useful conceptualization of 'context' in engineering studies and thus guide investigations in paying attention to how practices are normatively ordered according to the general and practical understandings, rules, and teleoaffective structures. Further, I have argued that engineering studies could benefit from the methodological resources of practice theories. Here I suggest that engineering studies employ the plurality of methods made available by practice theories in accordance with the specificities of the particular site of study and the perspective of the researcher. Finally, I have exemplified how the adoption of a practice perspective in research could suggest new avenues for structuring engineering studies that have 'practical' implications. All in all, I have made an argument for engineering studies to consider adopting the practice theoretical lens in developing the research field and for developing an adequate conception of context to understand engineering practice.

In closing this chapter I would like to point to the critical potentials of practice theory. Charles Taylor has discussed the development of the practice theoretical perspective through the work of Heidegger and Wittgenstein (Taylor 1995). His discussion shows that Heidegger's account of the 'finitude' of human existence (Dasein) and Wittgeinstein's account of 'meaning' as an unfolding 'form of life' (Lebensform) both aim to contextualize human understanding in relational and situated ways. Taylor sees Heidegger's and Wittgenstein's accounts as significantly counter-cultural and critical in the sense that they oppose the western cultural ideals of human intelligibility as disengaged and atomistic. Heidegger and Wittgenstein thus confronted the western intellectual legacy by criticizing the mentalist, rationalistic, individualistic, and disengaged ideals and conceptions that have informed science and technology in our culture. In drawing upon the insights of Heidegger and Wittgenstein, practice theory thereby installs a fundamental critique of the ontological and epistemological foundation of prevailing western scientific and technological enterprise. I think the critical perspective of practice theory would be an appropriate stance in the study of engineering practice and expert cultures – although, judged by the standards of the field under study, properly a rather awkward one.

References

Barley, S. R. (2005). What we know (and mostly don't know) about technical work. In S. Ackroyd, R. Batt, P. Thompson, & P. S. Tolbert (Eds.), *The Oxford handbook of work and organization*. Oxford: Oxford University Press.

Barley, S. R., & Kunda, G. (2004). *Gurus, hired guns, and warm bodies, itinerant experts in a knowledge economy*. Princeton: Princeton University Press.

Bourdieu, P. (1990). *The logic of practice*. Oxford: Polity Press.

Bucciarelli, L. (1994). *Designing engineers*. Cambridge, MA: MIT Press.

Bucciarelli, L., & Kuhn, S. (1997). Engineering education and engineering practice: Improving the fit. In S. R. Barley & J. Orr (Eds.), *Between craft and science. Technical work in U.S. settings*. Ithaca: IRL Press/Cornell University Press.

Buch, A. (2002). *Social læringsteori*. Frederiusberg: Roskilde Universitetsforlag.

Buch, A. (2012). Governing engineering. In S. H. Christensen, C. Mitcham, B. Li, & Y. An (Eds.), *Engineering, development and philosophy: American, Chinese, and European perspectives* (pp. 169–182). Dordrecht: Springer.

Buch, A., & Christensen, T. (1998). *Ingeniørliv – otte fortællinger*. København: Ingeniørforbundet i Danmark.

Cohen, M. J., Brown, H. S., & Vergragt, P. J. (2013). *Innovation in sustainable consumption: New economics, social-technical transitions and social practices*. Cheltenham: Edward Elgar Publishing.

Corradi, G., Gherardi, S., & Verzelloni, L. (2010). Through the practice lens: Where is the bandwagon of practice-studies heading? *Management Learning, 41*(3), 265–283.

Downey, G. L. (1998). *The machine in me: An anthropologist sits among computer engineers*. New York: Routledge.

Downey, G. L., & Lucena, J. (1997). Hiring in to the contested field of education. In G. L. Downey & J. Dumit (Eds.), *Cyborgs & citadels. Anthropological interventions in emerging sciences and technologies*. Santa Fe: SAR Press.

Downey, G. L., & Lucena, J. (2005). National identities in multinational worlds: Engineers and 'engineering cultures'. *International Journal of Continuing Engineering Education and Life Long Learning, 15*(3/6), 252–260.

Engeström, Y. (1999). Activity theory and individual and social transformation. In Y. Engeström, R. Miettinen, & R.-L. Punamäki (Eds.), *Perspectives on activity theory*. Cambridge, UK: Cambridge University Press.
Fairclough, N. (1995). *Critical discourse analysis: Papers in the critical study of language*. London: Longman.
Feldman, M. S., & Orlikowski, W. J. (2011). Theorizing practice and practicing theory. *Organization Science, 22*(5), 1240–1253.
Garfinkel, H. (1967). *Studies in ethnomethodology*. Englewood Cliffs: Prentice Hall.
Gherardi, S. (2006). *Organizational knowledge: The texture of workplace learning*. Malden: Blackwell.
Gherardi, S. (2012). *How to conduct a practice-based study. Problems and methods*. Cheltenham: Edward Elgar Publishers.
Goodwin, C. (1994). Professional vision. *American Anthropologist, 96*(3), 606–633.
Hager, P., Lee, A., & Reich, A. (Eds.). (2012). *Practice, learning and change. Theory perspectives on professional learning*. Dordrecht: Springer.
Henderson, K. (1999). *On line and on paper – Visual representations, visual culture, and computer graphics in design engineering*. Cambridge, MA: MIT Press.
Johri, A. (2010). Situated engineering in the workplace. *Editorial to Theme Issue of Engineering Studies, 2*(3), 150–151.
Knorr-Cetina, K. (1985). *The manufacture of knowledge: Essays on constructivist and contextual nature of science*. Oxford: Pergamon.
Knorr-Cetina, K. (1999). *Epistemic cultures: How the sciences makes knowledge*. Cambridge, MA: Harvard University Press.
Kunda, G. (2006). *Engineering culture, control and commitment in a high-tech corporation* (2nd ed.). Philadelphia: Temple University Press.
Lave, J. (1988). *Cognition in practice*. Cambridge, UK: Cambridge University Press.
Lave, J. (2011). *Apprenticeship in critical ethnographic practice*. Chicago: Chicago University Press.
Lave, J., & Wenger, E. (1991). *Situated learning. Legitimate peripheral participation*. Cambridge, UK: Cambridge University Press.
Marcus, G. (1998). *Ethnography through thick and thin*. Princeton: Princeton University Press.
Mellström, U. (1995). *Engineering lives, technology, time and space in a male-centered world*. Lindköping: Lindköping University.
Nicolini, D. (2009). Zooming in and zooming out: A package of method and theory to study work practices. In S. Ybema, D. Yanow, H. Wels, & F. H. Kamsteeg (Eds.), *Organizational ethnography. Studying the complexities of everyday life*. Los Angeles: Sage Publications.
Nicolini, D. (2013). *Practice theory, work, & organization. An introduction*. Oxford: Oxford University Press.
Orlikowski, W. (2000). Using technology and constituting structures. *Organization Science, 11*(4), 404–428.
Orlikowksi, W. (2002). Knowing in practice: Enacting a collective capability in distributed organizing. *Organization Science, 13*(3), 249–273.
Pickering, A. (1995). *The mangle of practice. Time, agency and science*. Chicago: Chicago University Press.
Quine, W.V.O. (1961). *From a logical point of view* (2nd ed.). Cambridge, MA: Harvard University Press.
Rabinow, P., & Bennett, G. (2012). *Designing human practices. An experiment with synthetic biology*. Chicago: Chicago University Press.
Rabinow, P., & Bennett, G. (2013). *Contemporary equipment: A diagnostic*. http://anthropos-lab.net/sites/default/files/book/Diagnostic_Digital_Editions/ContemporaryEquipment_Diagnostic-RabinowBennett.pdf. Accessed Apr 2013.
Reckwitz, A. (2002a). Toward a theory of social practices. A development in culturalist theorizing. *European Journal of Social Theory, 5*(2), 243–263.
Reckwitz, A. (2002b). The status of the "Material" in theories of culture: From "Social Structure" to "Artefact". *Journal for the Theory of Social Behaviour, 32*(2), 195–217.

Richards, D. (2001). Talking sense: Ethnomethodology, postmodernism and practical action. In R. Westwood & S. Linstead (Eds.), *The language of organization*. London: Sage Publications.
Rouse, J. (1996). *Engaging science. How to understand its practices philosophically*. Ithaca: Cornell University Press.
Rouse, J. (2002). *How scientific practices matter. Reclaiming philosophical naturalism*. Chicago: Chicago University Press.
Rouse, J. (2007). *Practice theory*. Division I Faculty Publications. Paper 43. http://wesscholar.wesleyan.edu/div1facpubs/43. Accessed Apr 2013.
Schatzki, T. (1996). *Social practices: A Wittgensteinian approach to human activity and the social*. Cambridge, UK: Cambridge University Press.
Schatzki, T. (2002). *The site of the social. A philosophical account of the constitution of social life and change*. University Park: Penn State Press.
Schatzki, T. (2003). A new societist social ontology. *Philosophy of the Social Sciences, 33*(2), 174–202.
Schatzki, T., Knorr-Cetina, K., & von Savigny, E. (Eds.). (2001). *The practice turn in contemporary theory*. London/New York: Routledge.
Shove, E., & Spurling, N. (2013). *Sustainable practices: Social theory and climate change*. Abingdon: Routledge.
Shove, E., Pantzar, M., & Watson, M. (2012). *The dynamics of social practice. Everyday life and how it changes*. Los Angeles: Sage Publications.
Taylor, C. (1995). Lichtung or Lebensform: Parallels between Heidegger and Wittgenstein. In C. Taylor (Ed.), *Philosophical arguments* (pp. 61–78). Cambridge, MA: Harvard University Press.
Tonso, K. L. (2007). *On the outskirts of engineering. Learning identity, gender, and power via engineering practice*. Rotterdam: Sense Publishers.
Turner, S. (1994). *The social theory of practices. Tradition, tacit knowledge, and presuppositions*. Chicago: University of Chicago Press.
Van de Poel, I., & Royakkers, L. (2011). *Ethics, technology, and engineering. An introduction*. Malden: Wiley-Blackwell.
Vinck, D. (2003). *Everyday engineering: An ethnography of design and innovation*. Cambridge, MA: MIT Press.
Warde, A. (2005). Consumption and theories of practice. *Journal of Consumer Culture, 5*(2), 131–153.
Wenger, E. (1998). *Communities of practice. Learning, meaning, and identity*. Cambridge, UK: Cambridge University Press.
Wittgenstein, L. (1958). *Philosophical investigations*. Oxford: Basil Blackwell.

Anders Buch M.A. in Philosophy, University of Copenhagen, and Ph.D. in Educational Studies, Roskilde University. He holds an associate professorship in Techno-Anthropology at Aalborg University Copenhagen at the Department for Learning and Philosophy and he is affiliated to the Centre for Design, Innovation and Sustainable Transitions (DIST). He has published articles and books on knowledge, learning, education, and the professional development of engineers. He is presently involved in the strategic research alliance PROCEED: "Program of Research on Opportunities and challenges in engineering education in Denmark."

Chapter 8
Design Methodology and Engineering Design

From Technical Problem Solving to Social Exploration

Pieter E. Vermaas

Abstract In this chapter I conceptually characterize the development of design methodology and analyze the changes this development has induced in the role engineers play in design. First, three successive types of design methods are described, from traditional engineering methods to current design thinking methods. Second, I show that this succession of methods has shifted the role of designing engineers from that of an assistant supplying technical solutions to customers, to an autonomous role of exploring and addressing user and societal needs independently of customers. I argue that these changes in their role may give engineers a more independent position and a broader grasp of design practices. Yet engineers also have to share this position with designers from disciplines other than engineering, and engineers may even lose their new role. According to current design methods, innovative design involves more than applying technology to address needs. Hence, if engineering remains to be seen as the discipline that provides technology, design becomes a discipline different to engineering. Engineers will in that case be forced back into their assistant role and become suppliers of technical solutions to other designers.

Keywords Development of engineering • Engineering design • Design methods • Reframing • User-centered design • Design thinking • Innovation through design • Sociology of engineering

P.E. Vermaas (✉)
Philosophy Department, Delft University of Technology,
Jaffalaan 5, Delft 2628 BX, The Netherlands
e-mail: p.e.vermaas@tudelft.nl

S.H. Christensen et al. (eds.), *Engineering Identities, Epistemologies and Values*, Philosophy of Engineering and Technology 21,
DOI 10.1007/978-3-319-16172-3_8

Introduction

When design methodology is regarded as being indicative of practice, it can be argued that the role of engineers in design has been broadening from the technical solving of problems to realizing goals, needs, and problems of customers, users, companies and society. According to the design methods of the second half of the twentieth century, engineering design is primarily about describing material products that solve problems as defined by customers. These methods were later supplemented with ones by which engineers can reformulate the problems presented by their customers and can improve the described products to better suit the envisaged users. Finally, at the beginning of the twenty-first century, methods of innovative design thinking were introduced, by which designers themselves can determine and propose the problems to be solved.

This development in design methodology suggests that engineering is broadening from a discipline that creates and applies technology in order to solve problems to a discipline in which people and society are probed to identify and address a variety of challenges, ranging from commercial innovations that are market 'game changers' to products that meet people's needs and resolve enduring societal and environmental issues. Moreover, this development may more effectively integrate engineering and its practices into society and give engineers a broader task and responsibility in design. Paradoxically it may, however, not be the discipline of engineering that will reap these benefits. The novel methods of design thinking in particular distance themselves from engineering practices. In these methods technology is seen as just one of the different sources for innovation, with a distinction introduced between the role of the design thinker and the role of the supplier of technology. From a conceptual point of view this split in roles leads to a choice in specifying what is meant by engineering design given the reported development: will engineering design be defined in the future as a practice aimed at supplying technical solutions, or will it be more broadly taken as the general practice of solving problems? From a sociological point of view the split is more involved, and reflects a struggle between groups about their disciplinary identities and the differences between these identities.

In this chapter, I start by giving a conceptual characterization of the development in design methodology in the last five decades in Section "Developments in Design Methodology". In Section "The Changing Role of Designing Engineers", I consider how this development has changed the role of engineers in design, and argue that this role can evolve into a broader role of realizing user and societal needs, or into a limited role of supplying technology.

Developments in Design Methodology

Design researchers may present their domain simultaneously in a broad and specific manner. They may introduce design as a broad and everyday practice, referring sometimes to Herbert Simon's (1996, p. 111) characterization that "everyone designs who devises courses of action aimed at changing existing situations into preferred ones" (Cross 2000; Lawson and Dorst 2009). In a second step design researchers may, however, limit design to more specific practices, by referring to what professional designers actually do (Lawson and Dorst 2009) or by defining design as the description of products that solve design problems (Cross 2000). Or design researchers may more swiftly introduce this more limited conception by taking design as the development and documentation of products for fulfilling perceived needs (Blessing and Chakrabarti 2009), or as the development of plans for the creation of products to assist users in attaining goals (Dorst and Van Overveld 2009).

The broad and limited conceptions of design in design methodology may be reconciled by relating them. The description of products may be regarded as a practice that is subsidiary to devising courses of action, turning design into a two-tiered practice: first it is determined by what actions users can take to achieve a particular goal, and second the material products used in these actions are described insofar as these products are not readily available to users (Hubka and Eder 1988; Houkes and Vermaas 2010). This reconciliation may be conceptually pleasing by capturing that design nowadays consists of different types of practices, and by pulling in practices that are traditionally not seen as design. The limited conception of design has indeed ceased to be apt, since the concept of a product has expanded over the years (Buchanan 2009) and may now refer to material products as well as processes, services, innovation strategies and planned actions. And consultancy, taken as a practice of recommending actions to achieve defined goals with existing means, is then also to be understood as design. Yet the limited conception of design still dominates design methodology and has its roots in earlier work in methodology for engineering design. Hence, in order to conceptually understand the development of design in design methodology, and the role of engineers therein, the starting point is this limited conception of design.

According to the earlier engineering design methods of the second half of the twentieth century (e.g., VDI 1993; Pahl et al. 2007), design is defined as finding a technical solution to a design problem. That problem is formulated as a set of physical, technical, and financial requirements that have to be met, and the solution is typically a description of a material product. The problem is fixed, and the source of the problem – goals or needs of agents, or ideas for products and services within commercial companies – is considered to be a matter of the customer ordering the design, and a matter that lies outside the realm of engineering design. In Fig. 8.1 this traditional understanding of engineering design is depicted schematically as a practice – the solid box in the figure – with a problem formulated by a customer as input, and a product that solves the problem as output. Engineering design itself,

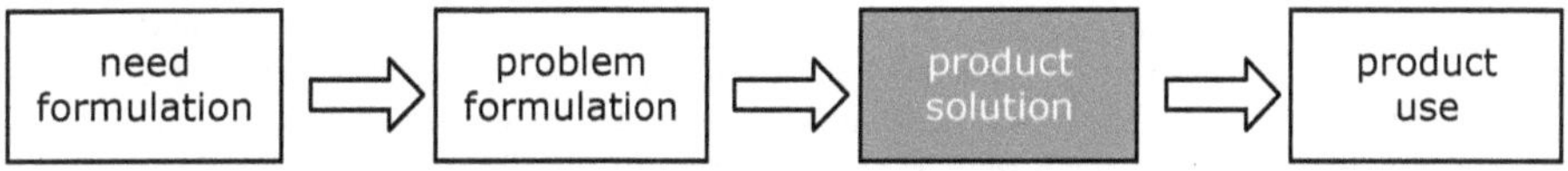

Fig. 8.1 Traditional engineering design (*the solid box*)

i.e., finding the description of the product that solves the problem, is typically divided into different phases, such as conceptual design, embodiment design, and detailed design. And engineering design may involve different iterative steps between these phases, meaning that the findings of a later phase may provide information that brings the design process back to reconsidering the decisions made in an earlier phase. Yet, in traditional engineering design such iteration or reverse feedback generally does not take place from the practice of use to the description of the product, or from the description of the product to the formulation of the customer's problem. This 'uni-directionality' in traditional engineering design limits the scope of engineering design to solving by technical means a given problem in terms of a material product.[1]

The scope of engineering design broadened with the methodological analysis of design by Donald Schön (1983). Although this work originated in part in architecture, engineering design also came to be seen as a practice in which the formulation of the design problem is to be analyzed and, if necessary, changed. A designer explores the problem and this exploration may yield reasons to *reframe* the problem. Designing starts with an initial interpretation of the problem and an initial solution or solution direction for finding a product that might solve it. The exploration of this solution direction provides the designer with new insights about the problem and these insights enable the designer to change and improve on the initial interpretation of the problem, thus allowing the designer to choose and explore alternative solutions or solution directions. This reframing of the problem can be seen as a characteristic that sets design apart from regular problem solving in, for example, science and mathematics. Moreover, reframing is sometimes even regarded as necessary in design. Scientific or mathematical problems are said to be well-structured by providing in the way in which they are formulated information and success criteria about the kind of solution that is required. Formulations of design problems may in contrast be *ill-structured* (Simon 1984), *wicked* (Rittel and Webber 1984) or *paradoxical* (Dorst 2006). Hence, in design it is sometimes necessary to transform the original formulation of the problem or to transform the design problem itself, in order to make it solvable.

This possibility of reframing led to design becoming methodologically seen as a practice in which the problem may evolve with the search for its solutions, and which ends when a satisfactory pair is found consisting of a reformulated design problem and a solution to it (Dorst and Cross 2001; Cross 2006). Figure 8.2 represents this new way of understanding engineering design, in which design – the two solid boxes in the figure – now concerns the problem formulation *and* the description of the product that solves the problem thus formulated. And since finding that description may reframe the problem formulation, there is now a reverse feedback

[1] Let me acknowledge that this characterization of traditional engineering design is somewhat simplifying matters. Design methodologists associated with the traditional understanding do recognize that engineering design is embedded in specific contexts and that engineering designers should mind the goals of their customers and be aware that much designing is instrumental to economic activities of companies (e.g., Roozenburg and Eekels 1995).

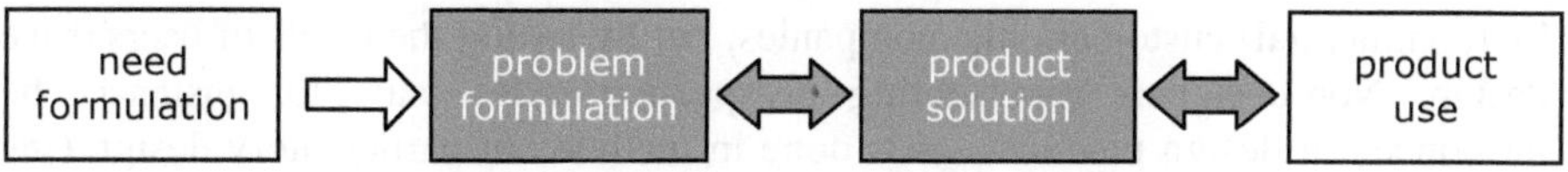

Fig. 8.2 Engineering design with reframing (*the two solid boxes*) and with user-centeredness

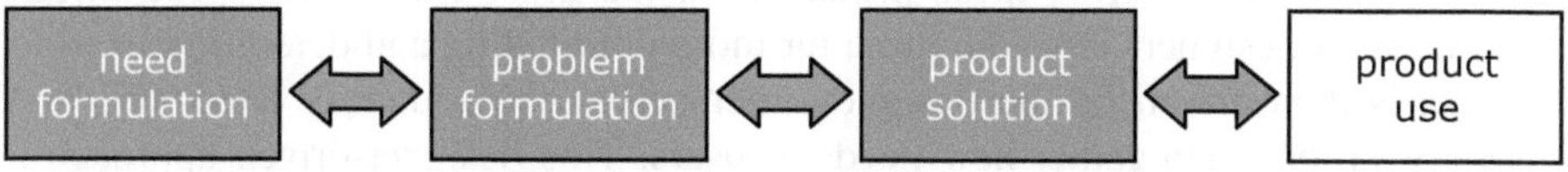

Fig. 8.3 Autonomous design (*the three solid boxes*) with user-centeredness

from product description to problem formulation, depicted in Fig. 8.2 by a two-sided arrow between these two activities.

This broadening in the characterization of design practices may be seen as a result of what design researchers call the descriptive phase in their field. In this phase design practices were being analyzed empirically rather than prescribed theoretically, and this analysis revealed that designers may change design problems. Similarly research in design started to focus on use, but now for more normative purposes of correcting design practices and the products they produce. This user-centered phase (see e.g., Koskinen et al. 2011, pp. 15–22) introduced research on usability and intelligibility of products and their interfaces for users, and added ergonomics and ethnography as design tools to understand users and their interactions with products in their daily life context. This user-centeredness corrected a neglect of user concerns in design, and created a reverse feedback between the design practice of describing products and the actual use of the products, depicted in Fig. 8.2 by a second two-sided arrow between these activities, and consisting of checks by designers to determine whether their product solutions actually achieve the goals for which they are designed.

Reframing design problems and the inclusion of feedback from users are now standard elements of design practices. Hence, designers, including engineering designers, no longer merely play an assisting role toward the customers for whom they design, but also a role of correcting customers. Moreover, the ability to explore and understand how users respond to products gave designers the means to design products more autonomously from customers, further broadening the scope of design and enabling a second development in design methodology. Novel design methods emerged by which designers can themselves determine the needs they aim to meet. This broadening of the scope of design may have its roots in the back end of product use, yet implies that designers appropriate the very front end of design by themselves formulating the needs that are to be designed for (Fig. 8.3).

This inclusion of the formulation of the needs in design practice can take different forms and creates a spectrum of design practices. At one end of the spectrum designers cash out their autonomy by choosing not to take up the problems defined by (commercial) customers like companies, but by letting the needs of users drive design. Non-designers representing prospective users may, for instance, be introduced in design practices, as is done in methods of participatory design (see e.g., Sanders and Stappers 2008). These non-designers bring in the needs of users and the conditions under which the resulting products are usable and intelligible for users, leaving designers with the intermediate role of making their skills and knowledge available to describe these products. At the other end of the spectrum the autonomy of designers is leading to a far more dominating and determinative role. Design is then driven by the designers themselves, who identify latent needs of users or propose altogether new needs to users. This designer-driven approach is taken particularly in the *design thinking* methods of the early twenty-first century. In this approach, average users are regarded as conservative, and market research on what such users want as leading to only incremental improvements in existing products. Designers use their acquired autonomy to break away from such incremental design, and aim at innovative design that is to result in products that are commercial or sociocultural 'game changers'. Designers do so by focusing their explorations on *non*-average users, such as 'future-focused persona' and 'cultural innovators' who hold the values and beliefs of the next area (Gardien 2006), or 'extreme' users such as novice or expert users who have unorthodox relationships with products (Brown 2009). And designers should engage with 'key interpreters' in society who explore changes in the context of users and in the meaning products have for these users (Verganti 2009). Design then becomes driven by designers who *propose* products to users on the basis of the designers' own insights and explorations with (non-average) users. Steve Jobs and his successes are often mentioned to illustrate this new role of designers, and the designers' autonomy is justified by Jobs' quote that "[a] lot of times, people don't know what they want until you show it to them" (Young and Simon 2005, p. 262).

The spectrum of current design practices is richer than the two extremes of user-driven participatory design and designer-driven propositional design. In other design methods, sometimes also presented under the label of design thinking (e.g., Brown 2009; Plattner et al. 2009; D.School 2011), designers actively research users in their daily context to determine user needs or to better articulate these needs, meaning that these users, including the average ones, remain central in design. Designers, for instance, place themselves in the position of users and attempt to identify and understand the experiences users have or will have with products in daily life. Designers show *empathy* with users, and carry out ethnographic studies. Designers then develop products to meet the needs and problems of users thus identified, and, moreover, use prototypes and mock-ups to check whether the envisaged products indeed help users (e.g., Koskinen et al. 2011). Designers may in this way be said to be guiding design by using their autonomy for arriving at product solutions, innovative or not, that are better fitting users. Designers still take their own responsibility to determine users' needs and problems for them, how these needs and problems can be addressed, and what impact products have for users in the long run. Hence, designer-guided empathic design also has a clear propositional flavor (e.g., Hekkert and van Dijk 2011) as compared to participatory design, yet it contrasts to the extreme of designer-driven propositional design by keeping (average) users in focus.

The autonomy of designers has also enabled the emergence of design practices aimed at meeting needs different to those typically considered in engineering design. It has made possible social design for the needs of people in developing countries, for societal needs, and for environmental issues, leading to an even richer spectrum of design practices. These practices can be found within user-driven, designer-guided and designer-driven design, and may have their separate design methods. The further distinctions in engineering design that can be made with these methods will not be considered in this chapter, although they have arguably led to further independence of designers, specifically from commercial customers such as companies.

The Changing Role of Designing Engineers

The development of engineering design as sketched in the previous section and the resulting changes in the role of engineers in design may be captured conceptually by two shifts. First, in moving from traditional engineering design to reframing (from I to II in Fig. 8.4), design has shifted from a practice in which engineers assist customers, to a practice in which engineers can correct customers with respect to the formulation of design problems and allowing the designed products to more closely fit user concerns. Second, in moving from engineering design with reframing to current practices (from II to III in Fig. 8.4), design has shifted to a practice in which designers can, autonomously from customers and with a focus on usability, define the needs and problems for which they design.

This development of the role of engineers in design has given them a more independent position specifically from customers, and a broader grasp on design practices. It has to be acknowledged that much of engineering design is still done for or within commercial companies, actually keeping many designing engineers in

relationship to users / relationship to customers:	User-insensitive	User-centered
Assistant	I: traditional engineering design	*IV: future engineering design (?)*
Corrective		II: reframing engineering design
Autonomous: - user-driven - designer-guided - designer-driven		III: - participatory design - emphatic design - propositional design

Fig. 8.4 The development of the role of designing engineers in relation to customers and users

their traditional assisting role of providing technical solutions to problems as defined by these companies and their managers. On a pessimistic reading of this situation, designing engineers may therefore still be seen as serving the decisions and values of companies and managers (e.g., Goldman 1984), but now in a way by which the resulting products are made more acceptable to users. Yet the development also means that designing engineers have an increased contribution to and responsibility for these products. On a more optimistic reading, designing engineers can now become involved in companies' exploration and development of new product ideas, thus taking over some of the tasks and responsibilities of their managers rather than serving them.[2] Moreover, engineers can start doing design outside the immediate context of commercial companies, for users, for developing countries, for society or for the environment. These new possibilities beyond traditional engineering design give designing engineers a role more connected with the needs and values of people and society; this means that engineers can combine their technological skills with social exploration through interactions with users and society, including the average and 'extreme' users, the 'future-focused persona', 'cultural innovators', and societal 'key interpreters'.

The new possibilities for engineers in design do not result in a role by which engineers gain full control over their technological skills in the sense of becoming the (sole) agents who fix design problems that are to be solved technically, and the (sole) agents who determine which solutions are to be adopted (Downey 2005). As sketched above, current design methods locate designing engineers in webs of interactions with agents within society, and these interactions result in design problems and the products that solve them being co-determined by the agents in society. Moreover, some current design methods (e.g., Plattner et al. 2009; D.School 2011) explicitly endorse design being carried out by multidisciplinary teams in which engineering is just one of the several disciplines involved. The argument for this multidisciplinarity is that designers on a team who originate from a single discipline will approach customer needs and their solutions by the received and standard ways of thinking within that single discipline, leading to incremental improvements in products. Conversely, in multidisciplinary teams, members cannot simply opt for such a received approach from a single discipline, but force each other to arrive at new perspectives and solutions to customer needs, thus leading to innovation. This propagated multidisciplinarity goes beyond the multidisciplinarity in design as analyzed by Louis Bucciarelli (1994); it is not just different engineering disciplines that have to collaborate in design. Finally, in other current design methods (e.g., Verganti 2009), a focus on technology in design is seen as limiting the potential of innovation through design by missing opportunities to find new potential product meanings for users.

---

[2] It may be argued that since engineers can themselves be managers in commercial companies, engineers were already earlier involved in the exploration and development of new product ideas. More precisely, the point made here is that nowadays designing engineers are also increasingly becoming part of this exploration and development (private communication with Byron Newberry).

	horizon 3 (create viable options)	horizon 2 (build emerging business)	horizon 1 (extend & defend core business)
communicate value	aspirational promise	concept car	specific campaigns
develop value	innovative debate (probes)	collaborative innovation	incremental innovation
identify value	social cultural trends & narratives	future focused persona research	people & market research

Fig. 8.5 The Philips design methodological matrix for innovation through design

Due to the development of design methodology, engineers are thus nowadays sharing the control over design with other agents by collaborating with those agents in formulating needs and design problems, and in finding their solutions. In fact, in current methods the agents involved in design practices are not called engineers but have labels like ‘product designer’, ‘industrial designer’, or simply ‘designer’. Engineering is no longer the only discipline involved in current design, and may in a grimmer scenario actually end up being confined to only particular parts of design.

To illustrate this possible fate of engineers one can consider the methodological description of innovation through design in commercial companies as proposed by Paul Gardien (2006). According to this description, which is based on experiences at Philips Design, innovation through design in companies can be organized by nine phases ordered in a matrix, see Fig. 8.5. This matrix ordering expresses that innovation through design typically does not follow a single linear series of design phases; particular innovative design efforts can take any path made up of (adjacent) phases in the matrix. With this methodological description of innovation through design it can now be made plausible that engineers need not be involved in each of its different phases.

The matrix for innovation through design superimposes two models of innovation. The first model advances that companies operate with three innovation horizons simultaneously (Baghai et al. 2000): companies are extending and defending their core business (a short perspective *horizon 1*); companies are developing new business (a longer perspective *horizon 2*); and companies are creating viable options for new business (the longest perspective *horizon 3*). The second model breaks up innovation up into three steps (Lanning and Michaels 2000): companies are identifying values for customers; companies are developing these values; and companies are communicating the developed values to customers. Each pair made up of one innovation horizon and one value-process step defines a cell in the matrix, and each of these cells corresponds with a possible phase in innovation through design. For instance: the matrix cell in row 2 and column 3 corresponds with the improvement of existing products for current users; the cell in row 2, column 2 corresponds with an innovative design project aimed at finding new products for future users; and the cell in row 1, column 2 corresponds with building an emerging business and communicating the associated values to customers, which is often done in the car industry when concept cars are created. It may be safely assumed that engineers are involved in each of these three possible phases in innovation through design, for instance because the design practices suitable for these phases resemble traditional engineering design or engineering design with reframing. Yet the matrix also defines phases which are less about the application of technology and more about sociological research, marketing research, cultural exploration and communication. For instance, technical skills are less relevant for spotting social cultural trends (represented by the cell in row 3, column 1) or exploring how users or society respond to design probes (the cell in row 2, column 1; probes are concept products meant to initiate and collect responses by users to possible new types of products (e.g., Koskinen et al. 2011)). So, if engineering design remains a discipline primarily associated with technology, these phases of innovation through design can be carried out without engineers.

Ultimately, the possibility that the role of engineers in current design practices will again become limited to only particular parts of design is, from a conceptual point of view, just a definition issue. One may associate engineering with technology, and then some of the agents participating in contemporary design are being labeled engineers while others are not. Or one may broaden the meaning of engineering, and call all agents involved in current design practices engineers. In either case the current design practices are just what they are: agents collaborating in exploring problems and finding solutions. Still, when engineering indeed remains primarily associated with technology, only the agents who contribute technical knowledge and expertise are regarded as engineers. Engineers then still participate in design, but design tasks such as the formulation of needs and problems are in this case carried out by other designing agents, pushing those agents called engineers back into the traditional engineering role of finding technical solutions to problems defined by others. Hence, if engineering remains primarily associated with technology, there will be another shift (from III to IV in Fig. 8.4) by which engineers again assume the limited role in design of assisting others, as given in Fig. 8.6. Yet if the understanding of engineering evolves in pace with design methods, such that the meaning of engineering broadens to societal exploration, the agents that carry out current design practices can all be called engineers. In the latter case engineers will indeed be given the role of finding the needs and problems for which they provide technical solutions, as in Fig. 8.3.

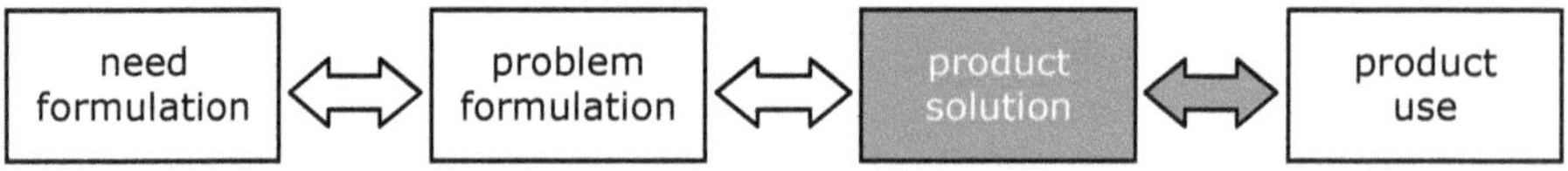

Fig. 8.6 A possible limited future of engineering design in autonomous design (*the solid box*)

There is no principled reason to assume that engineering has to continue to be associated with only the application of technology. The autonomous designer-driven practices associated with Steve Jobs resemble the designer-driven practices by Henry Ford, as the abovementioned quote by Jobs resembles Ford's famous claim that if he "had asked people what they wanted, they would have said faster horses." Hence, if Ford can be called an engineer, so can Jobs and many others involved in contemporary design. However, how engineering will become to be recognized in the near future is not simply a conceptual matter. Sociological mechanisms and practical issues may be more decisive. To establish a broader understanding of engineering, the current engineering community has to be capable and willing to accommodate the consequences of the associated change in their role, and be able to adjust the educational curricula in engineering accordingly (see, e.g., Williams 2003). And other groups of agents involved in designing must in turn be willing to be called engineers, even if their role consists of providing knowledge and skills other than technical expertise. The mechanisms that determine this willingness are beyond the realm of conceptual analysis of design methodology, yet are hopefully in the focus of other chapters of this volume.

Conclusions

In this chapter I conceptually characterized the development of design methodology in the last five decades and considered the changes this development has induced in the role of engineers in design. According to traditional design methods, engineers assist customers by providing technical solutions to the problems defined by customers. According to later design methods, engineers can correct the customers' formulation of design problems and can let the designed products to better fit user concerns. According to current design thinking methods, designers can define the needs and problems for which they design autonomously from customers. This development in design methodology has caused the role of engineers in design to shift from that of an assistant of supplying technical solutions to a corrective role of reframing customers' problems, and finally to an autonomous role of exploring and addressing user and societal needs independently of customers.

It was argued that this development of the role of engineers may give them a more independent position and a broader grasp of design practices, yet engineers have to share this position, and can even lose it if engineering remains associated with providing technology. Much contemporary engineering design still takes place in commercial settings of companies, keeping many engineers in their traditional role of assisting by providing technical solutions in design. Moreover, according to current design methods, designing engineers have to interact with agents within society in the exploration of user and societal needs, and collaborate in design practices with designers originating from disciplines other than engineering. Finally, innovative design practices involve more than applying technology for addressing user and societal needs. Hence, if engineering continues to be seen as the discipline that provides technology, design becomes a discipline different to engineering, and engineers will again be forced back into their assistant role by becoming suppliers of technical solutions to other designers.

References

Baghai, M., Coley, S., & White, D. (2000). *The alchemy of growth: Practical insights for building the enduring enterprise*. New York: Perseus Publishing.
Blessing, L. T. M., & Chakrabarti, A. (2009). *DRM: A design research methodology*. London: Springer.
Brown, T. (2009). *Change by design: How design thinking transforms organizations and inspires innovation*. New York: Harper Business.
Bucciarelli, L. L. (1994). *Designing engineers*. Cambridge, MA: MIT Press.
Buchanan, R. (2009). Thinking about design: An historical perspective. In A. W. M. Meijers (Ed.), *Philosophy of technology and engineering sciences* (pp. 409–453). Amsterdam: Elsevier.
Cross, N. (2000). *Engineering design methods: Strategies for product design* (3rd ed.). Chichester: Wiley.
Cross, N. (2006). *Designerly ways of knowing*. London: Springer.
D.School. (2011). D.School Bootcamp Bootleg. http://dschool.stanford.edu/wp-content/uploads/2011/03/BootcampBootleg2010v2SLIM.pdf
Dorst, K. (2006). Design problems and design paradoxes. *Design Issues, 22*(3), 4–17.
Dorst, K., & Cross, N. (2001). Creativity in the design process: Co-evolution of the problem-solution. *Design Studies, 22*(5), 425–437.
Dorst, K., & van Overveld, K. (2009). Typologies of design practices. In A. W. M. Meijers (Ed.), *Philosophy of technology and engineering sciences* (pp. 455–487). Amsterdam: Elsevier.
Downey, G. (2005). Are engineers losing control of technology? From 'problem solving' to 'problem definition and solution' in engineering education. *Chemical Engineering Research and Design, 83*(A6), 583–595.
Gardien, P. (2006). *Breathing life into delicate ideas: developing a network of options to increase the chance of innovative success*. Position paper, Philips Design. http://www.design.philips.com/philips/shared/assets/design_assets/downloads/news/Breathing_life_into_delicate_ideas.pdf
Goldman, S. L. (1984). The techne of philosophy and the philosophy of technology. *Research in Philosophy and Technology, 7*, 115–144.
Hekkert, P., & van Dijk, M. (2011). *Vision in design: A guidebook for innovators*. Amsterdam: BIS Publishers.
Houkes, W., & Vermaas, P. E. (2010). *Technical functions: on the use and design of artefacts*. Dordrecht: Springer.
Hubka, V., & Eder, W. E. (1988). *Theory of technical systems*. Berlin: Springer.
Koskinen, I., Zimmerman, J., Binder, T., Redström, J., & Wensveen, S. (2011). *Design research through practice: From the lab, field, and showroom*. Waltham: Morgan Kaufmann.
Lanning, M. J., & Michaels, E. G. (2000). *A business is a value delivery system*. A McKinsey staff paper dated June 1988, (pp 53–57). Available from *McKinsey Quarterly*: http://www.mckinsey.com/insights/strategy/delivering_value_to_customers
Lawson, B., & Dorst, K. (2009). *Design expertise*. Oxford: Architectural Press.
Pahl, G., Beitz, W., Feldhusen, J., & Grote, K.-H. (2007). *Engineering design: A systematic approach* (3rd ed.). London: Springer.
Plattner, H., Meinel, C., & Weinberg, U. (2009). *Design thinking: innovation Lernen – Ideenwelten Öffnen*. Munich: mi-Wirtschaftsbuch.

Rittel, H. W. J., & Webber, M. M. (1984). Planning problems are wicked problems. In N. Cross (Ed.), *Developments in design methodology* (pp. 135–144). Chichester: Wiley.

Roozenburg, N. F. M., & Eekels, J. (1995). *Product design: fundamentals and methods*. Chichester: Wiley.

Sanders, E. B.-N., & Stappers, P. J. (2008). Co-creation and the new landscapes of design. *CoDesign, International Journal of CoCreation in Design and the Arts, 4*(1), 5–18.

Schön, D. A. (1983). *The reflective practitioner: how professionals think in action*. London: Temple Smith.

Simon, H. A. (1984). The structure of ill-structured problems. In N. Cross (Ed.), *Developments in design methodology* (pp. 145–166). Chichester: Wiley.

Simon, H. A. (1996). *The sciences of the artificial* (3rd ed.). Cambridge, MA: MIT Press.

VDI (Verein Deutscher Ingenieure), (1993). *VDI Guideline 2221: Methodik zum Entwickeln und Konstruieren technischer Systeme und Produkte*. Düsseldorf: Verein Deutscher Ingenieure.

Verganti, R. (2009). *Design driven innovation: Changing the rules of competition by radically innovating what things mean*. Boston: Harvard Business Press.

Williams, R. (2003). Education for the profession formerly known as engineering. *Chronicle of Higher Education, 49*(20), B12–B13.

Young, J. S., & Simon, W. L. (2005). *iCon: Steve Jobs, the greatest second act in the history of business*. Hoboken: Wiley.

Pieter E. Vermaas M.S. in Theoretical Physics from University of Amsterdam, and Ph.D. in Philosophy of Physics from University Utrecht. Currently Senior Researcher with the Philosophy Department of Delft University of Technology, the Netherlands. Research interests: the concept of technical function; technical artefacts in engineering and metaphysics; design methodology; design for values. Editing for the *NanoEthics and Philosophy and Technology* journals. Editor-in-Chief of the Springer *Philosophy of Engineering and Technology* book series.

Chapter 9
The Epistemological Basis of Engineering, and Its Reflection in the Modern Engineering Curriculum

William Grimson and Mike Murphy

Abstract Perhaps unlike other professions, engineering is strangely difficult to define or describe. This is nowhere as evident as when an attempt is made to articulate its epistemological basis. Engineering has a rich and complex 'gene pool' which goes back to when people first built shelters and shaped implements for agricultural purposes. Throughout the ages one constant characteristic of engineering has been its readiness to avail of whatever material is on hand together with whatever knowledge or skill is available to meet the challenge of enhancing an object or making something which never previously existed. On occasion engineers have created new knowledge but for the most part they have been users of knowledge: borrowing from nature, science, mathematics, arts in order to meet their requirements to solve specific problems. The art of engineering is in the appropriate selection of knowledge coupled with an ability to use that knowledge in achieving an objective. A three-layer model is proposed to describe the epistemological basis of engineering. This layer consists of a foundational layer containing subject material such as mathematics and science, above which is a middle-layer largely populated by domain knowledge associated with engineering program learning outcomes, and with the final top layer acting as a capstone and expressed in terms of professional competences.

Keywords Knowledge • Engineering education • Learning outcomes • Competences • Professional engineers • Epistemology • Heuristics

W. Grimson (✉)
Dublin Institute of Technology, 143 Lower Rathmines Road, Dublin 6, Ireland
e-mail: william.grimson@dit.ie

M. Murphy
Technological University for Dublin Programme Team,
Dublin Institute of Technology, 143 Lower Rathmines Road, Dublin 6, Ireland
e-mail: mike.murphy@dit.ie

S.H. Christensen et al. (eds.), *Engineering Identities, Epistemologies and Values*, Philosophy of Engineering and Technology 21,
DOI 10.1007/978-3-319-16172-3_9

Introduction

> The ideal engineer is a composite … He is not a scientist, he is not a mathematician, he is not a sociologist or a writer; but he may use the knowledge and techniques of any or all of these disciplines in solving engineering problems. (N.W. Dougherty)

> Engineering is not merely knowing and being knowledgeable, like a walking encyclopedia; engineering is not merely analysis; engineering is not merely the possession of the capacity to get elegant solutions to non-existent engineering problems; engineering is practicing the art of the organized forcing of technological change… Engineers operate at the interface between science and society… (Dean Gordon Brown)

These two quotations point to the character of engineering that makes it so difficult to draw boundaries around both its fundamental nature and, as a consequence, its epistemological foundations. On the one hand engineering uses whatever knowledge is relevant, whatever its origin, to address a particular challenge. In that sense the totality of available knowledge (the body of knowledge) is in principle the epistemological basis of engineering. On the other hand that body of knowledge by virtue of its sheer extent is unknowable to or by the individual engineer. Described another way, if the duration of engineering degree programs matched the general explosion in technical knowledge over the last 50 years then the educational formation of engineers would far exceed the standard 4 or 5 years that is the norm. This means that an approach, other than a direct one, has to be adopted by engineers if a practical way of dealing with knowledge is to be realized within the practice of engineering. That in turns means choices have to be made constrained by limited resources – an intrinsically engineering activity – concerning what might be termed the epistemological problem in engineering.

Regarding engineers 'operating at the interface between science and society' this is now of growing importance when one considers problems of supplying clean water or the linked challenge associated with climate change and the generation of energy to meet the demands of an ever expanding population. Whilst the knowledge associated with these complex societal issues is very different in nature to purely technical issues it is not of less importance. Further the responsibility, and that is what it is, to be able to enter into meaningful dialogue with society on technological change is not just a communications challenge but is itself an epistemological challenge bearing in mind the likely knowledge mismatch between the participants in any discussion. Providing society with an inadequate explanation of what is involved in technological change or indulging in an exercise of 'dumbing down' can only result ultimately in a loss of trust: and trust once lost is hard to re-gain. Hence there is a clear responsibility on engineers to meaningfully and accurately account for their understanding of the underlying knowledge and its related consequences involved in whatever technological-based discussion is taking place with and within society.

Formal engineering education has gone through a number of evolutions by which a craft based approach was in turn replaced by an empirical practice-based one, leading on first to the engineering science model and then followed by a systems oriented one. Further, as engineering split into a multiple of sub-disciplines from the initial mechanical, civil and electrical ones, to the wide range now in evidence across the world, each new area by choice or necessity adopted the style of education that seemed best to suit its needs. And all this is reflected in the range of approaches taken to the 'epistemological problem' – there being more relevant knowledge than can be absorbed. (See for example Bucciarelli et al. 2009, p. 105)

In a paper presented at the Royal Academy of Engineering Antonio Dias de Figueirdo proposed a decomposition of the epistemology of engineering into four categories as follows: Engineering as Basic Science; Engineering as Social and Business Activity; Engineering as Design; and Engineering as Doing. With respect to the Basic Science the key features being its application to engineering, rigour, logistics, analysis, research and discovery. For Social and Business Activity the author identifies as key aspects socio-economic realities, social complexity, social and economic value, and satisfaction for end-users. As regards Engineering as Design the features noted include systems thinking, context, integrated representation, compromise, alternatives, incomplete knowledge, and non-scientific modes of thinking. Finally Engineering as Doing which concerns essentially the art of getting things done, overcoming barriers, the need for flexibility and adaptation. Without doubt this all points to the need to consider a complex concoction of knowledge elements with little in the way of a priori guidance as to how the bits fit together into a model. Custom and practice however has allowed some models to evolve and even if no claim can be made as to any deep philosophical justification at least experience has shown what is practicable from the perspective of educating engineers (de Figueiredo 2008).

Some other aspects of knowledge need to recognized if proper use, as would be the intention in engineering, of the application of specific bodies of knowledge. First, the range and extent of the applicability of such knowledge. Second, the provenance of the knowledge. This is especially the case concerning the widespread use of the web (Fox and Huang 2003). Third, how the knowledge is coded or represented, stored, transmitted, maintained and updated. This last feature is relevant to knowledge that encapsulates design methods, for example, where experience of its use inevitably leads to an updating of knowledge. Lastly, and not unique to engineering, there is the matter of secrecy or privacy where knowledge is withheld from a general audience either for commercial or strategic purposes and which at the very least raise ethical issues.

This chapter presents the knowledge relevant to engineering in the form of a three-layer model. The bottom or foundational layer represents fundamental knowledge, both rational and empirical, and which is commonly encountered through senior second-level and undergraduate years. The middle layer, building on the foundational layer represents the knowledge associated with the learning outcomes that students are expected master in their primary engineering degree program (i.e., domain knowledge). The final and top layer represents knowledge related to the competences that a practicing engineer should have achieved in order to be eligible to become the holder of the title Chartered Engineer or Professional Engineer. It should be said here that the proposed 3-layer model is representative of the epistemological basis found generally in English-speaking countries (e.g., United Kingdom, Ireland, Canada, United States). In many other countries this system of

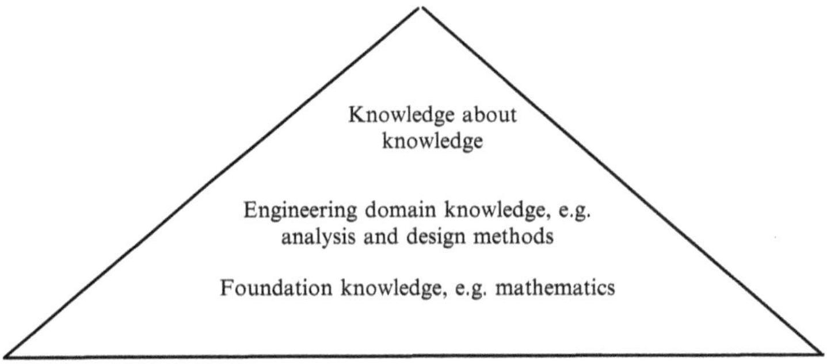

Fig. 9.1 Pyramid of engineering knowledge

becoming a chartered or professional engineer does not exist. In these countries the formal training of engineers generally ends with the completion of a second cycle engineering degree at Master's level. Figure 9.1 shows these layers as a pyramid of engineering knowledge.

Foundation Layer

Most educational models that reflect the various branches of knowledge found in engineering include the following subjects:

Mathematics
Science
Computer Science

Arts & Craft Practice
Engineering 'know-how' (e.g. design methodology)
Business & Economics
History of Science, Engineering and Technology
Ethics

As the opening remarks to this chapter imply, the list could easily be expanded, for example to include languages, communications (in the sense of written and oral interactions) and critical thinking, to name just three topics. Depending on how one views engineering, the selection of subjects together with their associated knowledge base will vary. So in one sense there is not a unique epistemological foundational basis for engineering unlike, say, mathematics. But it would be wrong to say there are numerous and widely diverse bases: rather a more 'correct' model is one of a fuzzy superset. Equally well a *mélange* might be considered an appropriate description.

We will return later to the *mélange* and examine how the design of an engineering curriculum is approached but first a few thoughts on the subjects listed above. First, the model is not simple because of multiple dependencies and linkages. For example to understand some parts of physics requires specific elements of mathematics. For example, vector calculus is necessary to understand the meaning of Maxwell's equations in an electrical engineering program. Likewise, elements of a business course will rely on an understanding and use of statistical analysis. Ethics taught to engineers without context would be sterile but fortunately relevant cases abound in the history of science, engineering and technology. Given the relentless pressure to cram more subjects into already crowded curricula, advantage is often taken of this interconnectivity. Thus some subjects can be taught as embedded topics within another subject.

The demands made on the use of mathematics as an analysis tool vary from engineering domain to domain; whilst the modern undergraduate mathematics syllabus generally is not that different to that of say 40 years ago. However the methods and modes of instruction have changed. In terms of foundation knowledge, one of the biggest changes in engineering curricula that has occurred is the inclusion of additional science subjects. From its earliest development, engineering education has generally based its foundational knowledge on the physical sciences of physics and chemistry, together with mathematics. With the rapid development of engineering disciplines such as biomedical engineering, the inclusion of life sciences in the curriculum is compelling.

Computer science was first routinely introduced into engineering programs about 40 years ago as an analysis tool. Today, application software is an indispensible tool in engineering analysis, design and graphical representation. One of the challenges for educators is that over-reliance on software without a proper understanding of the underlying processes can lead to undesirable or unexpected outcomes. Generally the approach adopted is that the engineering student should in principle be able to carry out a design without using a software package. This is perhaps akin to the merits of a pilot being able to manually fly an airliner as well as being confident that the auto-pilot can perform what is required in both routine and exceptional circumstances. However, computer systems and applications continue to become increasingly sophisticated, presenting continued challenges to curriculum designers. In such circumstances where lies the epistemological basis of engineering?

Mathias Heymann has written about the competing claims of 'art' and science, mainly in a German context, describing as a pendulum movement how in turn one and then the other contributed to engineering design methodology in the period 1850–2000. In some quarters as an engineering science approach was developed there was a tendency to downplay the role of arts and craft practice (Heymann 2009). Apprenticeships which were once a strong feature in engineering education in the UK served many purposes but certainly the mentoring role by which new recruits served with a 'master' to acquire craft skills was of great importance.

Michael Polanyi introduced into philosophy the term *tacit knowledge* and this concept applies to how some elements that are not explicit are an essential part of the engineering and engineering craft milieu (Polanyi 1958). Engineering 'know-how' is not confined to such tacit knowledge. For example, across many engineering disciplines much of the relevant working knowledge is codified (hence explicit) in one form or another often with customized software support. This enables both efficiency and the maintenance of minimum standards.

Economics and business are important not only for the sake of understanding how business and commerce work on a national and global scale but also because engineers are likely to move into senior management positions at a later stage in their careers. Younger engineers are also attracted into finance where the coupling of mathematics with a judgment of what makes sense (allegedly a characteristic of engineers) is a prized asset.

Ethics is given a high relevance amongst undergraduate engineering programmes not only by virtue of accreditation criteria but also with those regulating for practicing Professional Engineers or Chartered Engineers and who are members of professional bodies or institutions with Codes of Ethics.

Finally, history of science, engineering and technology is important, first because it helps undergraduate engineers formulate an identity. Because so much of engineering is hidden from view the understanding of what it means to be an engineer does not come easily to prospective or junior undergraduate engineers. The situation is very different in medicine for example where from an early age either through direct personal experience of healthcare systems or exposure to media (TV and film) dealing with how doctors and nurses work, children and then teenagers have a firm concept of what it means to be a doctor or nurse. That is not to say that medical student's identity is fully formed but it is well in advance of the situation amongst first year engineers.

The above overview is not sufficient however to describe engineering and its overall knowledge base but it does articulate the elements of the foundations. Put another way it is as if an orchestral piece of music work was defined by limiting discussion to the characteristics of the various instruments deployed. To an audience the music played depends on the selection of instruments used and the skill of the players, but above that there is the nature and quality of the music being played. The composer set out with an objective in mind and the degree to which that objective is judged to have been met is of course always an open question. Engineering has similar features in that the engineer's 'composition' might or might not be valued by society acting as an 'audience'.

The next section looks at the role of the material in the foundation layer in contributing to a set of learning outcomes which in turn have an associated engineering knowledge identity.

The Middle Layer of Model: Knowledge Associated with Engineering Program Learning Outcomes

A multi-nation European initiative resulted in the establishment of the European Network for Accreditation of Engineering Education (ENAEE) which authorizes accreditation and quality assurance agencies to award the EUR-ACE® label to accredited engineering degree programs. In addition to reviewing both the quality of the teaching facilities and the lecturing staff, importance is put on whether the program enables students to achieve a set of outcomes (ENAEE 2013). The six Program Outcomes are:

- Knowledge and Understanding;
- Engineering Analysis;
- Engineering Design;
- Investigations;
- Engineering Practice;
- Transferable Skills.

Currently Europe classifies programs as First Cycle, normally of 3 years duration, and Second Cycle normally together with the First Cycle of 5 years duration. The First Cycle is in essence a Bachelor program and the Second Cycle a Masters Program. In some cases the Master component is a 2-year 'add-on' to the Bachelor program, and in other cases the Master is an ab initio 5-year program. ENAEE specifies a range of competences under each of the six learning outcomes as follows:

Knowledge and Understanding
First Cycle graduates should have:

- knowledge and understanding of the scientific and mathematical principles underlying their branch of engineering;
- a systematic understanding of the key aspects and concepts of their branch of engineering;
- coherent knowledge of their branch of engineering including some at the forefront of the branch;
- awareness of the wider multidisciplinary context of engineering.

Second Cycle graduates should have:

- an in-depth knowledge and understanding of the principles of their branch of engineering;
- a critical awareness of the forefront of their branch.

Engineering Analysis
First Cycle graduates should have:

- the ability to apply their knowledge and understanding to identify, formulate and solve engineering problems using established methods;
- the ability to apply their knowledge and understanding to analyse engineering products, processes and methods;
- the ability to select and apply relevant analytic and modelling methods.

Second Cycle graduates should have:

- the ability to solve problems that are unfamiliar, incompletely defined, and have competing specifications;

- the ability to formulate and solve problems in new and emerging areas of their specialisation;
- the ability to use their knowledge and understanding to conceptualise engineering models, systems and processes;
- the ability to apply innovative methods in problem solving.

Engineering Design
First Cycle graduates should have:

- the ability to apply their knowledge and understanding to develop and realise designs to meet defined and specified requirements;
- an understanding of design methodologies, and an ability to use them.

Second Cycle graduates should have:

- an ability to use their knowledge and understanding to design solutions to unfamiliar problems, possibly involving other disciplines;
- an ability to use creativity to develop new and original ideas and methods;
- an ability to use their engineering judgement to work with complexity, technical uncertainty and incomplete information.

Investigations
First Cycle graduates should have:

- the ability to conduct searches of literature, and to use data bases and other sources of information;
- the ability to design and conduct appropriate experiments, interpret the data and draw conclusions;
- workshop and laboratory skills.

Second Cycle graduates should have:

- the ability to identify, locate and obtain required data;
- the ability to design and conduct analytic, modelling and experimental investigations;
- the ability to critically evaluate data and draw conclusions;
- the ability to investigate the application of new and emerging technologies in their branch of engineering.

Engineering Practice
First Cycle graduates should have:

- the ability to select and use appropriate equipment, tools and methods;
- the ability to combine theory and practice to solve engineering problems;
- an understanding of applicable techniques and methods, and of their limitations;
- an awareness of the non-technical implications of engineering practice.

Second Cycle graduates should have:

- the ability to integrate knowledge from different branches, and handle complexity;

- a comprehensive understanding of applicable techniques and methods, and of their limitations;
- a knowledge of the non-technical implications of engineering practice.

Transferable Skills
First Cycle graduates should be able to:

- function effectively as an individual and as a member of a team;
- use diverse methods to communicate effectively with the engineering community and with society at large;
- demonstrate awareness of the health, safety and legal issues and responsibilities of engineering practice, the impact of engineering solutions in a societal and environmental context, and commit to professional ethics, responsibilities and norms of engineering practice;
- demonstrate an awareness of project management and business practices, such as risk and change management, and understand their limitations;
- recognise the need for, and have the ability to engage in independent, life-long learning.

Second Cycle graduates should be able to:

- fulfil all the Transferable Skill requirements of a First Cycle graduate at the more demanding level of Second Cycle;
- function effectively as leader of a team that may be composed of different disciplines and levels;
- work and communicate effectively in national and international contexts.

Though the language used in setting out the learning outcomes is in the form of expressing an ability it is clear that behind each proficiency there is a knowledge that is to be applied. In some learning outcomes the knowledge is obvious but in others it is implicit. The first Learning Outcome (Knowledge and Understanding) has a straightforward link to the foundational layer through ‘scientific and mathematical principles’, but the added knowledge required for a ‘systematic understanding …’ is not so straightforward. Nor is it a simple matter to articulate exactly what this added knowledge consists of, even though its application might be perfectly clear. Nevertheless there is a specific form of knowledge required if a systemic understanding of an engineering task is to be carried out properly. With the second Learning Outcome (Engineering Analysis) there is another type of knowledge required in the ‘understanding to conceptualise engineering models, systems and processes’. In the third Learning Outcome (Engineering Design) a different knowledge base is necessitated for ‘an understanding of design methodologies’. To know the merits of potential design methodologies and then be sufficiently knowledgeable to choose the ‘right’ one is a valued knowledge based skill. (In passing it is noted that some authors make a distinction between knowledge and wisdom where the latter concerns the ability that comes with experience to wisely navigate through sets of knowledge.) The set of six learning outcomes have precursor knowledge starting points. And it should be clear that such knowledge is very different in character to the knowledge in the foundational layer. In this way the nature of engineering emerges but that is not the end of the matter.

The next section describes the third layer 'Competences', which, building on the knowledge of the two lower layers and added to relevant work experience, leads to the state of knowledge that an engineer must reach to be considered a Chartered Engineer or Professional Engineer. UK, Ireland and Australia amongst others use the term Chartered Engineer (CEng), whilst Canada and the United States amongst others use the term Professional Engineer (PE).

The Top Layer of the Model: Knowledge Associated with the Competences Required of Chartered or Professional Engineers

To set a context, professional bodies such as Engineers Ireland, consider that the formation of an engineer consists of two parts. The first part refers to the educational formation usually culminating in an accredited Bachelor or Master degree. To aid the mobility of engineers worldwide and ensure transparency of engineering qualifications, many countries have signed agreements or accords by which accredited degrees are mutually recognised. The Washington Accord is an international agreement entered into by Engineers Ireland with other professional bodies in the UK, USA, Canada, Australia, New Zealand, Hong Kong-China, South Africa, Japan, Korea, Singapore, Malaysia and Taiwan. Through this Accord, the signatories accept each other's accreditation decisions thereby enabling mutual recognition of each signatory's engineering degree programs. The second part refers to post degree experience gained in relevant engineering work situations. The total period of formation is a minimum of 8 years. The following is the definition of a professional engineer recognised by the Council of Engineers Ireland for the title Chartered Engineer and is the definition adopted in 1960 by the Conference of Engineering Societies of Western Europe and the United States of America (UESEC):

> A professional engineer is competent by virtue of his/her fundamental education and training to apply the scientific method and outlook to the analysis and solution of engineering problems. He/she is able to assume personal responsibility for the development and application of engineering science and knowledge, notably in research, design, construction, manufacturing, superintending, managing and in the education of the engineer. His/her work is predominantly intellectual and varied and not of a routine mental or physical character. It requires the exercise of original thought and judgement and the ability to supervise the technical and administrative work of others. His/her education will have been such as to make him/her capable of closely and continuously following progress in his/her branch of engineering science by consulting newly published works on a worldwide basis, assimilating such information and applying it independently. He/she is thus placed in a position to make contributions to the development of engineering science or its applications. His/her education and training will have been such that he/she will have acquired a broad and general appreciation of the engineering sciences as well as thorough insight into the special

features of his/her own branch. In due time he/she will be able to give authoritative technical advice and to assume responsibility for the direction of important tasks in his/her branch.

To become a Chartered Engineer, in most if not all relevant jurisdictions, an applicant submits a mandatory report which, if accepted, is followed up by an interview conducted by a panel of trained interviewers. The key areas explored are the five competences that the applicant must demonstrate they have attained (Engineers Ireland 2012).

- Competence 1: Use a combination of general and specialist engineering knowledge and understanding to optimise the application of existing and emerging technology.
- Competence 2: Apply appropriate theoretical and practical methods to the analysis and solution of engineering problems.
- Competence 3: Provide technical, commercial and managerial leadership.
- Competence 4: Use effective communication and interpersonal skills.
- Competence 5: Make a personal commitment to abide by the appropriate code of professional conduct, recognising obligations to society, the profession and the environment.

In Competence 1 it would be expected that a Chartered Engineer 'deepens their knowledge base systematically through research and experimentation', and 'extends knowledge of related disciplines or fields and fosters co-operation across discipline boundaries to identify future potential opportunities'. For Competence 2, Chartered Engineers should have the capacity to 'exercise original thought in synthesising satisfactory outcomes to engineering challenges'. Within Competence 5, Chartered Engineers are expected to 'give evidence, express opinions or make statements in an objective and truthful manner and on the basis of adequate knowledge'. This last requirement is a powerful one that should apply to all professionals as it demands that the individual operates within a knowledge space which they have the responsibility of judging to be adequate. In other words, this is an injunction not to operate outside one's competence (or knowledge boundary). As an aside there are those in both Science and Engineering who have proposed a Hippocratic Oath adapted to their respective professions (Grimson and Murphy 2013). Whilst few professional bodies have adopted such an oath it at least makes sense to discuss the underlying issues with students in their undergraduate classes and for codes of ethics to reflect the intentions involved.

At this stage a general picture should have emerged as to the epistemological web inhabited by engineers. But it is still very complex and difficult if not impossible for any one individual to master. This is why pragmatic approaches abound in engineering. One such approach is based on the use of heuristics and an example is presented later. But by far the most important approach encountered in engineering to control what otherwise would be an impossible situation – building from basics through to producing an optimum solution for any particular challenge – is the use of standardised (well-tested and widely adopted) methods. Another aspect of engineering is the positive role failure plays in developing new knowledge and refining methods. The next section discusses heuristics, standardised methods, and the role of failure.

Dealing with Complexity

Heuristics

Strategies that reflect constraints with respect to the acquisition of relevant knowledge to solving a particular problem include approaches that intentionally seek to obtain a solution through processes that are not in fact guaranteed to be valid. The use of a rule-of-thumb falls into this category. This might seem strange and anti-scientific but engineering puts more store on getting a result than proceeding elegantly without result. It is not that elegance is discounted, but more a reflection that elegance is not always possible. Billy Koen has written extensively about heuristics which he describes in essence as doing the best in an inadequately understood situation within available resources (Koen 2003). In fact Koen makes the claim that all engineering is heuristic. There is some truth to this claim but it should not be taken literally. Many engineering challenges can be addressed without recourse to heuristics, but they do have a place especially when applied intelligently. For a general introduction, not specific to engineering, the reader should consult Michalewicz Zbigniew, David Fogel's book 'How to Solve it: Modern Heuristics' (Zbigniew and Fogel 2004).

To illustrate the heuristic approach with a simple example and also to flavor why it is an attractive option when intelligently applied, consider the following problem. A very long ladder network consists of 1-Ω resistors as shown in Fig. 9.2. The problem is to determine the input resistance at terminals TT'. One line of attack is to truncate the network at AA' and calculate the input resistance when (a) a short-circuit is created at AA' and (b) when an open circuit is allowed at AA'.

Using the simple electricity laws of calculating the resistance of resistors in series and parallel quickly leads to the results:

$$R_{in}(\text{open circuit}) = 2\Omega \quad \text{and} \quad R_{in}(\text{short circuit}) = 1.5\Omega$$

Logically, the input resistance of the full ladder network must lie between these two values. An improved bracketing could be obtained by truncating the network at BB' in which case the results are as follows:

$$R_{in}(\text{open circuit}) = 1.666\Omega \quad \text{and} \quad R_{in}(\text{short circuit}) = 1.6\Omega$$

The heuristic 'solution' is then found from the upper and lower bounds by taking the geometric mean of these two values yielding 1.633. Two points can be made. First, adopting a truncation strategy allowed both the upper and lower bounds to be calculated with zero error. Second, no justification can be given for how the final result is obtained and the alternative of using an arithmetic mean would have been equally justified. However the solution obtained cannot be too far from the true value due to the approach used in the bracketing between the upper and lower bounds. As it transpires the true result is 1.618 Ω which can be found in a number of different ways. But that is not the point of the example; rather an intelligent approach using

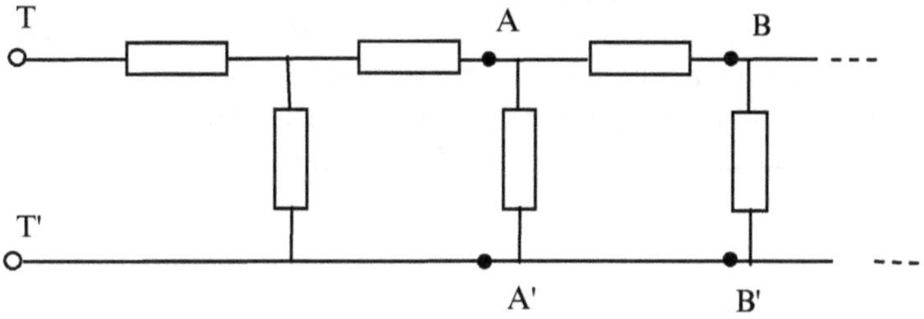

Fig. 9.2 Long resistor ladder network with equal components

easily available knowledge facilitated a sensible application of heuristics. Whether the bracketed solution is usable of course depends on the context in which the original problem arose.

Role of Failure in Engineering

Engineering is largely evolutionary wherein progress is incremental based on a selection process (see Grimson and Murphy 2009). The selection process is not just concerned with what is ‘good’, the ‘bad’ is equally important. In the extreme case the ‘bad’ can represent a total failure of a system or some vital component of that system. Henry Petroski has written much about *failure* in engineering and whilst failure is never intentional it is to some extent inevitable, but any failure comes with the benefit that post-failure analysis can result in new knowledge or at least information that can reduce future accidents (Petroski 1985). Unfortunately failure can be caused by poor management decision-making (see for example the Challenger Shuttle disaster) and negligence (Kansas City Hyatt Regency Hotel walkway collapse). But it is failure when neither poor management nor negligence is involved that generates the most useful technical insight leading to new engineering knowledge. One example is the well documented collapse of the Tacoma Narrows Bridge. What transpired in investigating this failure was that wind-induced instability (*aero-elastic flutter*) caused the bridge to collapse. The mode of failure was a new one and forced bridge designers armed with this new knowledge to take into account aero-elastic flutter.

Failure does not have to occur in live situations to generate new knowledge. In fact most failures occur under laboratory or testing conditions. The use of wind-tunnels and hydro-models for example are engineering substitutes for almost intractable mathematical calculations. And today the use of computers in modelling in almost all branches of engineering is so advanced that some educationalists are being forced to re-think how design is taught. Another challenge arises when engineering simply becomes an academic pursuit divorced from engineering practice with its rich experience of failure. Steen Christensen and Byron Newberry in reviewing academic drift in engineering note that 'the process whereby knowledge derived from practical engineering work experience and intended to be useful for industrial practice gradually loses its close ties to practice (Christensen and Newberry 2015). Instead engineering knowledge becomes increasingly theoretical and oriented toward engineering disciplines, including mathematics and natural science'. The danger here is analogous to a medical doctor losing sight of his/her patient!

The Role of Standardized Methods

There are multiple reasons for the use of standardized methods. Accumulated knowledge is encapsulated in such methods in the sense that they become evidence-based. Another reason is that safety concerns are addressed by restricting options that might with experience prove to be poor. The early history of ASME, founded in 1880, was inextricably linked with the problems of boilers exploding. The ASME website (see www.asme.org/engineering-topics/articles/boilers/the-history-of-asmes-boiler-and-pressure) provides the background.

The ASME Boiler and Pressure Vessel Code (B&PVC) was conceived in 1911 out of a need to protect the safety of the public. This need became apparent shortly after the conception of the steam engine in the late 18th century. In the nineteenth century there were literally thousands of boiler explosions in the United States and Europe, some of which resulted in many deaths. The consequences of these failures were locally focused and, other than one or two, received minimal national or international attention. Undoubtedly, one of the most important failures that proved the need for Boiler Laws was the boiler explosion that occurred at the Grover Shoe Factory in Brockton, Massachusetts on March 10, 1905. That incident resulted in 58 deaths and 117 injuries and completely levelled the factory. This catastrophe brought attention to the need to protect the public against such accidents with pressure-retaining equipment.

The outcome was the development of a code resulting in the publishing in 1915 of The Boiler and Pressure Vessel Code. This code set standards covering the rules for fabricating a component, materials that are to be used, welding, testing, and rules that permit the use of materials and alternative methods of construction. The work in devising this code must have involved many experts but the value to other engineers and society more than repaid the initial effort required. The central point is that knowledge became encoded in a particular and useful manner and then made available to others. This encoding might be thought of as creating second-order knowledge or knowledge about knowledge.

There are countless examples in engineering where knowledge is encapsulated in the form of codes, standards, guidelines, methods etc. and they all serve the purpose of engineering work proceeding in an efficient, effective, safe, and organised manner. Further, they serve the engineering community by providing a 'language' and a process by which enhancements can be made. Innovation it is claimed can be stifled by too rigorous an adherence to codes or standards. But in the longer term successful innovations yield their own standards, the internet for example. Finally, codes, standards, and approved methods allow engineers to proceed without having to revert to first principles every time a new task commences. Nor do engineers need to spend time and effort solving problems already solved many times over. Instead engineers are more readily freed to build on the work of previous generations.

Relevant Importance of Various Skills and Knowledge

It is one thing to identify what skills and knowledge need to be developed within an engineering program it is another matter entirely to assign corresponding weight in the design of a curriculum. In a report by the Center for the Advancement of Engineering Education data is presented indicating the importance of skills and knowledge in rank order as perceived by senior engineering students (Atman et al. 2010). In descending order the relative importance is ranked as follows: Problem solving (73 %); Communication; Teamwork (61 %); Engineering analysis; Ethics (40 %); Design; Creativity; Life-long learning; Math (19 %); Data analysis; Engineering tools; Leadership; Business knowledge; Science (13 %); Management skills; Professionalism (11 %); Conducting experiments; Global context; Societal context; Contemporary issues. Whether this ranking reflects what students have been offered or what they would like to have been offered is open to question. But the list and ranking is a good description of the challenges in designing a fit for purpose curriculum.

Design as a general engineering activity is one of the hardest to characterize. And the necessary knowledge underpinnings include much of what has been stated already (Science, Math, Analysis etc.). But there is much more to design. Another dimension is outlined in the above report, namely the most important Design Activities. These range from Understanding the problem (most important) to Identifying constraints, Testing, Modeling, Prototyping, Iterating to Abstracting (least important). The point that is emphasized here is that each activity must be supported by some form of knowledge if the engineer is to know how to proceed in a sensible manner rather than simply by trial and error. It follows that this spectrum of activities must be taken account of in the formation of an engineer (undergraduate, post-graduate and professional undertakings)

Conclusion

At first glance epistemology is the most easily understood of the five classical branches of philosophy (epistemology, metaphysics, logic, aesthetics and ethics). We are told that we live in a modern world that is knowledge driven and undoubtedly our ready access to knowledge is well supported through the use of information technology. We, it would seem, use, create and modify, store and transmit knowledge on a routine basis in our everyday lives, and so assume an easy familiarity with it. But you don't need to be a philosopher to realize that epistemology is in fact a most difficult and complex thing. From a very practicable perspective the issues surrounding determining provenance, the ability to authenticate, and understanding the limitations of knowledge, all have particular significance for engineering. With the possible exception of mathematics which since it is strictly rational can be excluded, all branches of knowledge on which engineering is based is to a greater or lesser extent a structure built on shifting sands. This is not to say engineering is useless or in some way defective. Rather it points to the need to constantly update its knowledge base knowing that circumstances change and as a consequence fresh evidence surfaces.

According to James Boswell, Samuel Johnson held that there are two kinds of knowledge, the first being that which we know ourselves, the second being knowing where to find what we wish to know (Boswell 1791). An engineer might well add a third kind, namely knowing when or why to search for new knowledge or update existing knowledge. It is not too surprising then that David Goldberg in a talk given at the Royal Academy of Engineering in London stated that engineering is epistemologically weak (see http://www.slideshare.net/deg511/engineering-in-context-the-professional-and-institutional-settingSlide 37). In the first place engineering is a net borrower (particularly mathematics and the sciences), and in the second place engineering knowledge is subject to gradual and occasionally abrupt change. One can accept therefore the 'weakness' assertion but this is not necessarily a negative attribute. At its simplest the demands made of engineering to solve problems, address challenges, and to create that which never previously existed means that the 'strength' of engineering epistemology is not of great importance. Instead the usability of the knowledge is the critical factor. Even a certain lack of consistency in the knowledge used can be tolerated in engineering provided the impact of such inconsistency is known and allowed for in the course of undertaking the work involved. This is analogous to the heuristic method discussed previously in this chapter where there is a need to do the best one can, in conditions that are not ideal.

Another issue centers on the definition of knowledge. The classical definition, going back to Plato, is that knowledge is 'justified true belief'. In an engineering context 'justified' is generally not a black or white attribute, in practice there would be degrees of justification depending on the context. This conditionality of engineering knowledge requires judgment to be exercised by engineers, a skill that comes with experience. It is tempting to refer to such judgment as being the application of wisdom; wisdom being the good or best use of available knowledge, and one could say it is a form of knowledge about knowledge.

Engineering relies on and is supported by the rational knowledge that is science and the empirical knowledge that is the sciences. As explained earlier there are other foundational areas too. The middle layer represents the core of what some would say is 'engineering' where the knowledge of how to carry out a wide range of activities is contained. The ability to determine requirements, analyze, design, test, evaluate, review – all these are required of an engineer in whatever sub-discipline of engineering they operate. The top layer of knowledge characterizes the deep understanding of the underlying layers by which engineering work can be carried out using appropriate methods and when necessary devising new approaches. The epistemological basis of engineering poses descriptive difficulties but there is an underlying structure that is both robust and resilient, largely determined by educationalists and practitioners who have built on a wealth of experience. New knowledge is created and old knowledge modified or discarded but the structure in which it fits remains constant. It is this structure that allows engineering to proceed without being submerged in the sea of knowledge.

References

Atman, C., Sheppard, S., Turns, J., Adams, R., Fleming, L., Stevens, R., Streveler, R., Smith, K., Miller, R., Leifer, J., Yasuhara, K., & Lund, D. (2010). *Enabling engineering student success – The final report for the Centre for the Advancement of Engineering Education*. San Rafael: Morgan & Claypool Publishers.

Boswell, J. (1791). *The life of Samuel Johnson*. http://www.gutenberg.org/ebooks/1564

Bucciarelli, L., Coyle, E., & McGrath, D. (2009). Engineering education in the US and the EU. In S. H. Christensen, B. Delahousse, & M. Meganck (Eds.), *Engineering in context* (Vol. 5). Aarhus: Academica.

Christensen, S. H., & Newberry, B. (2015). The role of research in academic drift processes in European and American professional engineering education outside the university sector. In S. H. Christensen, C. Didier, A. Jamison, M. Meganck, C. Mitcham, & B. Newberry (Eds.), *International perspectives on engineering education: Engineering education and practice in context* (Vol. 1). Dordrecht: Springer.

De Figueiredo, A. D. (2008). *Toward an epistemology of engineering*. Workshop on philosophy & engineering. London: Royal Academy of Engineering. http://www.slideshare.net/adfigueiredo/toward-an-epistemology-of-engineering-presentation

Engineers Ireland. (2012). *Chartered engineer.* Regulations for the title of chartered engineer (pp. 22–27). www.engineersireland.ie/EngineersIreland/media/SiteMedia/membership/professional-titles/Chartered-Engineer-Regulations-2012.pdf. Accessed 29 Aug 2013.

European Network for Accreditation of Engineering Education (ENAEE). (2013). *The EUR-ACE system*. http://www.enaee.eu/eur-ace-system. Accessed 29 Aug 2013.

Fox, M., & Huang, J. (2003). *Knowledge provenance: An approach to modelling and maintaining the evolution and validity of knowledge*. EIL Technical Report, University of Toronto.

Grimson, W., & Murphy, M. (2009). An evolutionary perspective on engineering design. In S. H. Christensen, B. Delahousse, & M. Meganck (Eds.), *Engineering in context* (Vol. 13). Aarhus: Academica.

Grimson, W., & Murphy, M. (2013). A Hippocratic oath for engineers. In D. Michelfelder, N. McCarthy, & D. Goldberg (Eds.), *Philosophy and engineering: Reflections on practice, principles, and process* (Vol. 15). Dordrecht: Springer.

Heymann, M. (2009). "Art" or science? Competing claims in the history of engineering design. In S. H. Christensen, B. Delahousse, & M. Meganck (Eds.), *Engineering in context* (Vol. 11). Aarhus: Academica.

Koen, B. V. (2003). *Discussion of the method: Conducting the engineer's approach to problem solving*. New York: Oxford University Press.

Petroski, H. (1985). *To engineer is human: The role of failure in successful design*. New York: St. Martin's Press.

Polanyi, M. (1958). *Personal knowledge: Towards a post-critical philosophy*. Chicago: University of Chicago Press.

Zbigniew, M., & Fogel, D. B. (2004). *How to solve it: Modern heuristics* (2nd ed.). Berlin: Springer.

William Grimson BAI, B.A., M.A.Sc., Chartered Engineer. An Electronic Engineering graduate of Trinity College, Dublin and the University of Toronto. He has recently retired but is still actively involved in re-thinking engineering education and the relevance of philosophy to engineering and engineering education. He first worked as a R&D engineer for Ferranti Ltd before joining the academic staff of the Dublin Institute of Technology (DIT) where he was at one time Head of the Department of Electrical Engineering. He ended his career as Registrar with overall responsibility for academic quality enhancement. His academic output was and remains eclectic ranging from publications in areas as diverse as plasma physics, clinical information systems, philosophy of engineering, and development issues. He is currently a member of the Executive of the Institution of Engineers of Ireland.

Mike Murphy H. Dip. from Dublin Institute of Technology, B.Sc. (Eng) in Electrical Engineering, Trinity College Dublin, M.Sc. and Ph.D. in Electrical Engineering, Stevens Institute of Technology. Fellow of Engineers Ireland. Member of the IEEE. He worked in Bell Labs and later at Bell Communications Research before returning to academia in 2002. Director of Dublin Institute of Technology and Dean of the College of Engineering & Built Environment. Member of SEFI Administrative Council, and Member of European Engineering Deans Council and current President. Area of special interest is engineering and technology education.

Chapter 10
The Tension Between Science and Engineering Design

Stig Andur Pedersen

Abstract Engineering design is an essential part of the process of constructing and maintaining modern complex systems as airplanes, power plants, and urban areas. As such engineering design must be based on scientific knowledge. But whereas it is the task of engineering design to assist in the realization of complex systems in their concrete real life context, it is the task of science and mathematics to find and justify new knowledge about the universal working of nature. In a science as physics mathematical structures and formalisms are developed and applied as means to identify and describe the form and nature of laws that govern the behavior of processes of very different scales in nature. This work requires comprehensive abstraction and idealization, and, as a consequence of that, advanced mathematical and physical theories are only valid in highly abstract and isolated systems. Consequently, these theories are far away from the concrete contexts that engineering design is about. In this paper we shall identify and discuss some of the epistemological problems that this tension between scientific idealization and engineering concretization may lead to.

Keywords Research and model objects • Model building skill • Tacit knowledge • Theoretical framework • Hierarchies of laws • Truth and reality

Introduction

Engineering design is an essential part of the process of constructing and maintaining modern complex systems as computers, smartphones, airplanes, power plants, and urban areas. All these artifacts have complex chemical, electrical, and mechanical features, which only can be understood and controlled by relying on scientific theories. In this sense, modern technology is science based. It is impossible to imagine how modern forms of artificial material, control systems, and structural systems would have been possible without advanced scientific mathematical modeling. At

S.A. Pedersen (✉)
Section for Philosophy and Science Studies, Roskilde University,
Box 260, Roskilde DK-4000, Denmark
e-mail: sap@ruc.dk

S.H. Christensen et al. (eds.), *Engineering Identities, Epistemologies and Values*, Philosophy of Engineering and Technology 21,
DOI 10.1007/978-3-319-16172-3_10

the same time it also true that modern mathematics and natural science depend heavily on technology. It is quite striking that some of the most theoretical questions in physics and cosmology, for instance, about the origin of life and the formation of matter, require the most advanced machines that ever have been built. The purpose of the Large Hadron Collider at CERN is to produce events which might shed light on the formation of matter. This machine is built with the aim of understanding the most fundamental aspects of the material world, all the technological benefits that flow from experiments with it are side effects. So, natural science and technology are intertwined and interdependent in a complex way.

In spite of the close interdependence between science and technology they are two different activities with their own specific logics. This has been forcefully documented by, among others, Derik de Solla Price in the essay *The Difference Between Science and Technology*:

> We have the position, then, that the normal growth, science begets more science and technology begets more technology. The pyramid like exponential growth parallel each other, and there exists what the modern physicist would call a weak interaction -at the educational level and the popular book and the Scientific American stage - that serves just to keep the two largely independent growths in phase (Price 1975, pp. 129–130).

The two exponential growth processes – science and technology – have the same formal structure and their own internal autonomous logic of development. Science develops in the following way:

> Science is a sort of growing jigsaw puzzle with a dozen sexes, and wherever there is a family of knowledge - an annual supply of knowledge - children are produced. Old knowledge gives rise to new at an exponential rate. From time to time new subdivisions of knowledge appear, but the general process goes on without let or hindrance, without fail in times of poverty and war, without hurting in times of need. There is, strangely enough, very little man can do to make knowledge come more or less quickly or to make it come in the directions we may wish (ibid., p. 129).

And technology has a similar growth pattern:

> Technology, the other twin, grows, I believe, in a very similar fashion. It is evident to any historian of technology that almost all innovations are produced from previous innovations rather than from an injection of any new scientific knowledge. We do not see it so well just because the technologists are keeping quiet rather that shouting from the rooftops as the scientists do (ibid., p. 129).

So, according to Price, science and technology constitute two parallel processes that interact weekly, but where the main driving forces in each process are internal problems. New scientific results lead to new scientific problems which are the basis for new scientific results, and new technical innovations leads to new technological challenges which result in new innovation. Although there are many other determining forces in the development and that science and technology today are more intertwined than ever before it seems to be a correct picture. Science has its own internal logic and so does engineering.

Price describes the situation from the outside. In this paper we will shed some more light on the interaction between science and engineering by having a closer look at the way in which knowledge is produced within the two fields.

The Subject Matter of Science and Engineering

The subject matter of natural science is not in general ordinary objects and phenomena as we see them in our daily lives. The objects and phenomena under scrutiny by scientists are carefully delimited from their environment. They are idealized and transformed to such a degree that many of them do not have natural counterparts. For instance, when molecular biologists or pharmacologists study how various drugs are transported over biological membranes, they do it by taking test samples of tissue and cultivating them in artificial environments and eventually decomposing them into molecular parts. This is a long and complex process which is difficult to control and interpret correctly. So, what the scientist eventually measures are variables, e.g. concentrations of enzymes, proteins, etc., which are related to the original biological tissue in a very indirect and complex way. The biologist cannot see the membrane directly, but only indirectly as constructed pictures made by advanced equipment like electron microscopes, MR scanners, ultra sound instruments and the like.

The situation in engineering research is similar. Studies of strength of various building components, for instance, are in many cases based on laboratory test samples which are isolated and manipulated in ways that make them appear very different from the way they would look like at real building sites.

The process of preparation and delimitation of objects so that they can be studied scientifically leads to the construction of what we are going to call *model objects*. These objects are abstract constructions. But they are the objects that scientific theories are about and construction of these objects is an essential part of scientific reasoning.

It is well-known that reasoning in natural and engineering science involves complex structures of hypothetico-deductive systems where principles, laws, and empirical generalizations are hierarchically ordered with respect to epistemological significance. We will briefly discuss such systems and their role in scientific reasoning.

As model objects in a sense are theoretical constructions we must face the problem of how scientific theories are related to reality. The paper ends with a few reflections on truth and reality.

It should be noted, that engineering science has as its main objects of research artifacts like pumps, industrial plants, electronic devices and buildings. It is impossible to define such artificial objects completely in naturalistic terms. They have an intentional side and they are defined and understood in terms of their operational principles in the sense of Polanyi (see Polanyi (1958), p. 176). One must consider the nature of the goals they serve. For instance, a water pump serves the goal of pumping water from a well, and such a machine is not well defined unless this goal is taken into consideration. As artifacts ultimately must serve certain functions in society, engineering science and work must include considerations of immaterial and intentional phenomena like production, planning, organization and other circumstances of general economical and societal significance. These aspects traditionally belong to the social sciences and in some cases the humanities. Hence, engineering science covers an extremely broad spectrum of activities. Many engineering fields of study cannot be characterized as only belonging to the natural sciences. However, in this paper we will only deal with those naturalistic aspects of engineering research which are epistemologically similar to what one finds in natural sciences. Intentional phenomena will not be taken up in this paper.

Theory, Research Objects and Model Objects

As already mentioned, scientists manipulate, idealize and transform their research objects in the laboratory so that the properties and partial aspects of the object, in which they are interested, are isolated and appear as pure as possible. They eliminate all kinds of irrelevant and disturbing properties so that they can be as certain as possible that the single aspect or property under study is present and not disturbed by irrelevant factors. For example, when solid-state physicists are studying magnetic properties of metals they abstract from the mechanical and chemical properties of these metals and they make sure that all possible disturbing magnetic fields are kept away. We call this process of delimiting a research object a preparation process.

It is evident that this process of preparation of the research object requires theoretical considerations. The solid state physicists must know that there are no interactions between gravitational and magnetic forces and they must know theoretically how to shield off other fields. Without such knowledge they would not be able to prepare a research object correctly and their experimental work would be useless. Consequently, the preparation process requires theory both as a tool for selecting relevant properties of the research object and as a tool for assessing the correctness of the preparation of the research object. It requires theoretical considerations to make sure that the research object is an adequate representation of reality.

By "reality" we mean the world as it looks for us as ordinary people. We do not mean anything metaphysically complex. What we want to stress is that natural sciences, and also many engineering research fields, are not directly about the world as it appears to us. They are about highly idealized and abstract features of this world, and even for engineering research it is often difficult to see how such research has anything relevant to say about the world as we know it, and it is not unproblematic to transfer results from the laboratory to the real world and use them for actual engineering design.

Preparation of Research Objects

The preparation process involves at least three partial processes. First, the part of the reality we want to study is delimited from its environment. We may, for instance, be interested in studying how a certain drug influences a membrane in the brain of a human being. The object under study is delimited to this membrane, and the membrane is so to speak isolated from the other parts of the human being.

Secondly, we ignore all properties of the membrane except some of its electrochemical properties which we want to measure. This is an *abstraction process*. We abstract from all properties except one or a few which we want to study experimentally. In this way we end up with a *generic object*. All relevant properties are lifted to this generic membrane. Consequently, when relevant properties of the concrete membrane are determined they are considered as properties of all membranes of that generic type. Research is not about properties of concrete tokens but about those properties that can be lifted to generic types of objects.

Thirdly, we also *idealize the object*. We consider the membrane to be an ideal, homogeneous and generic exemplar of the actual membrane. That is, the parameters we are measuring are supposed to be valid for any membrane of the correct type and not only valid for the actual membrane under consideration. All inhomogeneities and imperfections of the actual membrane are smoothed out. In praxis this is done by considering variations in measurements as noise which must be eliminated by, for instance, statistical analysis.

The result of the preparation process is an isolated object demarcated from its surroundings and processed such that it can be manipulated in a controlled way. It is considered as an instance of a perfect, generic object. Consequently, the *research object can be defined as an isolated and manipulated part of the world which is conceptualized as an instance of a generic object.*

We have given a biological example. But it is an easy matter to give engineering examples. Consider, for instance, the study of oscillations of steel bars. When we want to study eigenoscillations of steel bars we delimit the bar from disturbing parts of its environment, we abstract from all properties that are irrelevant for the study of its vibrations, and we idealize by considering it as a perfect, homogeneous bar. Although we of course are making measurements on real bars we consider these measurements as being properties of the perfect, generic bar. All variations and irregularities caused by the imperfections of the actual bar in the laboratory are smoothed out and considered as unintended disturbances. The bar is conceptualized as a generic bar.

Construction of Model Objects

The research object has been defined as the result of a preparation process. Although the object is physically real it is considered as an abstract entity. The preparation process involves at least three aspects: (i) delimitation of the object, (ii) abstraction from irrelevant properties, and (iii) idealization. This process leads to a complex artifact, the research object, which is an isolated and manipulated entity conceptualized as a perfect generic object. All basic properties measured and analyzed in the laboratory are comprehended as properties of the generic object. The actual object represents the generic object.

As an example consider measurements of the gravitational field of the Earth as part of precision surveying. In mountain areas like Greenland gravitational measurements are made on locations at different altitudes. Such measurements are not compatible unless they are transformed in such a way that they refer to the same altitude. That is done by introducing the *geoid.* Broadly speaking the geoid is defined to be the object bordered by the equipotential surface of the gravitational field at sea level. This surface is an abstract generic object used to define the form of the Earth as a research object. The geoid is an abstract planet which is as similar to the Earth as possible. It has the same mass, the same axis of rotation, the same angular velocity, etc. But it does not exist as a real object. It is a conceptual model. When geodesists talk about the gravitational field they usually refer to the gravitational field of the geoid. All kinds of geodesic measurements are planned and prepared in such a way that they can be construed as measurements of properties of the geoid. From a geodesic point of view the Earth is identified with the geoid. The actual Earth represents this generic object, the geoid.

Consequently, research objects are not only objects as they appear to us directly as part of the environment. They are delimited and modified in such a way that they can be regarded as instances of abstract generic objects. Such abstract generic objects are called *model objects*. Consequently, a research object is an entity which can be viewed as an instance of a model object. Model objects are abstract conceptual entities, and they are the proper targets of scientific theories. The geoid is a model object. The real earth is an empirical object. But when the earth is studied by geodesists it is represented by the geoid. All measurements are reduced in such a way that they appear as being made on the abstract model object: the geoid.

So far we have discussed two interrelated processes: Preparation of the research object and construction of a model object which conceptually represents the research object. The research object is prepared in such a way that it fits the model object as truthfully as possible. Usually further idealizations are made during the theoretical investigation of the research object and its properties. Thus, if one is studying vibrations of a homogeneous bar it is often assumed that these vibrations can be described by linear equations, or, if that is not sufficient, second order non-linearities are taken into account. In this way we construct even more abstract and idealized model objects. Whole series of model objects may be constructed during this further generalization process. In many cases this is a necessary condition for existing theories to be applicable. More and more abstract model objects are being constructed in such a way that it is possible to get through with numerical or analytical solutions of the fundamental equations which describe the problem under consideration.

As a specific example of such a model object construction in an engineering field we will consider a project in which certain dynamic and structural features of a human shin-bone are studied experimentally (Thomsen 1990). We want to know how such bones react when they are exposed to various kinds of loading conditions. Knowledge of this kind is important in many medical contexts. For instance, it is important as basis for design when various kinds of prostheses are being developed. The research object was in this case confined to mechanical properties of a well-defined carved piece of a shin-bone. Only structural eigenoscillations in a certain frequency domain were considered, and the object was prepared by cutting it out of a real human leg, gluing on sensors, freezing it down, etc. The model object that conceptually represented this research object was a perfect bar with definite physical properties. This generic object was conceived as the carrier of the oscillations. The laboratory measurements were made on the real piece of a shin-bone cut out of a human being and they were reduced in such a way that they could be considered as originating from the model object. That is, the research object was a technically prepared piece of shin-bone conceptualized as a perfect bar.

The theory relevant for the study of mechanical properties of such objects is taken from solid mechanics. This theory is based on, among other things, fundamental mechanics, and it contains theoretical analyses of dynamic properties of various kinds of objects: bars, plates and other geometric configurations with plastic or elastic properties. Within this theory we have massive knowledge about, say, stress–strain relations of various idealized objects. This knowledge is presented in a mathematical form and is about model objects which are idealized to such a degree that a mathematical analysis is feasible. The research object is represented by such a model object.

A real human shin-bone has a highly complex geometry. It consists of several types of substances, for instance, bone tissue and marrow, with very different mechanical properties. The substances are also inhomogeneous, anisotropic, and vast biological variations exist between different individuals. The inhomogeneities of the various substances cannot directly be eliminated in the research object. But when it comes to theoretical and mathematical considerations it is necessary to increase the level of idealization and construct a model object which fits into the theoretical framework. The inhomogeneities must in some way or other be reduced before it is possible to fit the shin-bone into a theoretical framework. In the example under consideration the research object was represented as a so-called Timoschenko beam. Only some of the constitutive properties of the shin-bone were then represented in the model object. They comprised, in the final definition, "dynamic and structural properties of a rectilinear, twisted, non uniform Timoschenko beam which was made up of two linearly elastic and transversally isotropic compounds and one perfectly flexible compound" (Thomsen 1990).

The actual piece of human shin-bone was in this case considered as a non-uniform Timoschenko beam. But other models might have been chosen. *Depending on available theory and the purpose of the analysis one might have been led to other choices.* In fact, an important part of the project consisted of deliberations of which other possible models might have been relevant for the analysis. The analysis actually showed that the structurally very complex human shin-bone had astonishing simple dynamic properties. Therefore, it was concluded that, for practical purposes, it might be possible to reproduce its dynamic properties with a simpler model. Other researchers had already suggested that a uniform Euler beam might be adequate. Consequently, several possible ways of reducing the complexity of the model object was analyzed. It was finally concluded that from a practical point of view it would be sufficient to reduce the complex model to a simple uniform bar model with concentrated inertia contributions.

Thus the representation of a research object by a model object involves an element of choice. Usually, there are several possible theoretical models that can be applied in a given situation, and the scientist must decide which one is best for the purpose. Such decisions are especially important in engineering science. They are based on deliberations of which goal the research is supposed to serve and of the theoretical and computational possibilities. *Complex models which represent the research object as faithfully as possible can be considered* as reference models *which form a theoretical framework from which more applicable simple models can be validated. Which simple model that eventually will be chosen as the most appropriate one depends heavily on which purpose the simple model is supposed to serve.*

There is also a kind of construction involved. *The model object is a conceptual object that the scientist must construct in such a way that it represents the research object as faithfully as possible.* In the actual case, it was argued that a specific Timoschenko beam constructed by the researcher was an adequate representation of a generic shin-bone. The Timoschenko beam representation would serve as a *reference model* and simpler models useful for design could be validated by comparing them with the Timoschenko beam model.

We shall illustrate the preparation process with another engineering example. In this project, methods for calculation of reinforced structural elements of concrete were developed (Andreasen 1988). When structural elements of concrete are reinforced, i.e. the concrete is deposited around reinforcing iron beams, the strength of the elements is increased. The increase of strength depends among other things on the anchorage. When the beams are ribbed, the strength of anchorage is greater. In the project methods for calculating the load carrying capacity of ribbed reinforcing beams were developed.

There exist two different main theories dealing with such structural calculations: One is about anchorage, i.e. the equilibrium of the loading and the resulting forces acting in the structural elements. The other is about how materials respond to the resulting "inner" forces. Consequently, two different, but interrelated, processes of preparation of the research object were undertaken, and this led to two different model objects.

In relation to the anchorage the problem was delimited in the following way. Only static load was considered. Movements between concrete and iron beams were considered to be unlikely and, therefore, excluded. Consequently, only failures between concrete holding on to the reinforcement and other parts of concrete were considered. Further idealizations about the geometry of anchorage were made. For example, it was assumed that the concrete surrounding the beams was axis-symmetrical to the beam axis, and that loads were evenly distributed. These delimitations, abstractions and idealizations resulted in an idealized model object of anchorage.

Similarly, a model object reflecting the dynamic behavior of the concrete was constructed. Properties like elastic effects, creep, and hysteresis were excluded, and the material was assumed to be homogeneous and perfectly plastic. The idealization into a perfectly plastic material is a very radical step. No material is perfectly plastic, and certainly not concrete. Nonetheless, this model object was chosen because calculation methods based on the theory of plasticity are simple and lead to relatively safe results when applied in an appropriate way.

Scientific Reasoning

In the last century philosophy of science has been dominated by a logical-linguistic view of scientific theories. Philosophers have been more interested in the final products of science, namely scientific theories, and not so much in the process of discovery. They concentrated on issues about justification and truth of scientific theories, and they required theories to be hypothetico-deductive systems expressed in an appropriate language. The kind of reasoning that led up to the construction of scientific theories was not of great interest. Consequently, most philosophy of science in that period overestimated the importance of well-established final scientific theories and underestimated the main body of scientific activity, namely, the reasoning involved in the development of new scientific ideas, concepts and theories. This attitude towards science studies was drastically changed by the appearance of Thomas Kuhn's theory of scientific paradigms where the role of communication and argumentation within the scientific community came to be of central importance for a proper understanding of science. However it took several decades before Kuhn's insights were fully appreciated.

It is, of course, true that many kinds of scientific analyses can be cast into a hypothetico-deductive form, and hypothetico-deductive reasoning is an important part of scientific rationality. However, it is not a complete description of the scientific rationality. It only gives a characterization of some very mature forms of scientific arguments. It is the form in which many result are presented in scientific journals. But it is not a form of representation that can be applied during the research process. During this process all kinds of reasoning are relevant, for instance, analogous reasoning, intuitive considerations, application of elucidating metaphors, and, especially in engineering science, praxis based reasonability considerations. Consequently, most scientific reasoning does not originally have a hypothetico-deductive form. In situations where one does not have a complete theory it might even be impossible to identify a deductive hierarchy. We only have a system of loosely related and vaguely defined conceptual models.

Another but related serious shortcoming of the logical-linguistic view of scientific theories concerns the fact that according to this view a theory is a general schematic framework. The basic laws are abstract schemata that do not have a concrete semantic meaning unless they are interpreted in concrete situations. That is, they only have concrete meaning when interpreted in connection with a concrete model object. But they do not by themselves give a method by which it is possible to identify the models which are essential for their own semantic interpretation. The way in which research objects and model objects are identified is an important aspect of the scientific activity. It is true, that knowledge of hypothetico-deductive systems in many situations is essential for the preparation process and for the definition of model objects. But it is not the only form of knowledge that is involved in this process. The ability to delimit concrete research objects and to identify model objects which represent them is a complex cognitive skill which is learned during scientific training. It involves both linguistically expressed knowledge and tacitly given kinds of knowledge (in the sense of Polanyi) concerning model identification. The schematic laws and propositions in hypothetico-deductive systems would not be semantically well-defined unless it was based on this knowledge.

The way in which implicit and tacitly given knowledge actually functions during scientific reasoning is very beautifully expressed by Heisenberg. He describes the way in which Niels Bohr reasoned when he was doing atomic physics:

> Bohr must surely know that he starts from contradictory assumptions which cannot be correct in their present form. But he has an unerring instinct for using these very assumptions to construct fairly convincing models of atomic processes. Bohr uses classical mechanics or quantum theory just as a painter uses his brushes and colours. Brushes do not determine the picture, and colours are never the full reality; but if he keeps the picture before his mind's eye, the artist can use the brush to convey, however inadequately, his own mental picture to others. Bohr knows precisely how atoms behave during light emission, in chemical processes and in many other phenomena, and this has helped him to form an intuitive picture of the structure of different atoms; a picture he can only convey to other physicists by such inadequate means as electron orbits and quantum conditions. It is not at all certain that Bohr himself believes that electrons revolve inside the atom. The fact that he cannot yet express it by adequate linguistic or mathematical techniques is no disaster. On the contrary, it is a great challenge. (Heisenberg 1972, pp. 36–37)

What Heisenberg is saying here is that Bohr, in his reasoning about the structure of atoms, operates with not very precisely defined model objects describing the structure of atoms. These model objects can be used to represent various atomic processes as seen in the laboratory, for instance, light emission phenomena (line spectra) and various aspects of chemical processes. These model conceptions can be communicated to other physicists by applying concepts from classical physics augmented with new quantum physical principles. But this augmented form of classical physics would be cryptic, senseless or even contradictory if not interpreted on the background of Bohr's conceptual models of the atom. To understand atomic physics at the beginning of this century would imply to be able to

1. understand the conceptual models developed by Bohr, Einstein, Heisenberg and many other physicists
2. understand in which way these models actually represented research objects, as, e.g., light emission phenomena,
3. relate these conceptual objects to selected theories, laws, and principles from classical physics and quantum theory.

Thus, the scientific activity consisted at that time in many other things than descriptions and deductions within hypothetico-deductive systems.

The situation in modern science is of course completely similar. A solid state physicist also knows various important model objects, he knows how to relate such objects to concrete research objects and he knows how to relate these objects to modern theories, laws and computational principles. This is not only true of basic science but also of engineering science. Thus, as we have seen, the study of reinforced concrete also require construction of model objects as representations of research objects and as objects of theoretical analysis.

Model Building Skills

The basic assumptions and skills of a scientific discipline are acquired by taking courses and participating in daily scientific work. Although the instrumental training in formal theories and laboratory techniques is important it only constitutes some significant aspects of the cognitive capabilities that are built up during the training to become a scientist. As we have seen the perhaps most important part of scientific work consists of establishing conceptual models in which problems under investigation can be represented in a way that makes solutions possible. Formal theories like fluid mechanics, thermodynamics and other fundamental physical formalisms are logical instruments which make it possible to describe model objects and research objects in a precise mathematical manner.

The ability to construct, select and elaborate conceptual models is a skill that must be established before one can claim to have acquired scientific competence. It is part of this skill to be able to identify and prepare research objects. It is also part of it to be able to construct and analyze model objects which represent research objects in a proper scientific way. These activities require conceptualizations of the world and as such the capability of constructing conceptual models.

This skill makes it possible to understand and apply formal scientific theories, and it makes it possible to interpret what we see in the laboratory. On the other hand, scientific theories and laws are important tools of this skill. They deliver the central conceptual tools that are necessary for a proper differentiation between adequate and inadequate models. Consequently, formal theoretical structures do not by themselves characterize a scientific discipline, but they are indispensable conceptual tools which scientists need in order to be able to prepare research objects and construct relevant model object representations.

Not all kinds of conceptualizations are allowed. A central part of establishing a scientific paradigm is to restrict the class of possible conceptualizations to those which fit into the ontological view that is characteristic of the science. At this point laws and formal theories play a central role. It is a main purpose of scientific principles to give descriptions of the scientific ontology.

Consider, for instance, classical electrodynamics. When Maxwell developed the classical field equations he had in his mind a rather concrete model of the ether (See for instance Nersessian 1992). The ether was considered as a fluid and magnetism was conceived as vortices in that fluid, and electric currents consisted of small particles that flowed between the vortices. This mechanical model of the ether was the basis and source of inspiration for his derivation of the equations. The full scope of his equations could, at that time, only be understood relative to this or similar models of the ether. Only much later when the special theory of relativity was introduced, it became possible to get rid of a concrete ether model.

During the process of developing this model building skill a certain ontological view of the world is established. This is not done in an explicit way but implicitly by learning which model objects are allowed and which are prohibited. For instance, when learning classical electrodynamics at the end of the nineteenth century model objects should be in accordance with the implicit and to some extent vague conception of the ether. Model constructions violating the established but tacit conception of the ether would be rejected as being too odd, too unrealistic, or too imaginative to be worth working on.

These assumptions define a definite scientific world view. It is established during a cognitive process by which the scientists actually learn the conceptual system of the research field, the basic laws, fundamental models objects, experimental methods, ways of preparing research objects, etc. All this is integrated into one specific way to comprehend the world in. We call this a *theoretical framework*.[1]

Hierarchical Levels in a Theoretical Framework

The system of laws within a theoretical framework is hierarchically organized. At the highest level one has abstract laws like energy conservation. They are very general structural descriptions of all kinds of systems and they can be considered as universal constraints that all kinds of physical systems must satisfy. We call these laws principles because they are part of the ontological characterization of the physical world. Our conception of nature requires that these principles are valid and that they govern all natural processes. Consequently, if they had to be changed or given up it would imply great changes in our scientific worldview.

At a lower level we have more concrete laws like, for instance, Newton's law of gravitation and Coulombs law of force between charged particles. Usually these laws are not considered as principles. They are less general although they are more than just empirical generalizations. The fact that both electrical and mechanical forces operate inversely with the square of the distance has far-reaching implications for the nature of physical phenomena. If these inverse square laws had to be modified it would be necessary to invoke radical changes in mechanics and electrodynamics to make these theories fit the phenomena. These laws are extremely well corroborated both theoretically and empirically.

When it comes to the study of more concrete phenomena like the strength of various materials, for instance steel and concrete, we find laws at a still lower level. It is not possible to deduce the strength properties of concrete within a well-defined hypothetico-deductive system. We are, in a way, in a similar situation as Bohr was in when he studied the nature of light emission. The laws which control say the

[1] A theoretical framework is one way of modifying Kuhn's concept of paradigm. Today there exist several other ways of generalizing or modifying Kuhn's idea. One is Bucciarelli's concept of object world in Bucciarelli (1994). Another one is the concept of a technical matrix in Hendricks et al. (2000).

process of rupture of blocks of concrete are not known and cannot be deduced from basic physics. Consequently, it has been necessary to develop a set of laws based on both empirical and theoretical considerations.

An Example from Engineering Science

To illustrate this we will again look at modern engineering methods for calculation of structural elements of concrete and reinforced concrete. Many materials are to some degree elastic. When they are exposed to loading (compressing, bending, tension, etc.) up to a certain point they follow Hooke's law, that is, they regain their shape when the loading ceases and deformations are proportional to the loading. But when materials are exposed to loadings beyond a certain point Hooke's law is no longer valid. The relation between load and deformation is no longer linear and the material may not regain its shape when loading ceases. The relation between load and deformation, which can also be expressed as a stress–strain relation, might look as shown on Fig. 10.1.

Hooke's law, which states the linear relationship between load and deformation, is a law of very low generality. It is an empirical generalization that only holds for deformations up to a certain limit. Genuine empirical generalizations may be empirically extremely well-corroborated, but they do not play a deep theoretical role and they would very easily be revised if observations required it.

Materials that follow Hooke's law are called elastic. No material is perfectly elastic, but many materials can be considered as being elastic within a certain range. A material is called perfectly plastic if the deformation continues without increasing the load, i.e. the stress–strain relation is horizontal. A material is perfectly elastic–plastic if it is perfectly elastic up to a point and thereafter perfectly plastic. The stress–strain relation of such materials is shown in Fig. 10.2.

They do not exist in nature, but they are model objects which give reasonable and approximate descriptions of many existing materials. Important examples are many types of steel for which these model objects have been used extensively in many years.

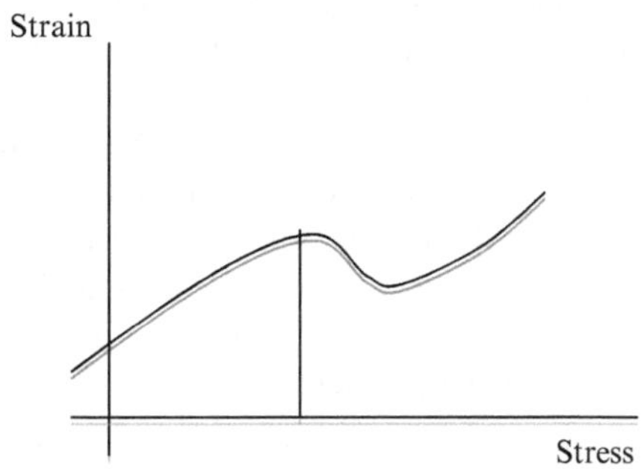

Fig. 10.1 General stress-stain relation

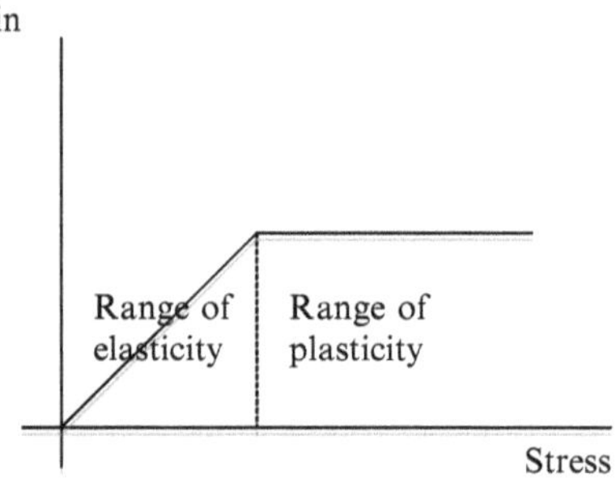

Fig. 10.2 Stress–strain relation of perfectly elastic–plastic material

When dimensioning a structural element one can, for a given load, calculate the necessary conditions for keeping the deformation inside permissible deformations. In this way it is possible to find safety conditions for collapse or yielding. Such calculations are based on the elastic properties of the material. That is, it is assumed that the material is perfectly elastic up to a certain point, and that the deformations are within the range of elasticity of the material. Calculations of structural elements of concrete and reinforced concrete have for a long time mainly been based on such elasticity properties. The fact that the stress–strain relation is not entirely linear in actual materials has been compensated for by using safety factors.

However, one can also try to calculate the necessary strength against yielding and eventually collapse. Such calculations are based on plasticity theory. A difficulty when doing so for concrete constructions is that it is not known in advance which part of the element will participate in the collapse. Consequently, it is not possible to reduce the problem to a given single set of differential equations. Another difficulty when studying concrete is that concrete is far from being perfectly plastic. It has a work curve similar to the one in Fig. 10.1. For small loads it is rather near to being elastic but for greater loads it only poorly resembles a plastic material. That is, the calculations are based on a model object that corresponds relatively badly to the research object. The development of modern plasticity theory has been an attempt to overcome these difficulties.

At the beginning of the twentieth century only minor works on structural elements of concrete based on plasticity considerations existed. However, in the thirties a new important and productive development was initiated. In 1931 the Danish engineering scientist K.W. Johansen proposed a practical method for calculation of certain types of homogeneously reinforced slabs. His method was an extension of a method suggested by another Danish engineer, Aage Ingerslev.

The method was based on the plastic properties of reinforced concrete using the often observed fact, that concrete structures, when collapsing, yield at certain lines, the so-called yield lines. In the beginning the method was primarily meant as a pragmatic way of getting results, and it was justified by empirical observations. But later on the idea of yield lines gained theoretical significance. In his dissertation from 1943 Johansen writes:

> In 1931 I gave an extended technical theory of yield lines. At that time I considered the theory as a practical approximation method, but a later review of the experiment convinced

> me about the reality and theoretical justification of the yield lines. This conception was further enhanced by my own experiments with small model plates, and I, therefore, began a more comprehensive theoretical investigation which in 1934 led to the mathematical theory of yield lines in slabs. (Johansen 1943)

The theory of yield lines meant an important step towards a method for determining where in the material yielding would occur, and as such it was a significant step in the direction of modern plasticity theory. However, this early theory of yield lines only made it possible to calculate safe upper bound solutions for load carrying. A more complete theory was developed independently by Russian and American researchers and published in the fifties. In fact, the Russian formulation coincided with the one by Johansen, but it was unknown to the Western World until the fifties. The complete theory of plasticity contains methods for calculating both upper and lower bound solutions.

As we have seen, it is possible to study structural elements of concrete from two different perspectives. On one hand, we can base the theory on elastic properties of materials. Within this view concrete is considered as behaving as an elastic material up to a load which leads to rupture. Determination of rupture conditions and other properties of the material are based on the assumption that the material up to rupture is elastic. That gives one theoretical framework on which both practical and theoretical analyses of structural elements can be founded. Structural elements are then construed as model objects which are within the range of elasticity.

On the other hand, one can assume a plasticity theoretical perspective on structural elements. Within this view concrete is considered as a rigid-plastic material which means that no deformations occur for stresses up to a certain limit, the yield point. For stresses at the yield point arbitrary large deformations are possible without any change in stresses. Although concrete is far from being a rigid-plastic material it is possible within this framework to develop a general theory of rupture which fits experimental data reasonably well.

The existence of these two very distinct theoretical views of concrete reflects nicely the situation in engineering science. As it is not possible to deduce material properties directly from basic physical theory we must develop theoretical views and model objects from experimental observations and those fundamental theories which seem to imply the best possible practical methods. We are free to comprehend the situation in any possible way as long as it leads to applicable results. Even contradictory views may be developed.

Both elasticity and plasticity theory can be developed within existing physical theory, and, even though they in many cases may lead to different results, both theories can be used to solve practical construction problems. However, they cannot both hold at the same time, and to some extent it is fair to say that none of them are right, as they both are based on highly idealized models of concrete. They are constructed by generalizing two different areas on the stress strain curve. But they both lead to results that are safe and technically applicable. As in the case of Bohr, a skilled expert has the capability to select those models and theories that are most adequate as tools for solving a given construction problem.

As an illustration of laws at an intermediate level between principles and empirical generalizations we will present some ideas from plasticity theory. A rigid-plastic object is characterized by a system of generalized stresses, $Q_1, \cdots, Q_n$ and strains, $q_1, \cdots, q_n$. The product

$$W = Q_1 q_1 + \cdots + Q_n q_n$$

represents the virtual work per unit volume.

There are two fundamental laws which govern objects of this kind, namely the *yield condition* and the *yield law*.

The yield condition gives information about which combinations of stresses can cause rupture. The yield law determines the properties of strains during yielding. It says that the strains $q_1, \ldots, q_n$ must be proportional to the outward directed normal to the yield surface which mathematically means

$$q_i = \lambda \frac{\partial f}{\partial Q_i}$$

where λ is a positive constant.

This law, which also is called *von Mises' flow law*, can be derived from a general variational principle introduced by Ludwig von Mises. von Mises introduced the hypothesis that stresses corresponding to a given strain field assume such values that the work W becomes as large as possible. That is, the material strives against deformation. From this hypothesis and the yield condition it is easy to derive the yield law.

These constitutive equations of plasticity theory are not expressions of universal basic physical laws like, say, energy conservation. They are empirical generalizations based on careful observations of failure properties of various kinds of materials. Although von Mises' flow rule can be derived from a principle of maximum work it is still a hypothesis that requires further justification from a more fundamental understanding of the structure of solids. Until such a deeper explanation is found von Mises' flow rule must be considered as a well-corroborated hypotheses which has important practical applications. But besides being empirical generalizations they also serve as a theoretical framework within which studies of structural elements can be organized. Therefore, they are intermediate level laws like Newton's law of gravitational force and Coulomb's law of force between charged particles.

It follows from these examples that there are many kinds of natural laws and that they can be organized hierarchically with respect to their generality. Laws like energy conservation are valid for all kinds of physical interactions, whereas laws like Newton's gravitation law only holds for mechanical interactions. At an even lower level of abstraction we have the laws of plasticity theory which are valid only for bodies which can be considered as approximately plastic, and, finally, there are laws which only hold for specific types of material like, for instance, concrete and steel. The most abstract laws are also those which are the most difficult to revise mainly because they form part of our scientific ontology. They are epistemologically basic and therefore impossible to revise without changing central parts for our scientific world view, whereas low level laws can be changed when required without changing our view of nature.

It is interesting to notice that in many cases a scientific law is first introduced into a scientific framework as a rather low level empirical generalization or as a heuristic tool to better calculations. An example of this is Johansen's theory of yield lines discussed above. When he introduced this theory in the early thirties he himself considered it as a practical method of calculation. But later on, when he studied the experimental results more carefully and further developed the idea, he realized that the idea of yield lines might have a deeper theoretical meaning. This was further elaborated by himself and other scientists around the world, and it finally led to the modern plastic theoretical analysis of concrete. Epistemologically, this development is similar to the development of the concept of a photon. The idea of light quanta was introduced by Einstein in 1905 as a heuristic, mathematical tool. Only many years later, in the 1920s, was it realized that light quanta was real particles, and only much later was the idea of light particles, i.e., photons, incorporated into modern physics. It had to wait until the idea of quantized fields was acceptable.

Truth and Reality

From the analysis above it follows that scientific statements are claims about model objects and not directly about the world as it exists independently of us. Usually scientific statements are true for model objects, but they may be false, or at least only approximately true, as characterizations of the world. For instance, when engineers calculate the strength of a shin-bone or of a construction element they have in mind a conceptual model which is used as a base for setting up their equations. The equations give a true description of the model object, and, if they can be solved, it is possible to produce true statements about the model object. But the model object is not the reality. It is an abstract, idealized, conceptual model of the research object which in turn has been transformed by the preparation process.

The test piece of shin-bone is not a human shin-bone. It is a manipulated piece which has been cut out of a real human body and has been modified to such an extent that it is possible to produce stable measurements on it. Furthermore, it has been modified in such a way that only certain important features of it, which are related to some of its mechanical properties, have been controlled. The test shin-bone, i.e. the research object, is a laboratory artifact. It is a non-trivial problem how this object is related both to the real human shin-bone, as it exists in a living human being, and to the model object, which is the object that theories are about. Data are produced by making measurements on the research object, i.e. the test shin-bone, but they are interpreted as claims about the idealized model object. They are used to "put blood and flesh" on the Timoschenko beam, that is, data are reduced in such a way that they can be considered as statements about the dimensions and oscillations of the Timoschenko beam. However, the results of this scientific analysis are regarded as giving information not only of the model object or the research object but also about the real human shin-bone. Consequently, data are not only measured on an artifact that is fabricated in the laboratory. They are interpreted as being about a highly abstract model object and, finally, they are believed to give real information about a piece of reality. This complex process where a piece of reality is being delimited, generalized, abstracted, idealized and finally identified as the object which scientific propositions are about must be reverted in order to give information about the original piece of reality.

If these considerations are true they raise serious questions about how scientific theories can be said to give true information about the reality. How can highly idealized knowledge about model objects which are only very remotely related to the part of the world they are supposed to represent lead to reliable knowledge about actual phenomena in the world? How can we be sure that statements about the strength of concrete building elements, based on calculations on highly idealized model objects, also hold true for real constructions? Fortunately, experience tells us that it, in fact, is possible in many cases to base real constructions on theoretical calculations. But we still have the epistemological problem of accounting for how that is possible.

Immanuel Kant introduced the distinction between the world in itself and the world as it appears to us. As we are finite beings and only have limited cognitive capacities we cannot know the world as it is in itself. All objects we identify and develop knowledge about are already shaped by our form of perception and by our conceptual system. Things in themselves are not accessible to us. Only things as they appear to us can be known. In science one goes even further. Only objects that are abstract and conceptual in nature – model objects – are accessible to scientific scrutiny. Hence, scientific statements about the world do not in any sense refer directly to objects in a world completely independently of us. They do neither refer to things in themselves nor to objects as they appear to us in practical life.

This strange situation has motivated some modern philosophers to claim that the objects of science – the scientific world – is a social construction. The model objects of science are social constructions based on our interests and social attitudes. Our theories about these objects are also social constructions. Consequently, science is a product of creative imagination that in a serious way is circular. Its conception of reality and its theories of this reality are constructions of our mind that are more governed by social values than by confrontation with an objective reality.

The social constructivist view has some good points as it is true that science is only able to develop true knowledge about model objects and as the construction of these abstract objects is inevitably based on our interests and epistemological possibilities. Furthermore, experiments are in many cases developed in such a way that they manipulate objects and processes in our environment with the intention of approximating the abstract model objects as closely as possible. When we force Nature to "fit our ideas" in this way we very often work with technological constructions which are at the borderline of what is technically possible. Usually, the experimental set-up is so complex and badly understood that it is nearly impossible to differentiate between real effects stemming from the research object and unexpected properties of the experimental set-up. Where to stop an experiment and how to interpret the outcome is often a matter of choice. The scientific community makes this choice. If we want to avoid social constructivism we must in some way explain how "nature strikes back" on our conceptual constructions.

This problem requires a deeper analysis. But let us conclude by suggesting a possible answer. A scientific theory is rendered true if it holds that (i) its statements are true for the model objects (in a correspondence sense of true), and that (ii) the model objects sufficiently approximate the research objects. The fit between abstract model objects and laboratory produced research objects is difficult to estimate. It requires that both kinds of objects are modified, and that involves both conceptual reconstructions and engineering of physical objects in the laboratory. These modifications cannot be done arbitrarily. The conceptual reconstructions must comply with consistency and other epistemic requirements and engineering of laboratory objects is limited by practical and physical constraints. Consequently, we cannot arbitrarily force the fitting process to converge; it may easily diverge and develop in a direction that does not serve our interests. If this process diverges or does not stabilize, aspects of the theory under scrutiny will be overthrown not by arbitrary decisions but by being unable to comply with constraints given by Nature.

It is true that scientific experiments always allow several interpretations and it is up to us – the scientific community – to choose the one that fits best into our scientific world. Therefore, especially experiments that are at the borderline of what is technically possible are not acceptable standards for deciding between truth and falsity. They admit several interpretations and our choice must be constrained by other norms and standards of the scientific community in order to be uniquely determined.

However, sometimes the experimental praxis leads to anomalous situations where new qualitative properties of Nature show themselves. Such situations constitute natural non-social conditions that often require a reorganization of the theoretical framework. The history of science delivers examples of that abundantly. Descriptions of these originally anomalous phenomena appear in textbooks, often referred to by the scientists involved in their discovery: Newton's rings, the photoelectric effect, the Compton effect, the Zeemann effect, the Hall effect, etc. Such situations, when they appear within a scientific discipline, first of all indicate that a phenomenon has appeared which cannot be reduced to irregularities of the equipment and the experimental and theoretical techniques involved. Furthermore, the adjustment of the theoretical framework must take the new phenomenon into account. This cannot be done in a sociologically free way; it may even lead to changes in the social structure of the scientific community.

When these various constraints are respected, the stability and convergence between conceptual constructions and laboratory manipulations may lead to a worldview which cannot in any sense claim to be a true picture of the reality as it is in itself. This claim of metaphysical realism must be given up. But at least the convergence results in a view that respects the constraints that the world puts on us. We do not know what the reality is in itself but we know that it constrains us as just described.

This view leads to special problems for engineering design. Usually, new forms of design involve processes that are badly understood. We may be able to model the processes and construct devices that fit our models. But it is still an open question how well the models fit the practical reality, and, therefore, it is often unknown how the devices will behave when they no longer are under controlled laboratory conditions. Consequently, engineers face further problems. As scientists they are able to build models that to some extent describe and explain natural processes under abstract and idealized conditions. But the devices that they design and construct must live outside the controlled laboratory conditions. The abstract, idealized conditions may not hold out there and the scope of our scientific theories is too limited to cover these circumstances. New technologies must cope with the unknown. Their ultimate test is historical. Their success will eventually follow from how well they are adapted to our practical life. Luckily, it is an incontestable fact that they by and large do adapt.

References

Andreasen, B. S. (1988). *Anchorage of ribbed reinforced bars*. Lyngby: Department of Structural Engineering, Technical University of Denmark.

Bucciarelli, L. L. (1994). *Designing engineers*. Cambridge, MA: The MIT Press.

de Solla Price, D. J. (1975). *Science since Babylon*. New Haven: Yale University Press.

Heisenberg, W. (1972). *Physics and beyond*. New York: Harper Torchbooks.

Hendricks, V., Jakobsen, A., & Pedersen, S. A. (2000). Identification of matrixes in science and engineering. *Journal for General Philosophy of Science, 31*(2), 277–305.

Johansen, K. W. (1943). *Brudlinieteorier*. Technical University of Denmark, Denmark.

Nersessian, N. (1992). How do scientists think? Capturing the dynamics of conceptual change in science. In R. N. Giere (Ed.), *Cognitive models in science* (pp. 3–44). Minneapolis: University of Minnesota Press.

Polanyi, M. (1958). *Personal knowledge*. Chicago: Chicago University Press.

Thomsen, J. J. (1990). Modelling human tibia structural vibrations. *Journal of Biomechanics, 23*(3), 215–228.

Stig Andur Pedersen M.Sc. in Mathematics, Physics and Philosophy, all from University of Copenhagen. Professor of Philosophy of Science at Roskilde University. He has published articles and books on mathematical logic, epistemology, philosophy of mathematics, philosophy of natural sciences, philosophy of medicine, philosophy of technology and engineering science. He is involved in the research alliance PROCEED, "Program of research on Opportunities and Challenges in Engineering Education in Denmark".

Chapter 11
Efficiency Animals: Efficiency as an Engineering Value

Byron Newberry

Abstract In the literature on the study of engineers and engineering practice, the pursuit of *efficiency* is often claimed to be a prime directive for engineers. The objective of this chapter is to examine that claim. It starts with an exploration of the concept of efficiency, which has a multitude of meanings, some very technical and precise, and some more broad and equivocal. Some important philosophical distinctions are made, such as between the notions of efficiency as an instrument of conservation and efficiency as an instrument of growth, between efficiency at a micro-scale and efficiency at a macro-scale, and between efficiency and effectiveness. Questions of how the concept of efficiency relates among the arenas of technology, nature, and economics are also addressed.

Keywords Engineering • Efficiency • Effectiveness • Economics • Optimization • Nature

Introduction

As an engineer who in recent years has become engaged with the philosophy of technology and engineering, I am particularly intrigued by the ideas that scholars in this field – often non-engineers – have with respect to the methods, motives, or values attributable to engineers. One such value – which serves as the topic of this chapter – is the engineer's perceived regard for *efficiency* as a core engineering design value.

There are many works that discuss the role of efficiency in technological activity in a broad sense – at a societal level – such as in Jacques Ellul's classic critique of *technological society* (1964). Ellul famously defines his all-encompassing notion of *technique* as, "the totality of methods rationally arrived at and having absolute

B. Newberry (✉)
Department of Mechanical Engineering, Baylor University,
One Bear Place 97356, Waco, TX 76798, USA
e-mail: Byron_Newberry@Baylor.edu

S.H. Christensen et al. (eds.), *Engineering Identities, Epistemologies and Values*, Philosophy of Engineering and Technology 21,
DOI 10.1007/978-3-319-16172-3_11

efficiency in every field of human activity." In technological society, according to Ellul, efficiency has become the end-in-itself. In a recent homage to Ellul's lasting intellectual influence, Wha-Chul Son (2013) suggests that even in cases where technological actions are really motivated by values other than efficiency, such actions are often nonetheless justified "in the name of efficiency," as if the very invocation of the name provides the imprimatur of necessity.

More specifically for the engineering profession, many authors have suggested that efficiency is a foundational, if not *the* foundational, design value for engineers. Carl Mitcham wrote (1991), "The thesis here is that the internal value constitutive of and operating in engineering is the ideal of efficiency." Eugene Ferguson (1979) ranked efficiency as a primary "imperative of engineering." Stanley Carpenter (1983) wrote, quite strongly, that, "Technical efficiency, with its quantified precision, is introduced early in engineering education, and its pursuit thereafter by each engineer tends to take on a character of a quest for the 'holy grail'." More recently, Joe Pitt (2011) wrote that, "For engineers, the design of an artifact or a system is approached with questions of utility and efficiency foremost in mind." Finally, I once heard philosopher of technology Peter Kroes refer to engineers as *efficiency animals*, an appellation I subsequently adopted for the title of this chapter. So at both the level of technological society in general, as well as at the level of the engineering profession in particular, strong claims are made that give efficiency a telic character.

Is this portrayal of the centrality of efficiency accurate with respect to engineers and engineering? For engineering, is efficiency really a 'constitutive value', a 'primary imperative', or the 'holy grail'? The goal of this chapter is to explore these questions.

A simple-minded starting point for this exploration was to peruse a bookshelf loaded with standard engineering design textbooks. Out of 12 contemporary texts, there was not a single entry for the word efficiency in any index. This does not guarantee that the word does not appear somewhere in each of these texts; in fact, it was found in all of them. But it was always narrowly construed in the context of a specific problem or example, and was not itself the focus of the ideas being explicated. Though not conclusive of anything, this absence from these texts seems at odds with the notion that the idea is the *holy grail* of engineering design. This, however, is congruent with my own experience as a teacher of design for more than two decades. Never has efficiency been a focus of my course in any general or explicit way. The subject does get discussed at times, but, again, in fairly narrow, ad hoc, and context-specific ways.

This discrepancy about the perceived role of efficiency in engineering suggests two possibilities. Either the premise which claims that efficiency is a quintessential value of engineering – what might be called the *holy grail* thesis – is wrong, or, if it is correct, perhaps efficiency is so elemental to engineering, insinuating itself into the very fabric of what engineers do, that the tacit centrality of its role has largely become transparent to practitioners. In what follows, I will argue that both of these statements are correct in their proper contexts. That is, I will argue that efficiency is not a unique, distinctive, or signature value of engineering per se, that it is not generally pursued by engineers in any macroscopic sense connoted by the *holy grail* metaphor. However, it is elemental and pervasive to engineering in a microscopic sense, in a way that transcends engineering inasmuch as engineering is an economic activity.

What Is Efficiency?

What is efficiency? The question may seem trivial since most people have a relatively common understanding of what efficiency is. Efficiency has to do with avoiding unnecessary time and effort, with not being wasteful, with getting things done in a clever and intelligent way, and with saving energy. But these commonplaces notwithstanding, efficiency is an elusive concept. The fact that engineers have some very precise and quantitative definitions of it does not necessarily help avoid this elusiveness. Returning for a moment to Ellul, Peter Fitzgerald-Moore (1997) criticizes him for having too non-specific a definition of efficiency. "Ellul's proposition conceals…fundamental difficulties," writes Fitzgerald-Moore, "…there are very many distinct kinds of efficiency and the concept of a generalized *efficiency* is not very useful." Likewise, Jennifer Alexander (2008, Kindle Locations 134–135) writes,

> Efficiency is a slippery concept. As the previous examples illustrate, it has taken on not only a variety of technical configurations but also a bewildering array of more common meanings.

In the context of engineering science, technical efficiency is often a precisely-defined parameter – typically the ratio of output energy or power to input energy or power, which provides a measure of how much loss occurs in an energy transformation process. This type of efficiency corresponds to what Alexander calls *bounded* efficiency. That is, the values are bounded between zero and one, or 0 % and 100%. This might also be called an *absolute* efficiency. A particular numerical value *is* the efficiency in an absolute sense, independent of the observer. And 60 % is more efficient than 50 %, regardless of the perspective. This type of bounded or absolute efficiency can be extrapolated to other situations besides energy conversion. For example, efficiency in a quality control process intended to catch defective parts can be defined as the ratio of defective parts identified to total defective parts. Again this ratio is bounded by 0–100 % and the efficiency can be referred to as a number, such as a defect-catching efficiency of 98 %.

Such measures of efficiency pervade engineering science textbooks. Interestingly, while engineering science texts typically define efficiencies and then utilize them in a variety of quantitative problems, very little is actually said about efficiency in a qualitative or value sense. That is, problems are stated in forms such as, "For the given system, calculate the efficiency," or, "If the efficiency of the given system is X, calculate its output work." But it is difficult to find in engineering science texts many explicit normative statements. For example, if a higher efficiency is thought to be better than a lower one, that belief is not typically expressed in the text. Perhaps the most forceful articulation I found comes from a thermodynamics text, which states, "Heat engines are built for the purpose of converting heat to work, and engineers are constantly trying to improve the efficiencies of these devices since increased efficiency means less fuel consumption and thus lower fuel bills and less pollution" (Çengel and Boles 2002, p. 386). This quote appears more or less as a throwaway line buried deep into a text devoted to energy conversion. Thus, in the absence of other clues, the overall conclusion one might draw from a collection of engineering science and design texts is that engineering students are either expected to enter school already having a belief about the purpose and value of efficiency – absorbed perhaps from the wider culture – or to develop such beliefs through informal professional socialization mechanisms in school and beyond. At first glance, the formal mechanisms of engineering education appear to simply teach the proper ways for measuring or calculating various efficiencies without elaborating much on their significance.

In addition to the energy conversion type of definition, there is a more generalized set of technical efficiency measures, a set that comprises a relatively open-ended variety of ratios that relate the value of a performance characteristic – or output – of an engineered device or system to the value of one of the system inputs or resources. Thus we speak of cost efficiency, weight efficiency, energy use efficiency, materials efficiency, time efficiency, or labor efficiency. These indicate an achievement of some performance objective – strength, say, or speed – with a minimum of, or at least a savings of, respectively, cost, weight, energy, materials consumption, time, or labor. In effect, these are manifestations of a class of technical efficiency measures that derive from an economics-like definition of efficiency. Unlike energy conversion efficiency, which conveys fundamental scientific information about a physical system or process, these types of efficiency measures are arbitrarily defined for the purpose of gauging performance relative to human wants; that is, they allow us the opportunity to maximize outputs for a given set of inputs, or to minimize inputs for a given set of outputs, in order to further some technological or organizational objectives. Also, in contrast to the definition of energy conversion efficiency, which is a dimensionless ratio having a range of zero to one, with *one* being an unobtainable perfection, efficiency in this sense is often a ratio between incommensurables, and its numerical value is only meaningful relative to alternatives. For instance, aircraft structural components might be judged on their ratios of strength-to-weight, with higher values generally considered more desirable, all other things being equal, but with the absolute values being effectively meaningless. This is consistent with what Alexander calls *arbitrary* efficiency measures. They might also be called *relative* efficiency measures.

The phraseology is different for these arbitrary or relative efficiency measures. If the tensile-strength-to-weight ratio of a particular steel is 264,000 psi/(lb/in^3), one typically does not say that the efficiency of the steel is 264,000. One would say that with respect to strength-to-weight, this steel is more efficient than another type of steel that has a ratio of 220,000. That is, one thing is more or less efficient than another by comparison of the values, but a thing does not have an efficiency value in-and-of itself. Also, in this example the implicit assumption is that a higher strength-to-weight ratio is more efficient than a lower one. This is true in airplane design, for example, but it may be just the opposite dam design. Or, to borrow an example from Sunny Auyang (2004), a low drag coefficient is efficient for designing a wing, but it is inefficient for designing a parachute. For arbitrary efficiency measures, *more* or *less* efficient is relative to the objective.

Optimization as Efficiency

For most engineering designs, however, there exist multiple performance-based relationships of critical importance, which leads to the related concept of optimization. Roughly speaking, optimization is the process of trying to achieve the most desirable balance of all relevant quantities – "minimizing the most significant undesirable effects and/or maximizing the most significant desirable effects" (Ertas and Jones 1996, p. 292). In the case of the aircraft structural components, we might want to optimize – or find the best balance of – relationships between strength, stiffness, weight, machinability, environmental tolerance, availability, and cost. The typical result is a situation in which none of the individual efficiency relationships – such as strength-to-weight, for example – achieves the level it has the potential to achieve independent of other considerations. Rather, analogous to economic Pareto-optimality, a compromise is sought such that no relationship can be further improved without an unacceptable degradation in another. In biological evolution, this is known as the *principle of frustration*: "This principle captures the notion that different needs will often have (partially) conflicting solutions, so that the overall optimal design for an organism will rarely be optimal for any of the specific tasks it needs to perform" (Marshall 2006).

Regarding optimization, and lending credence to the *holy grail* thesis, engineer Walter Vincenti writes that optimization is "a constant element, implicitly or explicitly, in engineering thinking. For the engineer optimization has the nature of an ethos" (1990, p. 165). Optimization methods, from relatively simple heuristic approaches, to more sophisticated mathematical computations, are discussed to varying extents in many engineering design texts. As Herbert Simon (1996) points out, however, such methods result not in real-world optimal solutions, but rather in satisfactory ones – ones that are deemed acceptable given the finite bounds on our knowledge and resources. And, importantly, whether in economics or in engineering, formal optimization efforts typically occur at lower, rather than higher, levels; that is, optimization methods are most likely to be applied to arranging deck chairs in the most efficient way, whereas the ship itself was likely set sailing on the basis of a top-level, experienced-based, exercise of judgment and/or speculation. Put another way, formal optimization methods always presuppose a generic design that has been articulated well enough to create a mathematical representation of it, to define an *objective function* that rationalizes the balance of desirables and undesirables. Thus the optimization is local in the sense that we might find the "best" solution within the bounds of our assumed optimization space, which includes a generic design morphology, along with constraint factors that we have identified as important. But there is no guarantee that our generic design is at or near the optimum of the set of all possible generic designs, or that the constraint factors that we have included represent the most relevant or most complete set we could have defined. The initial – and higher-level – choice of the generic design and constraints might often be made on the much more informal basis of experience, socio-cultural preferences, habit, and so forth.

Common in design texts are quantitative optimization example problems, such as finding the optimum spacing and positioning of towers along a route for an electrical transmission line. Presupposed in the problem, however, are the route and the design of the towers themselves, both of which could of course be different, not to mention that the higher-level decisions that led to the choice of having an electrical power transmission line in the first place are taken for granted. Further, it is difficult to include in such optimization procedures factors such as social and environmental impacts of the transmission line, including aesthetic concerns, habitat alteration, and so forth. Thus, the results of an optimization calculation may ultimately be modified or overruled altogether by other, more intangible, considerations. This is not to say that optimization procedures are not useful, just that they are never optimum in some macroscopic sense, and are typically most useful in defining lower-level details once higher-level decisions have been made. Auyang (2004) writes, "To assert what is the best is difficult….Engineers are too hardheaded to dream about the universal best, which anyone sensible knows to be infeasible. Optimization theories aim at the best relative to a specific objective criterion and a range of options and constraints." Likewise, engineer Billy Vaughn Koen (2003) says that engineering optimums do not "pretend to be the absolute best in the sense of Plato," only best relative to the specific objectives people have.

But what objectives ought people to have. Just because something can technically be made more efficient according to a specific criterion, or optimized relative to a specific objective function, it does not follow that we have done anything meaningful, useful, or good. "In a society of cannibals," writes Koen colorfully, "the engineer will try to design the most efficient kettle" (2003).

Judging by the example problems in textbooks, we might guess that maximizing a ratio of just about any desirable thing relative to cost is universally good, while, as we have seen, maximizing strength-to-weight ratio, for example, is good in certain applications, but is indifferent – or even undesirable – in others. But once again, engineering textbooks largely seem to suggest, via their silence, that beliefs about the kinds of relationships among variables that should be valued are either patently obvious, are context dependent, or are otherwise externally acquired.

Efficiencies or optimums are ostensibly pursued in a utilitarian fashion, for their effectiveness in furthering the pursuit of our current goals. Efficiency "cannot be said to be an intrinsic value if it is primarily an instrument for implementing the other values that define the context of its use" (Alexander 2008, Kindle Locations 150–151). But when efficiency targets are set, or objective functions for optimization defined for a specific end in a specific context, they risk being undeservedly reified if conceptually divorced from those ends and contexts. As a result, achieving some optimum or some measure of efficiency can potentially seduce one into thinking that such an achievement is objectively good, or that one has advanced toward the positive end of some absolute scale. In reality, the achievement is such only relative to the limited criteria that were set in the beginning. Someone else with different criteria can just as easily define and pursue different efficiencies or optimums for purposes that may be orthogonal. As Fitzgerald-Moore (1997) writes, "efficiencies are often in conflict or contradiction with one another."

This observation provides a good entrée into another meaning of efficiency. Sometimes phrases such as "the efficient use of resources" mean more than simply saving money on the inputs to a process; rather, a concern with the long-term preservation of some set of resources is literally implied. Efficiency in this sense is related to the notions of sustainability and conservation. The normative content of this meaning of efficiency is fairly clear; a high value is placed on maximizing the long-term stability of either some specific resources or of the biospheric system in general. A distinction can be made between couching efficiency in the language of scarcity and couching it in the language of plenty. In cases of plenty, efficiency promises more, bigger, and better. In cases of scarcity, such as is implied by the conservational meaning of efficiency, the efficiency is required to preserve what exists, if even that is possible. It is fairly easy to illustrate the conflicting nature of efficiencies using this meaning. Sustainability-related efficiencies are frequently perceived to be at odds with economic efficiencies defined in terms of profit and growth by individuals, corporations, or societies. This distinction roughly parallels Alexander's (2008) distinction between *static* and *dynamic* efficiencies, or the efficiencies of equilibrium and growth. Distilled to their essence, definitions of efficiencies define things upon which people place value, and provide a quantitative metric for assessing that value. But different people, or the same people at different times and in different circumstances, value different things. "Subsuming efficiency in context suggests that it is a shell, ready and waiting to take on the values and objectives of whoever uses it but with little content or character of its own" (Alexander 2008, Kindle Locations 142–143).

When put this way, efficiency appears less of an *end* and more of a *means*. This is the point Alexander makes when she writes that efficiency is a tool that allows us to bring things under our control. Efficiency can be put in the engineering toolkit alongside scientific understanding, methods of visual representation, methods of reduction and simplification, and other forms of knowledge and technique that engineers routinely utilize to achieve their aims. Auyang writes, "It is important to note that the choice and evaluation [of objectives and constraints] are made by engineers, who use optimization theories as tools and stay outside the theories" (2004).

But saying that efficiency is more properly a *means* than an *end* does not necessarily make the case that efficiency is not somehow an overarching goal of engineering; it does not in itself guarantee that it has not taken on the character of an end. As the old saying goes, *to a man with a hammer, everything looks like a nail.* The implication is that while a hammer is clearly a tool, a person who has a hammer may make finding things to hammer his or her end objective. Or more generally, those who possess what they consider to be a powerful tool or technique, may come to extol the virtues of that tool or technique to the point of making the application of it an end in itself, an elixir to cure all ills. And this effect is likely a factor in why people – including engineers – seem to have a very definite popular notion of efficiency as being something inherently desirable, despite the fact that the popular meaning is rather vague and ambiguous.

One way this can manifest itself is when someone argues *this* instead of *that* should be done because *this* is more efficient, without a clear definition of efficient. Others may be confounded, at a loss as to how to counter that argument. This is the point made by Son (2013), that often efficiency is blindly invoked as justification. Two subtle things have happened in such an argument. First, a *specific* or *limited* definition of efficiency has been conflated with the more *general* notion, which preys on prevailing perceptions that to disregard efficiency would be some sort of sacrilege. Second, efficiency has been slipped into the role of primary decision-making criterion, the trump card. And these are likely the effects that bothered Ellul in his wide-ranging criticism of the pursuit of efficiency: (i) that considerations of efficiency had become *the* litmus test for decision-making, rather than simply informing the decision-making process, and (ii) that the specific definitions of efficiency upon which decisions are made are mistakenly viewed as essential rather than arbitrary, and might most often be ones favorable to specific interests, rather than to society at large.

Engineering, Nature, and Economy

It is instructive to note that while Vincenti (1990) asserts that optimization is a pervasive element of the engineering ethos, he first qualifies that assertion by recognizing that optimization is in no way unique to engineering – that optimization alone cannot be used as a basis for differentiating engineering from some other types of activities. In fact, my own thinking about this topic was initiated from a non-engineering source – upon rereading Darwin's *The Origin of Species* in preparation for a teaching a history of science course. Darwin writes,

> [N]atural selection is continually trying to economize in every part of the organisation. If under changed conditions of life a structure before useful becomes less useful, any diminution, however slight, in its development, will be seized on by natural selection, for it will profit the individual not to have its nutriment wasted in building up a useless structure. (1985 Penguin Edition)

What Darwin calls economizing, might now be called *fitness* optimization, an efficient allocation of resources that increases the organism's chance of success in the face of competition and scarcity. Care must be taken, though, with the terminology. Nature, through evolution, does not strictly optimize. Optimization implies a conscious evaluation of a range of options followed by the selection of the one which best accomplishes an objective. Nature does not perform a predictive evaluation of options, and it knows nothing of *best*. Nor does it have objectives. Rather, nature proceeds by a trial-and-error method that can only build upon existing structures and which charts a historically contingent course that is guided by differential success in the presence of potentially fluctuating resources and constraints. People often tend to overlay their interpretations of that success post facto in terms of efficiencies. This must be kept in mind even though the language may be used loosely in what follows.

Darwin's choice of the word *economize* is instructive. An economy, defined as an arena in which resources are extracted, produced, allocated, and consumed by entities competing for their own welfare and propagation, holds the key to the notions of efficiency and optimization, whether that arena is a natural ecology, a household, an organization, or a society. "Economics," writes Auyang, "features decision makers, households and business firms, each intentionally trying to find what is best for itself. They are engineers except they are designing different sorts of things" (2004). And when discussing optimization and satisficing, Simon draws frequent parallels between processes in biological evolution, market economics, and technological development. Technological development is itself wholly subsumed under the umbrella of socioeconomic activity more generally conceived. "The 'demand' for technology," writes Joel Mokyr (1990, p. 151), "is a derived demand, that is, it depends ultimately on the demand for the goods and services that technology helps produce; there is little or no demand for technology for its own sake."

But when we talk about efficiency and optimization in real-world economies, whether natural or social, we are invariably talking about *localized* efficiency and optimization. Like falling drops of water molded into teardrops by the pull of earth and the push of wind, economic entities continuously streamline, other things being equal, in response to the pressures of their respective arenas. But such streamlining is myopic; for nature, strictly so, and to varying degrees for human-constructed entities. The water drop knows nothing of the ground upon which it will soon fall, the organism knows nothing of the invasive species against which it will soon compete, and the company may know little of the incipient technological development that will soon render its products obsolete. Yet all continue to optimize – to be molded by their environment, consciously or unconsciously, into more efficient forms or behaviors. But that environment is potentially an ephemeral one, so what is efficient today may in fact be terribly inefficient tomorrow.

This might be called *micro-efficiency*, the ineluctable tendency of a given economic entity to pursue efficiency within a context that pits its current internal morphology against its short-range (both spatially and temporally) external environment. With respect to technological change in particular, Mokyr (1991) distinguishes between *microinventions* and *macroinventions*, where microinvention denotes the stream of improvements that a technology, once developed, undergoes with time in an effort to make it more and more efficient. This is congruent with the meaning of micro-efficiency intended here. It is also consistent with what is called in the economics literature *process innovation* (e.g., Adner and Levinthal 2001; Quintanilla 1998). Process innovation is the streamlining over time of both the basic product design and the production process for that product for the sake of economic efficiency. Interestingly, and germane to an important point to which we shall return, process innovation starts out relatively slowly at the birth of a new technological product type, reaches a peak of activity sometime in the midlife of that product type, and then decreases as the product type ages. In this timeline, which is subject to great variation, a newly conceived product type exploits some opening in the economic arena and may enjoy a period of relatively unfettered success. Only after competition arises does the need to optimize fully assert itself. This need intensifies in concert with the competition until eventually a point of diminishing returns may be reached and further improvements are exhausted.

But as is well known in both biology and economics, this pursuit of micro-efficiency, in and of itself, holds no guarantee for efficiency or optimization on a more global scale; entities often condemn themselves to paths of extinction, or else marginal existence in backwaters of the economic arena as substantial changes occur in the external conditions of that arena. One needs only to think of horse-drawn carriages and mechanical watches. Entities sometimes can even enjoy various degrees of long-term success having locally optimized what globally is recognized as a non-optimal situation. This phenomenon has been explored in-depth in the literature on the economics of technological change under the headings of *path dependence* and *lock-in* (e.g., Cowan and Gunby 1996; Stack and Gartland 2003; Arthur 1989; Liebowitz and Margolis 1995). In these cases, the development of a technology along a globally non-optimal path (as a result of the inevitable myopia that exists) results in the non-optimal technology becoming permanently entrenched because of the daunting economic costs of later switching to a better path once that better path becomes known. But these persistent technological inefficiencies are in a real sense still economically efficient when factoring in the transaction costs associated with correcting them.

Gains in micro-efficiency are not guarantees of any long-term or long-range efficiency or optimization; i.e., they do not guarantee *macro-efficiency*. They are not leading monotonically toward some social, economic, technological, or natural optimum. They lead instead toward some local maximum, which could be a stagnation point, or could be wiped out altogether with changes in the landscape. So the whole notion of pursuing efficiency is a localized, not a generalized, concept. Earlier, I suggested optimization occurs more often at lower, rather than higher levels of abstraction. Similarly, it occurs in engineering and socioeconomic systems more often in the shorter-range, rather than in the longer-range. Human-constructed systems can be optimized, or made more efficient, with intentionality, as opposed to natural systems that cannot. But when human socio-technological systems are taken in larger, more complex aggregates, and as they operate over longer time spans, the intentionality becomes noisy and the types of efficiency that prevail degenerate closer to those of nature. Intentional optimization depends on prediction and control, both of which become more difficult with increases in scale and complexity. In short, while we can make a product or manufacturing process more efficient, and likewise with a company's financial operation, the overall long-range development of economies and technologies are historically contingent and largely unpredictable.

The pursuit of micro-efficiency is an imperative of economic entities inasmuch as competition is a fact of economic arenas. Engineers, as agents of economic entities and designers of economic goods, are unquestionably immersed in and contribute to that pursuit, and quite often do so consciously and deliberately. This fact undergirds the suggestion that efficiency is elemental to engineering, and in two main ways. First, many of the problems to which engineers get assigned are problems of optimizing technologies for the sake of incremental gains in efficiency. And second, engineers, just as much as other people, are possessed by the ghost of *homo economicus*, and therefore have an intrinsic understanding of, and susceptibility to, the imperatives of economic efficiency.

The other half of the equation is exemplified by Mokyr's *macroinventions*, or economists' *product innovation* – development with the primary aim of *effectiveness*, in contrast to *efficiency*. My thinking about these notions was influenced by the work of biologist Geerat Vermeij. Vermeij writes,

> I believe the emphasis on efficiency is misplaced. Economic success depends on absolute performance, and very often – in human-economic contexts as well as in the evolutionary marketplace – high levels of performance go hand in hand with reduced efficiency. (Vermeij 2004, p. 124)

There are basically two ways to increase power or performance. One is to become more efficient; that is, to find ways to increase the system output for a given set of inputs. The other is the brute force method of simply consuming more inputs, regardless of how efficiently they are used. In fact, with sufficiently elevated quantities of inputs, losses in efficiency can still correlate to gains in output. Which path an economic entity takes, according to Vermeij, will depend on the competitive nature of the environment and the availability of resources. "Although supply of resources dictates what level of metabolism can be achieved, it is demand – imposed by consumption and by competition – which drives some entities toward higher metabolic rates" (Vermeij 2004, p. 136).

A good example of this comes from the work of Anders G. Finstad and colleagues (2011) on energy efficiency of salmonid fishes. They found that Arctic char convert input resources (food) into body mass twice as efficiently as brown trout, which might seem like a significant advantage for the char. But the more docile char only outcompete the trout in colder, resource-poor environments. The char might be said to adopt a static, equilibrium, or conservative efficiency strategy, a strategy that enables them to extract the most benefit from meager resources. In contrast to the char, in warmer, more resource-rich environments, the less efficient (in terms of food-mass conversion) trout grow dominant by aggressively consuming resources in such high quantities that they more than offset their greater "wastefulness" in converting those resources. They might be said to adopt a dynamic, growth-oriented efficiency strategy. But the trout's strategy is dependent upon sustained high levels of resources.

Vermeij provides other detailed data on the various types and classes of organisms with respect to energy consumption and biomass production. Higher performance organisms, ones typically larger, more complex, and higher in the food chain – ones we might say have superior natural technology – have a much higher ratio of energy consumed to biomass produced than lower performance organisms. That is, in the overall natural economy their biomass energy conversion output-to-input ratio is relatively poor. But that overall inefficiency is secondary to the absolute economic power conferred by the high metabolic rate. Flying organisms, for example, comprise the class of organisms with the highest energy consumption per unit of body mass. But the benefits for evolutionary success of exploiting air as a transport medium, which include speed of fleeing and/or pursuing, range of forage, and seasonal migration to avoid weather extremes, are worth the investment, provided the resources are available to sustain consumption levels. Cold-blooded animals, according to Vermeij, absorb energy passively from the ambient environment and are thus at the mercy of environmental conditions for their physiological performance. Warm-blooded animals have invested in their own internal power-generation mechanisms, which provide them with more autonomy from the current thermal conditions of the environment, but at the cost of committing them to dependency on a consistently abundant food supply. But this in turn requires the existence of an overall large economy. Large, high-performing mammals, for example, aren't generally found on smaller islands due the inability of the economy to supply the requisite resources. And warm-blooded animals have energy conversion efficiencies, for example, an order of magnitude below those of cold-blooded animals. "In fact," writes Vermeij,

> ...a problem for active warm-blooded animals is disposing of excess heat produced by an inefficient engine. This is why we sweat, dogs and birds pant, and bees and termites ventilate their nests. In our technological world, internal combustion engines and atomic power plants give off vast amounts of unused heat, but their power yield is so great and provides such clear economic advantages that their inefficiency is tolerated, much as it is in warm-blooded animals.
>
> ...In all economies...efficiency becomes important when power is low and output cannot be increased in absolute terms. This occurs when energy or raw materials are sufficiently scarce that reducing the cost of acquiring them is the only way of not losing ground. Increases in power, however, are sufficiently beneficial that considerations of efficiency are secondary, especially if productivity also benefits the supply of raw necessities. In such cases, absolute performance is far more important than efficiency. Thus it pays to be efficient for subordinate members of an economy, and it pays to increase in performance for those in power. (2004, p. 125)

In concert with this development of macroscopic inefficiency – justified on the basis of exploiting an open avenue for economic power – is a parallel process of micro-efficiency. For example, over time flying organisms evolve lighter weight tissues, more streamlined shapes, and more precisely tailored flight mechanisms (just as human flight technology has). This is the process described earlier as process innovation or micro-invention. But such trends in micro-efficiency, while increasing an organism's competitive advantage incrementally, will never completely reverse the overall commitment to very high levels of consumption inherent in the base technological adaptation.

Inefficient Specifications

In a similar way, technologically advanced societies have levels of per capita resource consumption many times greater than those of most less developed countries. They also have greater absolute levels of economic power. But they are not necessarily more efficient in terms of the ratio of economic output to resources consumed, and often less so. Technologically advanced societies are analogous to organisms with high metabolic rates. The investment in energy intensive technology confers power in the form of better control over wider ranges of resources, the ability to specialize and decentralize functions, the storage of vast reserves, quicker response times, and more flexibility in adaptation, all of which tend to buffer the entity against the uncertainties of short-to-medium-range spatial and temporal fluctuations in the environment of the economic arena. The Faustian bargain for this economic stability is the dependence on long-term, sustained, and abundant inputs. The book *Collapse* by Jared Diamond (2005) chronicles past societies whose technologically-driven, rapid metabolisms exhausted their input resources, leading to precipitous societal failures. This is the ultimate in macro-inefficiency, and it parallels the extinction of higher-performing species in the wake of major environmental changes, species which have staked their economic power, stability, and success on high levels of resource consumption.

Consider the example of the *sport utility vehicle* (SUV) type automobile, which is quite popular in the United States. For the automobile manufacturers they are economically efficient in that they produce large profits due to high demand. But why are they in high demand? They are expensive both to buy and to maintain. From a technological point of view, it could be argued that they are inefficient. But they do confer absolute economic power, provided the resources are available to sustain them. They provide safety to the passengers in collisions. They enable the transport of many persons at once. They can be used to pull a trailer. They can go into four-wheel-drive mode and drive through mud. They confer prestige. For all these reasons they provide a much wider range of performance than many other vehicles, and so confer economic power. But for most owners the vast majority of miles driven by SUVs are not driven in the mud, nor with the trailer, nor carrying the soccer team. The vast majority of miles are driven by a single person to the store to get milk, or some such. Because of that, the powerful, low gas mileage engine, the large body size and corresponding heavy materials consumption, the expensive four-wheel-drive systems, and so on, are grossly underutilized and hence effectively wasted except for occasional events. The economic power that the vehicle confers with its wide range of abilities is at the expense of tremendous resource consumption, resources that are largely held in reserve due to the extremely low duty cycle of extreme requirements on the vehicle. We might compare it to a lion, a large powerful animal with a high metabolic rate, but which spends large amounts of time inactive, burning resources on idle, and only utilizing its high performance characteristics sparingly. And like the lion, its continued success is dependent on the abundance of resources, whether large herds of game, or large herds of deep-pocketed consumers coupled with large deposits of oil and iron.

So in many respects SUVs are not efficient from a technological point of view. They are favored because they increase power and performance, at least so long as resources are abundant (and in fact their favored status fluctuates with resource availability in terms of fuel prices, wages, etc.). But we could also consider them efficient in other respects, such as relative to personal convenience or corporate economics. And engineers are constantly engaged in making the finer details of both the vehicle designs and their production processes more efficient. Thus, economic entities are engaged in parallel processes of pursuing micro-efficiency and macro-power. In the former, an entity seeks to find ways to get more for less from its current technological paradigm, and this is compatible with our typical notions of efficiency. In the latter, an entity seeks to develop new technological paradigms that confer greater economic power by either exploiting new resources or exploiting old resources in a new way, and this represents an absolute increase in performance made possible by the availability of resources that may in fact be used inefficiently. Engineers are instrumental in the achievement of both objectives – pursuing savings through efficiency and exploiting resources for power-enhancing innovation. And while engineering considerations constrain and influence both objectives, the objectives themselves are socioeconomically mediated. For the engineer, the common denominator for both kinds of work is the notion of *specifications*, and if efficiency is ever a design value in any macroscopic sense, it is only so because it was specified to be so. If there is a *holy grail* of engineering it is surely the *meeting of specifications*.

I have previously worked as a structural analyst for a large aircraft modification company, and in the course of that work had the occasion to inspect work being done to a 747 that belonged to the monarch of an oil-rich country. Since the monarch was himself a pilot who occasionally liked to fly his own plane, the aircraft's throttle was quite literally gilded. That is just one example of what might be considered excesses that were part of the aircraft's design, excesses that seemingly violated the rules an aerospace engineer would consider part of good, efficient aircraft design. For another example, a different customer desired to install a granite conference table in an airplane. Not only was the weight of a granite tabletop contrary to the aircraft design principle of choosing the lightest weight materials, but its susceptibility to cracking required engineers to design an elaborate mounting system that would allow the table to float stress free while the underlying structure to which it was attached flexed during flight. Counterintuitive (with respect to efficiency) designs such as these happened because they were specified by a customer possessing the resources to afford such inefficiencies. The engineers worked to meet those specifications in as efficient or optimal a way as practical. But such efficiency or optimality was only local, defined within the context of what might be considered globally inefficient or even outrageous specifications.

In describing the phases of matter to students, science teachers will often define a gas as having neither definite shape nor volume, but rather expanding to fill the limits of the space that ultimately constrains it. Similarly, the work of engineers will often expand to fill the limits of the specifications that constrain it. In teaching engineering design, I have the students work on projects in which they must design, build, and test an electromechanical (mechatronic) device. The specifications given to the students define performance objectives, and constrain the design with respect to such factors as budget, time, physical size, and energy sources. These constraints are imposed with a general rationale related to efficiency (i.e., practical resource limitations), but the particular values chosen for the projects are somewhat arbitrary and can reasonably vary by an order of magnitude or more. While there may be a practical lower limit on the constraint levels (that is, a particular set of performance objectives may not be practically achievable below certain levels of cost or size or energy, say) there is not as clearly an upper limit (that is, more resources could always be invested in the design).

My anecdotal experience is that when constraints are set near the lower limit, issues of efficiency come to the forefront and significantly influence solution approaches. If constraints are set much more loosely, a highly efficient design is still very much a possible solution, but it will generally not be realized in the presence of excess resources. Questions of efficiency typically fade into the background, and the designers will absorb those excess resources into enhancing the power of the device or system; that is, making it more robust, more flexible, more accurate, or more elegant. In fact, the particular *value* which excess resources are employed to enhance varies greatly with individuals and teams.

One of my engineering students once wrote in an essay on engineering, "An engineer as an individual is not interested in adhering to tight schedules, nor to minimizing cost." His suggestion was that efficiency, rather than being the *holy grail* of engineers, is more appropriately the *bane* of engineers because it constrains what they truly want to do, which is to make things that are bigger, faster, more powerful, and more sophisticated. This sentiment complements the *macroinvention*, or *innovation*, side of engineering. But engineers also operate in the *microinvention/process innovation/micro-efficiency* world in which technologies and products are continuously refined and improved for efficiency's sake. But in these cases, issues of efficiency take on the explicit role of performance objectives rather than constraints. That is, the problem may be to find a way to manufacture a given product for ten percent less cost. In that case, that gain in efficiency itself becomes the technological challenge, and as such can become the focus of the engineer's drive for technical achievement and satisfaction, not because it is efficiency per se, but simply because it is now the performance specification to be conquered. And conquering performance specifications is what engineers like to do.

References

Adner, R., & Levinthal, D. (2001). Demand heterogeneity and technology evolution: Implications for product and process innovation. *Management Science, 47*(5), 611–628.

Alexander, J. K. (2008). *The mantra of efficiency: From waterwheel to social control*. Baltimore: Johns Hopkins University Press.

Arthur, W. B. (1989). Competing technologies, increasing returns, and lock-in by historical events. *The Economic Journal, 99*, 116–131.

Auyang, S. (2004). *Engineering: An endless frontier*. Cambridge, MA: Harvard University Press.

Carpenter, S. (1983). Alternative technology and the norm of efficiency. In P. Durbin (Ed.), *Research in philosophy & technology* (Vol. 6, pp. 65–76). Greenwich: JAI Press.
Çengel, Y., & Boles, M. (2002). *Thermodynamics: An engineering approach* (4th ed.). New York: McGraw Hill.
Cowan, R., & Gunby, P. (1996). Sprayed to death: Path dependence, lock-in, and pest control strategies. *The Economic Journal, 106*(436), 521–542.
Darwin, C. (1985). *The origin of species*. New York: Penguin.
Diamond, J. (2005). *Collapse: How societies choose to fail or succeed.* New York: Viking.
Ellul, J. (1964). *The technological society*. New York: Vintage Books.
Ertas, A., & Jones, J. C. (1996). *The engineering design process* (2nd ed.). New York: Wiley.
Ferguson, E. (1979). The imperatives of engineering. In J. Burke et al. (Eds.), *Connections: Technology and change* (pp. 29–31). San Francisco: Bond and Fraser.
Finstad, A., Forseth, T., Jonsson, B., Bellier, E., Hesthagen, T., Jensen, A., Hessen, D., & Foldvik, A. (2011). Competitive exclusion along climate gradients: Energy efficiency influences the distribution of two salmonid fishes. *Global Change Biology, 17*(4), 1703–1711.
Fitzgerald-Moore, P. (1997). Efficiency. Unpublished lectures on technology and society. http://people.ucalgary.ca/~pfitzger/. Accessed 19 Oct 2013.
Koen, B. V. (2003). *Discussion of the method: Conducting the engineer's approach to problem solving*. New York: Oxford University Press.
Liebowitz, S. J., & Margolis, S. E. (1995). Path dependence, lock-in, and history. *Journal of Law, Economics, and Organization, 11*(1), 205–226.
Marshall, C. R. (2006). Explaining the Cambrian 'explosion' of animals. *Annual Review of Earth and Planetary Sciences, 34*, 355–384.
Mitcham, C. (1991). Engineering as productive activity. In P. T. Durbin (Ed.), *Critical perspectives on nonacademic science and engineering* (pp. 80–117). Bethlehem: Lehigh University Press.
Mokyr, J. (1990). *The lever of riches: Technological creativity and economic progress*. New York: Oxford University Press.
Mokyr, J. (1991). Evolutionary biology, technological change and economic history. *Bulletin of Economic Research, 43*(2), 127–149.
Pitt, J. (2011). *Doing philosophy of technology: Essays in pragmatist spirit*. Dordrecht: Springer.
Quintanilla, M. A. (1998). Technical systems and technical progress: A conceptual framework. *Techne, 4*, 1.
Simon, H. A. (1996). *The sciences of the artificial* (3rd ed.). Cambridge, MA: MIT Press.
Son, W.-C. (2013). Are we still pursuing efficiency? Interpreting Jacques Ellul's efficiency principle. In H. M. Jerónimo, J. L. Garcia, & C. Mitcham (Eds.), *Jacques Ellul and the technological society in the 21st century* (pp. 49–62). Dordrecht: Springer.
Stack, M., & Gartland, M. P. (2003). Path creation, path dependency, and alternative theories of the firm. *The Journal of Economic Issues, 37*(2), 487–494.
Vermeij, G. (2004). *Nature: An economic history*. Princeton: Princeton University Press.
Vincenti, W. G. (1990). *What engineers know and how they know it.* Baltimore: Johns Hopkins Press.

Byron Newberry B.S. in Aerospace Engineering, University of Alabama. M.S. in Aerospace Engineering and Ph.D. in Engineering Mechanics, both from Iowa State University. Professor of Mechanical Engineering, Baylor University. Baylor Fellow for teaching. Professional Engineer (PE) Texas, USA. Research interests: engineering design, engineering ethics, philosophy of engineering and technology. Aircraft structural engineering consultant. Executive board member, National Institute for Engineering Ethics. Editor of the Springer *Philosophy of Engineering and Technology* book series.

Part III
Engineering Values and Normativities

Introduction

Jen Schneider and Wayne Ambler

Entitled "Engineering Values and Normativities," this part features chapters united by their concern to explore the values that engineers do or should act upon in their professional work, as well as the way their work influences the values – and, indeed, the entire lives – of those who live in a world that is increasingly engineered. As Winston Churchill put it, in a quotation we borrow from Chap. 14 by Sylvain Lavelle, "We shape our dwellings; thereafter, our dwellings shape us." In other words, the interplay between the values engineers share, and the way engineering in turn shapes the values of us all, our physical and metaphorical "dwellings," is of special interest in these chapters.

Since the values pursued by engineers will often be influenced by their educations, the professional associations to which they belong, the society of their fellow engineers, and the larger historical and social contexts in which they live and work, the articles in this part take up the question of engineering values in diverse ways. One question several chapters ask, however, is whether the engineering profession makes engineers more ethical, or whether it in fact narrows their vision in some way. Or is it both? Martin Meganck in Chap. 12 "On the Normativity of Professionalism" suggests that it is perhaps neither: "professionalism" for engineers differs wildly across historical and cultural contexts and, in the final analysis, the

J. Schneider
Department of Public Policy and Administration, Environmental Research
Building, Office 5135, Boise State University, 1910 University Drive, Boise, ID 83725, USA
e-mail: Jen.Schneider@mines.edu; jenschneider@boisestate.edu

W. Ambler
Herbst Program of the Humanities, College of Engineering,
University of Colorado Boulder, Boulder, CO 80309, USA
e-mail: wayne.ambler@colorado.edu

foundation for professional ethics may rest most basically on one's "ordinary morality," rather than on any professional ethics that might be imparted simply by dint of being an engineer.

Other work in this part suggests that while the philosophical foundations of engineering ethics may rest on one's "ordinary morality," the presence of engineering cultures shapes and is shaped by engineering attitudes toward the political. Christelle Didier and Kristoff Talin in Chap. 13 argue that political valences such as a tendency to discount or underestimate ecological challenges are endemic to French engineering, thus putting engineers out of step with general French sentiment. Jen Schneider, Abraham Tidwell and Savannah Fitzwater in Chap. 15 demonstrate this mechanism in detail in the case of nuclear engineers, who remain largely apolitical on the question of climate change, while the industry around them constructs contradictory narratives and silences about the connection between global warming and nuclear power.

In short, the relationship between engineers and neoliberal economic systems, which tend to exclude non-economic concerns from public deliberation or consideration, is one that is explored elsewhere in this edited collection, and serves as an implicit backdrop to many of the arguments in this part. Indeed, Chaps. 17 and 18 look more to the larger societies in which engineers work in order to track how the social shapes engineering. Chapter 17, by Carl Mitcham and Wang Nan, traces the historical development of engineering ethics, beginning with its early emergence in the German context. This early history would surprise those who believe engineering ethics began in the U.S.; they might also be surprised to see China included alongside the Netherlands and Denmark as key countries contributing to engineering ethics as we know it today. This decentering of American and European hegemony adds nuance to our understandings of history.

Wayne Ambler's Chap. 18 "Guiding Gulliver: Challenges for Ethical Engineering" directs our attention to the ways in which politics and economics shape the kinds of questions engineers are called to address. Engineering is powerful, Ambler argues, yet we do not often enough ask, "What ends should this power serve?" How is "Gulliver" to be guided in a political age in which corporations wield outsize power, the philosophical zeitgeist is one of moral relativism, and we must heed national imperatives? This leads to a complex landscape indeed for determining what "good" engineering is for.

Both of these chapters are perhaps surprisingly inclined to stress the philosophical context or zeitgeist that envelops practicing engineers. While several of the articles included here also note the influence of engineering on the larger society, be it direct or indirect, it is especially Chap. 14, Sylvain Lavelle's study of engineering as a way of world-making, that is focused on this. All recognize that engineering produces technological artifacts, but Lavelle shows how this can be tantamount to designing and producing a world for those who use them. As both this chapter as a whole and the Churchill quote above suggest, engineering is simultaneously influenced and influencing, even at the level of general societal and philosophical values.

Several chapters in this part, therefore, speak globally about challenges engineering must face. Others, however, are more closely devoted to particular issues that are of decisive importance today. They offer, respectively, insightful introductions to the relationship between engineering and environmental issues, nuclear power and climate change, and the coming smart grid. Returning to the influences that act upon engineers, it is no small matter that engineers are professionals and belong to professional associations, at least in most countries. Consequently, they have special codes of ethics to follow, and they have an influential group identity, one they would like to protect and enhance. But to say that this is important is not yet to say what effect this influence has in complex cases. To explore the role of this professional culture, several articles in this part take up the following questions: differences between the professional associations for law, medicine, and engineering; the dilution of the term "professional"; the risk that a narrow understanding of one's profession may blind one to larger ethical questions; and the extent to which membership in the profession influences such judgments as the importance or unimportance of the environment, efficiency or optimization in general, and the need to take climate change into account. To simplify these complex chapters for the sake of a general introduction, we may say they share a concern that even as the profession seeks higher standards of ethics, it may end up fostering more insular ones.

That engineering is a profession, in other words, has consequences. Having sketched some of these with a broad brush, let us touch upon some particular points of special note. No one who reads these chapters will be able to forget that Albert Speer, Hitler's architect, was a professional. That he was one will serve as a caution against assuming that professionalism narrowly defined is sufficient to ensure ethics in engineering. If professionalism in the narrow sense requires expertise put in action on behalf of clients, a deeper goal is the wellbeing of people. Even when clients are found not in a Nazi regime but in a mostly free market, Meganck's Chap. 12 helps one wonder whether serving one's clients always entails serving the common good. Such tensions are only exacerbated in the age of neoliberalism, in which the question of "doing good" becomes harder to imagine outside of economic rationales. This said, it is admitted that an unfailing understanding of human welfare is not readily available, either for engineering students or for others. Ambler's Chap. 18 illuminates this problem and traces its philosophical pedigree; Meganck stresses that "everyday ethics" is a necessary supplement to a narrower professional view.

We must take up the complex question of what constitutes human welfare, and we will need tools such as everyday ethics in order to have such a conversation. This is because the shaping and communication of engineering values happens both implicitly and explicitly, and these values can be invoked and inculcated in contradictory ways. The engineering profession may encourage certain values, for example, even if it does not insist upon them in explicit codes of ethics. Possible examples include eco-skepticism as discussed by Didier and Talin in Chap. 12, and the effects of climate change on the standards used in planning and building nuclear power plants by Schneider, Tidwell, and Fitzwater in Chap. 15.

The first of these three chapters reports on surveys conducted concerning environmental issues. The responses of graduate engineers turned out to be rather different from those of others, even when they shared many demographic similarities. Nor did the engineers' opinions vary much in relation to other individual traits. While Didier and Talin hesitate to give a causal account of the lack of pluralism they detected, they do raise concerns about subtle influences operating within the profession and about a surprising lack of environmental consciousness.

The effort to reduce the emission of greenhouse gasses in power production has reopened and strengthened the case for nuclear power, but increasing climate change should also be taken into account when designing and operating a power plant, as the events at Fukushima demonstrated with special clarity. Research presented by Schneider et al. in Chap. 15 shows, however, that there is very little professional discourse on this issue within the engineering community. An important observation in itself, for it cries out for remedy, it also suggests the profession's blind spot may be the result of an insular culture that would affect other issues as well. There may be significant "talk" about climate change in relation to nuclear power, for example, but the quality of the talk – and who is engaged in it, and to what ends – matters.

The smart grid is coming, and engineers will have much to do in determining the final form it takes. High expectations await it, not only as a way of fixing the problems that grow with the aging of the current grid, but especially as a way of improving our energy efficiency and increasing the fraction of our energy that comes from renewable sources. But while these great expectations may prove justified, the new grid will not only have to overcome substantial technical problems but will also need to face social and ethical challenges, including ones involving privacy, security, and equity. In Chap. 16 Joe Herkert and Timothy Kostyk clarify these challenges by attention to discussion in both the EU and the USA and with reference to similar problems affecting other technologies, such as the vulnerability of Iranian nuclear development to US and Israeli cyber warfare. Since these challenges are not merely technical in nature, the article concludes with some consideration of ways their resolution will depend on reform in engineering education.

While all the articles in this part reflect on engineering education and the general need for it to reach beyond narrow technical training, Mitcham and Nan make it their focus in Chap. 17 to trace the way engineering education has become a subject of philosophical focus. This emergence of interest in the philosophical study of engineering ethics – or one might say in the relationship between responsibility and the ever-growing power of technology – has taken somewhat different shape depending on time and place. To cite but a single example, the role of engineers in the German war effort was bound to limit the extent to which engineering could be celebrated as contributing to some grand philosophical project or the unfolding of History. Notwithstanding its focus on the philosophical study of engineering and its proper ethics, the article concludes with Aristotle's reminder that without politics to give it force, ethics remains but words, a sobering but necessary reflection for all authors, and engineers, who wish to make the world a better place.

Jen Schneider Ph.D. in Cultural Studies from Claremont Graduate University. Associate Professor, Public Policy and Administration, Boise State University, USA. Co-author with Juan Lucena and Jon Leydens of *Engineering and Sustainable Community Development.* Past projects have addressed the role communication plays in environmental crises such as climate change and engineering for development work. Current projects address the role of rhetoric, media, and communication in energy controversies in the United States, with a special focus on fossil fuel production and consumption. Jen is completing a co-authored book titled *Under Pressure: Coal Industry Rhetoric in the Age of Neoliberalism.*

Wayne Ambler BA in Government, Cornell University. M.A. in Political Economy, University of Toronto, M.A. in Classics, Boston College, and Ph.D. in Political Philosophy, Boston College. Associate Professor and Co-Director of the Herbst Program of Humanities for Engineers, University of Colorado at Boulder. Scholarly articles and books on Aristotle, Xenophon, and Aristophanes, as well as on the challenges facing modernization through science, technology, and democracy.

Chapter 12
On the Normativity of Professionalism

Martin Meganck

It definitely was professional, and it was definitely smart, if you can call it that, but it was very conservative, very risk-averse, very aware of what mattered.

(Lance Armstrong, Interview with Oprah Winfrey, January 2013)

Abstract Why should engineers behave ethically? Often, this question is answered by qualifying engineering as a "profession", and professional organizations have codes of ethics that members should comply with. In many countries however, engineering is organized differently. The present chapter explores conceptions of "professionalism", inspired by evolutions in different occupational areas. A second part questions the idea that professionalism encompasses ethical responsibilities "beyond ordinary morality". The thesis will be defended that, although there may be specific rules for "professionals", the philosophical foundation of professional ethics yet rests on ordinary morality.

Keywords Professionalism • Engineering ethics

Introduction

> Why should engineers be ethical? Well: engineers are professionals. And professionals are members of professional organisations. And these professional organisations have codes of ethics that regulate their activity. That's why engineers should be ethical!

The passage just cited, was the intervention of an authoritative senior member of the international engineering scene, during a convention on philosophy of engineering and technology a few years ago. If stated that way, the foundations of engineering

M. Meganck (✉)
KU Leuven, Faculty of Engineering Technology, Technologiecampus Ghent, Gebroeders De Smetstraat 1, Ghent B-9000, Belgium
e-mail: martin.meganck@kuleuven.be

S.H. Christensen et al. (eds.), *Engineering Identities, Epistemologies and Values*, Philosophy of Engineering and Technology 21,
DOI 10.1007/978-3-319-16172-3_12

ethics are simple: ethical is what the code of ethics declares to be ethical, and as an engineer you are bound to comply with the rules of your association.

Founding engineering ethics on this rationale, stumbles upon two problems: a pragmatic one and a fundamental one. The pragmatic question deals with the fact that in many countries a system of chartered engineers or of an engineering board or any similar organization is absent: countries where membership of an engineering organization is not compulsory to work as an engineer, or to carry the title. The tendency to institutionalize professions rather strictly is – according to Julia Evetts (2003, p. 398) – typical for countries with an Anglo-American background. The European-continental approach by contrast would focus more on expertise and employment questions, and less on organizational and regulatory questions. Looking at the situation of engineering associations in Europe, Evetts' analysis seems to be confirmed, with the exception of some countries in the Mediterranean area (Spain, Portugal, Italy) where the organizations do have a more stringent status (Didier 2015). If professional ethics finds its foundation mainly or exclusively in membership of professional organizations and in the regulatory activity thereof, many engineers will find themselves ethical orphans. In a more nuanced formulation, this view would imply that engineers who work under the rule of such a regulated, membership-based engineering profession may have more or other ethical obligations than those who live in a country where such a system does not exist – even if they have the same capacities and are doing essentially the same kind of job.

The second, and more fundamental problem with the aforementioned rationale, has to do with the assumption that being ethical or behaving ethically equals rule compliance. This seems to presuppose that:

- either the act of compliance constitutes ethical quality as such, whatever the rules are;
- or that the ethical qualification of the rules is accepted: e.g. because they emanate from a recognized authority, or after an evaluation of their contents.

The present chapter will dwell upon two questions: if professionalism cannot just be equaled with membership of a professional organization, what can be contemporarily valid interpretations of this notion? And what are the foundations of an ethics for professionals?

On Professionalism

Discussions about professionalism are stained by the reminiscence of traditional learned professions, with medicine and law as paradigmatic examples. These occupations require an extended and exclusive body of theoretical knowledge and of related practical expertise. Their services are of the highest importance, as well for individual clients/patients as for society as a whole. Because of the importance and the exclusivity of the competences, there is a strict control over the practice of the profession: the exercise of the profession requires membership of a professional body, that is also responsible for monitoring the delivered professional services. A code of conduct serves as a tool for internal discipline, and as an external pledge soliciting the trust of society. In a very traditional view of these professions, they are further characterized by a set of attitudes, rituals, traditions and symbols that form together a professional culture (Greenwood 1957). Because of the importance of the field of activities and regarding the presumed incompetence of the clients or patients, the exercise of these traditional professions is not supposed to follow the logic of commercial supplier/client-relationships. Instead of (negotiable) prices on a free market, remuneration of professional work occurs through an (often fixed) fee or honorary. Commercial advertisements are usually judged unbefitting for a real professional.

Extending this view of professionalism to other occupational fields, stumbles on a few obstacles. In 1964 already, Harold Wilensky wrote that "nowadays everybody wants to become professional", and this tendency has not weakened since then (Noordegraaf 2007). The term *professional* can be used with almost any function or occupation: engineers, teachers, nurses, but also funeral undertakers, massage parlors, tanning centers, cleaning services and bicycle repair shops. It will be obvious that the degree to which this broad range of activities meets the rather strict set of attributes of traditional professionalism as described above, varies largely. For some of them, the importance of the delivered services, the necessary skills and knowledge, the modes of organization and the societal status may be similar to the corresponding qualities in the medical and juridical sector. But for some of the latter few examples in the list, the necessary knowledge or skills seem rather limited, and the idea of "professionalism" goes little further than what one does for earning one's living, or refers mainly to the use of specialized products and equipment. It may also be limited to the promise of following standardized procedures. Anyway, these would-be-professions will have more difficulties in making themselves perceived or accepted as pure professionals. Yet they too strive for a label of professionalism that should inspire trust, thereby appealing to the original idea of professionalism: i.e. to warrant the quality of delivered services, by controlling the access to and the exercise of the activity, by establishing standards and quality control systems, etc. Legal sanction of this desire to protect customers and users may result in criteria that make small scale or artisanal practice of the activity virtually impossible, notwithstanding the sympathy and trust that the public often has for this mode of operation.

This broadening of the concept of professionalism can result in an erosion of its meaning. The idea of public service e.g., or that the mission of a profession is in the aspiration of some public good, may fade completely. The idea of sound organization and specialized competences may prevail, like in Lance Armstrong's interview with Oprah Winfrey (where he qualified the doping program in which he participated as "definitely professional"), or like in designations like *a professional killer,* where the idea of efficiency and competences is "enriched" by the pecuniary aspect. And there are other instances still where being (or calling oneself) a professional is used to claim privileges or exemptions, or to advocate attitudes or traditions that would otherwise be rejected: e.g. the plea that traffic rules would be applied differently for people whose profession necessitates a lot of driving, or that fair play does not mean the same to a professional sportsman compared to a hobbyist.

Examples like these seem antagonistic to the original highbrow ideas of professionalism, or can make a caricature of it. But even without such extreme examples, several phenomena question the "old style", "pure" conception of professionalism. One is that traditional professions like medicine cannot be exerted without taking into account a wide range of influences and contexts that put pressure on the idea of professional autonomy. The passive, incompetent patient who was unable to define her own needs, has in some cases evolved to a client or partner with whom one has to negotiate about the presence and nature of a problem, about the desirability of different scenarios, and about the terms, circumstances and prices of the treatment (Stapel 2013). The individually operating professional with his individual client has in many cases been transformed into a member of a team or organization, directed and supervised by managers. These managers often do not belong to the same professional groups as their collaborators. Management may even be considered as a profession in itself, with its own set of competences and its own claims of autonomy. Professional managers should be able to run a school as well as a hospital, a chemical factory as well as a chain of supermarkets. Besides the manager, there are other colleagues in the organization with different occupational backgrounds, the expertise and operating modes of which have to be taken into account. And besides the immediate client or patient, external third parties may interfere in decisions about the financing or execution modes of projects and operations. This may be a threat to the idea of professional autonomy, or even make that "the contexts of work actually undermine the profession's fundamental purposes and its standards of quality and ethical practice" (Colby and Sullivan 2008, p. 408).

Add to that the striving for a professional status by other occupations: some with a highly intellectual character, some more skills-based, and a quasi-continuous spectrum in between. The degree of exclusivity of their skills and knowledge may differ widely. The organizational context in which they work may be very diverse: as individual service providers with individual clients; as employees of industrial or commercial companies; as public servants; etc..... They may or may not have some degree of autonomy or protection towards their clients or employers. They may or may not have regular formal or informal contacts with colleagues having similar functions in other organizations. In short: the degree in which they meet the traditional criteria of the ideal type of professionalism, may vary largely.

Confronted with these deviations from the traditional ideal type of professionalism, different reactions are possible. Mirko Noordegraaf (2007) distinguishes three possible approaches:

- either one sticks to the traditional definition of professionalism. There can be no real professionalism outside these well-defined and well-organized traditional domains. Even if they work in contexts with interferences by management, consumers or other stakeholders, professionals must keep their autonomy. Noordegraaf refers to this as *purified professionalism*;
- or one accepts a form of *situated professionalism*: the organizational context in which many professionals work, is recognized and accepted. The traditional idea of elite professionalism has to be broadened to include experts;

- or one yields to the observation that the idea of professionalism is shifting, and that a strict definition has become impossible. Forms of *hybridized professionalism* arise, in which the relatedness to outside worlds is part of the professional identity. Interdisciplinarity, contextuality and interactivity become part of professional work. The professional's performance is multi-facetted; the awareness of this puts the professional in a network with other professionals stakeholders.

Several other scholars have made similar analyses, using a different terminology, but essentially dealing with the same questioning. Here are a few examples.

Donald Schön (2001) used the term *reflective practitioner* for describing the new professionals who are confronted with a *new epistemology of practice*. The traditional rigorous professional practice "depends on the use of describable, testable, replicable techniques derived from scientific research, based on knowledge that is objective, consensual, cumulative and convergent." But real world problems do not come well formed. New professionals do not just solve well defined problems: they are also involved in the process of constructing the problem to be solved. In this process, they must be open to "artistic, intuitive processes which bring to situations of uncertainty, instability, uniqueness and value conflict" (Noordegraaf 2007, p. 774). When facing a situation that is puzzling, unique or conflicted, reflective practitioners should be "able to turn thought back on itself, surfacing, criticizing and restructuring the thinking by which they have spontaneously tried to make the situation intelligible to themselves" (Schön 2001). This reflection-in-action puts pre-established paradigms under pressure, and confronts them with *tacit knowledge*; the input of third parties can in these instances be illuminating.

Edgar Burns (2007) focuses on the use of the term *post-professionalism*. The "post" in this term is an indication of the fact that the traditional conceptions of professionalism have become empirically inadequate; but it also reminds of movements like post-modernism, post-structuralism,... in philosophy and sociology: movements fundamentally questioning "narratives and discursive truths". Similar to Noordegraaf's notion of hybridized professionalism, the use of post-professionalism tends to include the different networks and relationships surrounding professional work: relationships with governments, business, and the voluntary or not-for-profit sector. On the one hand, this is a critique of the straightforward functionalist view on professionalism. On the other, there is also some sympathy with professionals: the "rules and expectations that are imposed on the professional by the public, governmental regimes and bureaucratic organizations, often entail huge compliance costs without – for the majority of practitioners – making a practical difference to the level of performance" (Burns 2007, p. 5). The "post"-thinking method is undoubtedly valuable in questioning paradigms and rhetorics, especially when they seem laden by suspicions of privileges and power. But after the "deconstruction" of the old concepts, the reconstruction of alternatives is difficult, fraught as it is by the uncertainties, precautions and auto-criticism that are inherent to the method itself.

Andrew Jamison et al. (2011) situate science and technology in cultural and historical perspectives, resulting in a form of *hybrid imagination* that is in line with Noordegraaf's idea of *hybridized professionalism*. Scientific and technological professionals are invited to give up isolationism and hubris, and to acknowledge and foster the contextualized knowledge that is necessary to function properly with a sense of cooperation and social responsibility.

A provisional conclusion can be that, due to the changes in the landscape of professionalism, the traditional idea of professionalism still figures at the background of the reflections, as a kind of ideal type. Attempts to advance the discussion, taking into account the multitude of new would-be professions and the changing circumstances in which even very traditional professions actually operate, will often refer to that ideal type. The ideal type had the advantage of conceptual clarity, at the expense of eventually getting alienated from real life practice. Other approaches may have more empirical support, or may be more philosophically nuanced and critical, thereby risking to become confused, or at least: to lose visibility as a clearly set standard. Moreover, ordinary language seldom bothers about the results of semantic discussions about the proper meaning of terms. If – to say it with Wittgenstein – "the meaning is in the use", a "declaration of invalidity" of certain uses of the word professionalism may appear in the end rather fruitless…

On Keeping Focused…

One of the key elements in discussions about professionalism is the idea that the aim of professions *as a group* (even if they are not formally organized) is the pursuit of some important public good: be it health, justice, knowledge, etc.… At the same time, professionals *as individuals* are often seen as experts, valued for their knowledge and skills that they can put at the service of their individual clients, and their needs, desires and preferences. At first sight, both pictures seem equivalent. Yet it deserves further inquiry whether the aggregated needs, desires and preferences of clients are compatible with, promote or constitute the common good. Similarly, it can be questioned whether the aggregated microrational actions of the experts result in an evolution towards the common goal on a macrorational level. And finally, who is best placed to judge about the common good: the professionals (individually or as a corps), the clients, or other instances?

An outspoken example of the tension between the exercise of the technical expertise, and the attainment of the public good, was Albert Speer's self-defense for his functioning as "Hitler's architect" and (1942–1945) Minister of Armaments and War Production. His plea was that he considered himself as a *pure technician*, solely concerned with his technical skills, and that he had mentally kept his functioning as a trained architect separate from the moral implications and political reflection about the overall goals and consequences of Nazism (Sammons 1993, p. 179). In a memory to Hitler, he wrote:

> The task I have to fulfill is an unpolitical one. I have felt at ease in my work only so long as my person and my work were evaluated solely by the standard of practical accomplishments (Sammons, p. 190).

It was only later, in prison, that he realized that "becoming a human being requires stories and images a good deal richer than professional ones" (Sammons, p. 193). Traditionally, this example leads to discussions about the fragmentation between *acting as an architect* and *acting with respect for human dignity*. Reasserting oneself as a moral being then may require a "rebellious ethics", flirting with the question whether one can be obliged to act heroically. Sammons however refuses this line of defense: Speer didn't even act as a good architect, because his very own architectural insights unmasked Hitler's "lack of concern for the social dimensions of architecture" (Sammons, p. 185). Being a good architect would include working according to the internal values of architecture, whereas Speer, even in his building activity, was guided mainly by external concerns. Architecture could have given him a morality that Speer failed to understand…

In the context of law, Bradley Wendel (2005) takes a stance against the view that professionals are mere expert agents acting in the name of their clients. Lawyers have "to apply the law to her client's situation with due regard to the meaning of legal norms, not merely their formal expression". As the internal *raison d'être* of law is to serve the functioning of good institutions, lawyers should not participate in operations destined to subvert this cause. Law is "worthy of being taken seriously, interpreted in good faith with due regard to its meaning" (Wendel 2005, pp. 1168–1169). The good professional is – according to Wendel – not just a person knowing all the tricks and maneuvers with which the clients' interests can be pursued, but an expert working at the service of justice. And probably law is an eminent example of an area where the aggregate of the interests of individual clients does not automatically coincide with "justice" as a societal value… Similar observations can be made in fields like accountancy, salespeople,…: even if in each individual situation optimal use is made of technicalities on a micro-scale, this does not necessarily guarantee optimal effects on the macro-level.

In his studies on social work, Harry Kunneman (2007; also Driessens and Geldof 2008) too takes a stance against professionalism being reduced to mere technical expertise. Social work is the domain par excellence where a merely cognitive-technical expertise may be blind to moral and existential dimensions of a situation. Where professionals experience this tension, they can either adapt to the rules and structures of the organization, thus shifting responsibility towards higher levels in the organization. A second strategy may consist in the professional – on her own initiative, and often unknowingly to the organization – adding a complement to the services she is officially supposed to deliver. In some instances, where such services seem to imply too personal a relationship, this may even be interpreted as "unprofessional". Ideally, forms of *normative professionalism* can be developed, where the gap between the "system world" (the way the situation is perceived and framed by the professional apparatus) and the "life world" (the way the situation is lived by the client) can be bridged. For the further development of his views, Kunneman refers to social theorists like Habermas, Giddens and Castells, especially concerning their critique on the narrow view on humans that reduces them to beings who are to be empowered to become production and consumption oriented autonomous individuals. Instead of that, moral involvement, respect for existential values and attention to relationships should become part of real professionalism, and thus help the profession to keep in touch with its proper objective: the well-being of people.

Taking fields like medicine and clergy as examples, Anne Colby and William Sullivan (2008, pp. 414–415) issue a warning against "misalignments": these may be caused when "extrinsic rewards [can] overwhelm and even actually undermine the ultimate purposes of the profession" or when there is a lack of "deep engagement with the profession's public purposes, along with a sense of meaning and satisfaction from one's work that is grounded in or aligned with those purposes." Evaluation and promotion criteria (and in general: criteria for recognition of the quality of work) may be diverse, and sometimes difficult to combine. When people experience that they are evaluated on criteria that differ from what they perceive as being adequate for the contents of their job, they may either adjust to these external criteria (sometimes at the expense of properly functioning in the job they were hired for), or be left behind with the feeling that their superiors are either incompetent or dishonest in judging what they do. Especially in competitive environments, *survival of the fittest* may lead to an adaptation of the individual's behavior to the survival criteria. Whether these criteria correspond to the ultimate purpose of the activity or the profession, is not always self-evident. In a context of financial management e.g., Ian Herbert refers to the possibility that knowledge workers may be driven "towards competitive advantage […] in which process controls and performance measures will ultimately win out over more subjective notions of 'doing a good job'" (Herbert et al. 2012, p. 54).

For engineering, threats for the over-all purpose of the profession by micro-rational or external influences can be found in cases of planned obsolescence: products may be designed to have a limited useful life, either by wear, by becoming unfashionable, or by being outdriven by newer products with which they are no longer compatible (Dannoritzer 2010). Products may also be unnecessarily complex, or equipped with a multitude of possibilities that the majority of clients do not need and will never use; yet finding old-fashioned or simpler apparatus may be virtually impossible. A market-driven industry may also focus more on luxury products for a wealthy group of customers, thereby neglecting more fundamental needs of larger and more needing populations.

Finally, if keeping focus on the ultimate purpose of the profession is a key element in the debate on professional ethics, a serious question is raised by Carl Mitcham (2009), especially in the context of engineering. Attempts to indicate the ultimate purpose of engineering can be found on different places. Adam Briggle and Carl Mitcham (2012, p. 294) refer to the traditional definition of engineering formulated by Thomas Tredgold (1828), when he was president of the Institution of Civil Engineers. He defined engineering as:

> "… the art of directing the great sources of power in nature for the *use and convenience of [human beings]*" (our italics)

IEEE, in its Mission Statement and Vision Statement, declares:

> "IEEE's core purpose is to foster technological innovation and excellence *for the benefit of humanity*," and "IEEE will be [...] universally recognized for the contributions of technology and of technical professionals *in improving global conditions*" (our italics)

In its Code of Ethics, IEEE wants engineers to "take responsibility by making decisions consistent with *safety, health and welfare of the public*". And also in other texts reflecting on the function of engineering, the ultimate purpose is usually defined in terms of human wellbeing, e.g., Allison Ross and Nafsika Athanassoulis in a study of the social nature of engineering in a context of risk taking:

> ...the chief good internal to the practice of engineering is safe efficient *innovation in the service of human wellbeing* and that this good can only be achieved where highly accurate, rational decisions are made about how to balance the values of safety, efficiency and ambition in particular cases (Ross and Athanassoulis 2010, p. 159 – our italics).

Be it with slightly different words, all these sources seem to agree on engineering having as its ultimate purpose some form of human wellbeing. There can be – of course – discussion among techno-optimists and techno-pessimists about whether engineering or technology, in their final consequences, actually promote the human good. But even without going into that fundamental debate, Mitcham (2009) points to a major difference between engineering and some of the other traditional professions. As "human health" is the core business of medicine, medical students will – together with their courses on anatomy, pharmacology etc.... – acquire a robust view on what constitutes "health". And a typical law school curriculum will include a sound introduction into procedural justice, which is the raison d'être of their profession. One can, of course, question whether this makes these professionals the sole adequate instances to judge about how health or justice can be pursued (actually, the implementation of their codes of conduct suggests that there are external boundary conditions within which their conception of their ultimate good should be framed). Contrary to what seems the case for medicine and law however, engineering education often pays little attention to reflection on the alleged ultimate purposes of the occupation: what constitutes "safety, health and welfare", or how the "balance of values" should be reached or evaluated. The social construction of safety, health and welfare is a societal process that is difficult to grasp, and in which "engineers *qua* engineers are no more qualified to make such determinations than anyone else; they legitimately participate in making such determinations, but only as consumers, users, and citizens" (Mitcham 2009, p. 349). Definitions of the problems to solve, and decisions about how to solve them, do not belong to the jurisdiction of engineers alone, and it may be good to ask if they would be the most adequate judges. One can here compare with Nicolas Rescher's remark about the activity of scientists: "As war is too important to be left to generals, so knowledge is too important to be left to scientists and scholars without, at any rate, moral checks and balances" (Rescher 1987).

Beyond Ordinary Morality?

In Chap. 4 of this book, Michael Davis defines a profession as

> a number of individuals in the same occupation voluntarily organized to earn their living by openly serving a moral ideal in a morally permissible way beyond what law, market, morality, and public opinion would otherwise require (Davis 2015).

The question of the "moral ideal" in professions has been commented on in the previous section of this chapter. In this last section, the last part of Davis' definition will be dealt with: "... beyond what law, market, morality, and public opinion would otherwise require." This expression seems to presuppose the idea that the ethical evaluation of certain behaviors of professionals is founded by the idea of "professionalism" itself.

Of course, one can find many examples of types of behavior that would be evaluated differently, depending on whether the person in question is considered as a professional or not. In an attempt to circumscribe professionalism by listing a series of behaviors that would be deemed unprofessional, Erde (2008, pp. 14–15) indicates that some of these instances of unprofessional conduct cannot really be called unethical: among them misplaced forms of humor, or e.g. a doctor who would stand smoking at the entrance door of a hospital. Expectations about confidentiality may be different depending on the professional status. Boucher (2007) even refers to cases where compliance to rules of professional deontology may lead to decisions that would go against ordinary moral intuition.

Pragmatically and empirically speaking, there seems to be a professional ethics that may differ from ordinary morality. The foundation of this specificity however is less obvious. Starting from a distinction made by Boucher (2007), the following lines of argumentation can be seen:

Contractualist Arguments

When an individual voluntarily accepts to take up a professional role, she also accepts the complex of obligations, benefices and privileges that go with that role. This is clearly visible in contexts where professionalism is strictly regulated, e.g. in professional organizations with a code of conduct, or where an oath is part of the membership rituals (in religious orders, the "oath" sealing the membership of candidates, is called "profession"!). Such an oath or code of conduct may moreover be part of the social contract by which the professional group acquires its status in society.

Where professionalism is not organized in this way, the commitment of the practitioner may be formalized in a contract (employment contract, business contract,...), or simply by accepting to carry out a task in an environment where certain rules prevail. But also informally, creating the perception that one can deliver services "on a professional level" (or failing to adjust false perceptions in this regard) may entrain moral obligations: be it in cases that are not governed by formal regulations, or when one deliberately choses to keep things informal.

Boucher (2007) extends on some possible criticisms against this contractualist view (who are the contractants? and can one legitimately make a promise that may imply actions that would elsewhere be illegitimate?). Besides these, the idea of a "contract" presupposes that the contractants have equal liberty to accept or refuse it: a presupposition that can be severely questioned, especially when some of the stakeholders are dependent on the others' services (as can be the case for professions). Whereas contracts imply (free) mutual agreement among contractants, the mere (implicit or explicit) unilateral *promise* of a quality of service can generate moral obligations, even without formal acceptance by the other party.

Fiduciary Arguments

A *fiduciary* argument points to the fact that obligations can be generated by the trust that clients, patients, and even society put in professions and professionals. For domains that may be of vital importance, we rely on the capacities and good will of experts. In this argument, two poles appear: the *capacities* of the professional on the one hand, and on the other hand the *dependence* (or even vulnerability) of the receiving parties. Organized professions can be an answer to this dependence, but the dependence would exist also without these organizations. And similarly, the (intellectual, physical, skills-related) capacities can be present with as well as without organized professions; professions can make a difference where legal authority is concerned.

Following Levinas' philosophy, the confrontation with the dependence and vulnerability of "the other" is the *fait primitif* of existence and therewith also responsibility (Levinas 1982). Responsibility grows when the other cannot but trust in one's capacity and willingness to take care of a situation; in cases where trust is less based on dependence, the contractual or promise-based logic reemerges. In this view, when faced with questions of vital importance, the obligation to use one's capacities to deal with it cannot be brushed aside by the mere argument that one is not a professional. For a person with adequate capacities, the obligation to take care of an emergency situation also stands without her officially recognized professional status. In less urgent situations, a person with sufficient capacities can shift responsibility towards a "professional", if such a professional is at hands; this however would be based more on the contractual or promise-based argument, than on the fiduciary.

Teleological Arguments

Finally, specific professional ethics could be supported by *teleological* arguments: the important common good purpose that professions are to pursue. The goals of professions, justified as they are, would legitimate the means that are necessary to pursue that goal. Boucher (2007) dissects this argument into four steps: (1) the acts of a professional are justified by professional rules or obligations; (2) these rules or obligations are justified by the professional's role; (3) this role is justified by the (institutionalized) profession; and (4) the profession is justified by its "ultimate purpose": the common good that it is supposed to serve.

It deserves careful attention to examine whether, in the trickling down from the (undisputable?) good objectives to the very concrete behaviors that are expected from a real professional, the rules of necessity are followed. In the concretization of expected techniques and procedures that should contribute to the common good, rules may appear that can be unnecessary, counterproductive, or at least questionable. And besides sometimes being unnecessary, they may also be insufficient: see the comments in the previous section on how professional rules may fall short in attaining the good that professions aim at, on how nowadays professionalism has to deal with external expectations that may divert from the profession's goals, on how sticking to merely technical micro-rationality – even if it is highly dependent on professional expertise – may "miss the point" of the profession's ideal purpose, and on how the professionals themselves may or may not be good judges of what the ultimate purpose can be and imply.

A more fundamental remark on this line of legitimation is the question whether the aim justifies the means. Can one accept that otherwise illegitimate actions are undertaken, even if they are well-intended and covered by a rule-utilitarian logic? A second critique points to the fact that a profession may be a rather composite complex of different roles with different purposes, and that the teleological justification of professions is therefore underdetermined. And finally: if the norms and rules of a profession are justified by the profession's ultimate purpose, this ultimate purpose in itself can only be justified by ordinary morality (Boucher 2007).

Conclusion

The tendency of these reflections is that attempts to justify a professional ethics on other grounds than ordinary morality, do in the end often fall back on principles of everyday ethics. This is not to deny that there may be rules or expectations that are specific for professionals. However, in the cases where these rules or expectations differ from what would be used for lay people, the basis for these deviations seems to rest on principles that are accepted in very common ethics too: the obligations that are created by making a promise, the responsibility that results from having the capacities to help – especially in situations where people are dependent and vulnerable –, or the pursuit of an important societal good, where – it is true – a rule-utilitarian approach may lead to other conclusions than act-utilitarianism. Claims for very specific rights or duties that would be linked to the idea of professionalism, deserve to be examined very critically.

The self-evidence of choices that are made within a professional paradigm, may meet resistance in a larger audience. Lay people may have other preferences than professionals. The choices professionals have to make recurrently, may be once-only events for lay people; the emotions and resistances that lay people experience in these cases, may have worn off in the hearts and minds of the professionals. In such instances, communication can be very important; and this may result in rethinking practices that were undoubtedly well-intended, but the efficiency or adequacy of which can from time to time be questioned.

It can be helpful to see professions as instruments, developed and used for well-defined purposes. But like in all instrumental rationalities, the relationship between the means and the goals should be monitored carefully: are the means adequate and efficient? And what are the side-effects? Building wastewater treatment plants can be a good measure to protect environment, but is it sufficient and efficient to make this the automatic choice? People may feel safer if they possess and know how to handle a weapon; but does one get a safer society when everybody is equipped with weapons? What about the possibility of means-end-inversions? A professional security corps may be instrumental in dealing with situations where people feel insecure, but does the security corps (as a corps) have interest in creating a safe society? Organizing things professionally and generalizing this choice, may change society in a way that may be unintended and unforeseen.

Professions and professionals: we need them. We trust them. But we have to be careful to keep them on the right track…

Acknowledgments I express my gratitude to Christelle Didier, for her critical and constructive comments on a first draft for this chapter, and to Dr. Erik Van Achter, for his linguistic polishing of the text.

References

Armstrong, L. (2013). Interview with Oprah Winfrey. Transcript on http://www.bbc.co.uk/sport/0/cycling/21065539

Boucher, F. (2007). *Le problème de la moralité de rôle en éthique professionnelle*. http://www.intranet.idea.ulaval.ca/Documents/FB-role%20EP.rtf. Consulted 18 Aug 2012.

Briggle, A., & Mitcham, C. (2012). *Ethics and science. An introduction*. Cambridge, UK: Cambridge University Press.

Burns, E. (2007). Developing a post-professional perspective for studying contemporary professions and organisations. *Proceedings conference on critical management studies,* University of Manchester, July 2007. http://www.mngt.waikato.ac.nz/ejrot/cmsconference/2007/proceedings/newperspectives/burns.pdf

Colby, A., & Sullivan, W. M. (2008). Formation of professionalism and purpose: Perspectives from the preparation for the professions program. *University of St. Thomas Law Journal, 5*(2), 404–427.

Dannoritzer, C. (2010). The light bulb conspiracy. Documentary available at: http://dotsub.com/view/aed3b8b2-1889-4df5-ae63-ad85f5572f27

Davis, M. (2015). Chapter 4: Engineering as profession: Some methodological problems in its study. In S. H. Christensen, C. Didier, A. Jamison, M. Meganck, C. Mitcham, & B. Newberry (Eds.), *Engineering identities, epistemologies and values. Engineering education and practice in context*. Springer.

Didier, C. (2015). Engineering ethics: European perspectives. In J. Britt Holbrook and Carl Mitcham (Eds.), *Ethics, science, technology, and engineering: A global resource* (Vol. 2, pp. 87–90). Farmington Hills: Macmillan Reference USA.

Driessens, K., & Geldof, D. (2008). Normatieve professionaliteit in het sociaal werk. *Alert, 34*(2), 66–75.

Erde, E. L. (2008). Professionalism's facets: Ambiguity, ambivalence, and nostalgia. *Journal of Medicine and Philosophy, 33*, 6–26.

Evetts, J. (2003). The sociological analysis of professionalism: Occupational change in the modern world. *International Sociology, 18*(2), 395–415.
Greenwood, E. (1957). Attributes of a profession. *Social Work*, July 1957, 45–55. Reprinted in: Johnson, D. G. (1991). *Ethical issues in engineering* (pp. 67–77). Englewood Cliffs: Prentice Hall.
Herbert, I., Lambert, S., & Rothwell, A. (2012). Professionalism in the 21st century: Symbol or substance? *Financial management*, April, 54–55
IEEE (Institute of Electrical and Electronics Engineers). Code of ethics. http://www.ieee.org/about/ethics.html. Accessed Mar 2015.
IEEE (Institute of Electrical and Electronics Engineers). Mission and vision statement. http://www.ieee.org/about/vision_mission.html. Accessed Mar 2015.
Jamison, A., Christensen, S. H., & Botin, L. (2011). *A hybrid imagination. Science and technology in cultural perspective* (Synthesis lectures on engineers, technology, and society, Vol. 12). San Rafael: Morgan & Claypool Publishers.
Kunneman, H. (2007). Sociaal werk als laboratorium voor normatieve professionalisering. *Ethische Perspectieven, 17*(2), 92–107.
Levinas, E. (1982). *Éthique et infini*. Paris: Fayard.
Mitcham, C. (2009). A philosophical inadequacy of engineering. *The Monist, 92*(3), 339–356.
Noordegraaf, M. (2007). From "pure" to "hybrid" professionalism. Present-day professionalism in ambiguous public domains. *Administration and Society, 39*(6), 761–785.
Rescher, N. (1987). *Forbidden knowledge*: *Moral limits of scientific research*, as quoted in: Oger, E. Hoeveel waardevrijheid? In: P. Reynaert (Ed.). *Wetenschap en Waardevrijheid* (p. 35). Leuven: Garant, 1993.
Ross, A., & Athanassoulis, N. (2010). The social nature of engineering and its implications for risk taking. *Science and Engineering Ethics, 16*, 147–168.
Sammons, J. L. (1993). Rebellious ethics and Albert Speer. (Reprinted as: *Albert Speer: ethics, architecture and technology*). In: E. Katz (2006). *Death by design: Science, technology and engineering in Nazi Germany* (pp. 179–200). New York: Pearson Education.
Schön, D. (2001). The crisis of professional knowledge and the pursuit of an epistemology of practice. In: J. Raven, & J. Stephenson (Eds.). *Competence in the learning society* (pp. 185–207). New York: Peter Lang Publishing. http://www.heacademy.ac.uk/
Stapel, M. (2013). Beroepsethos versus patiëntenautonomie binnen de plastisch-esthetische chirurgie. *Filosofie en Praktijk, 34*(2), 25–40.
Tredgold, T. (1828). Description of a civil engineer. *Minutes of the Proceedings of the institution of civil engineers*, Vol. 2 (1827–1835), January 4, pp. 20–23, as quoted in Briggle & Mitcham (2012), p. 294.
Wendel, B. (2005). Professionalism as interpretation. *Northwestern University Law Review, 99*, 1167–1233.
Wilensky, H. L. (1964). The professionalization of everyone? *American Journal of Sociology, 70*, 137–158.

Martin Meganck M.Sc. in Chemical Engineering from Ghent University, Ph.D. in Chemical Engineering and M.A. Moral Theology both from KU Leuven. Lecturer in Philosophy and Ethics in the Faculty of Engineering Technology, and researcher at the Center for Science, Technology and Ethics and the Theological Faculty of the KU Leuven. Teaching areas: Philosophy of Science, Philosophy of Technology, Professional and Business Ethics. He was co-author and co-editor of *Philosophy in Engineering* (Academica 2007) and *Engineering in Context* (Academica 2009).

Chapter 13
Engineer's Ecoskepticism as an Ethical Problem

Christelle Didier and Kristoff Talin

Abstract The graduate engineers' attitude towards environmental issues differs profoundly from that of their fellow citizens. This is what we have found out when comparing the answers given by 27,000 graduates to an original survey we conducted in 2011 with those of a representative sample of French people who participated to the "European value survey". The engineers' attitude is also very different from those of business managers and executives. It also differs from those of other master's degree graduates. Contrary to our expectations, the demographic change observed in the profession (growth, place of women, development of new educational tracks) has little influence on the professionals' attitude. The engineers' attitudes toward environmental issues seem to depend more on their professional position than on their individual traits. While the younger generation seems a little bit more pro-environment than their seniors, females do not differ significantly from their male colleagues on that topic. By contrast, we found out that the engineers' attitude towards environment is strongly related to their attitude and values in general and their political, ethical and religious attitude in particular.

Keywords Engineers' attitude • Environment • Engineering ethics • Politics • Religion

C. Didier (✉)
Département des sciences de l'éducation UFR DECCID,
Université Charles de Gaulle-Lille3, 3 Rue du Barreau,
Villeneuve-d'Ascq, Lille F-59650, France
e-mail: christelle.didier@univ-lille3.fr

K. Talin
CNRS-Clersé, MESHS, 2 Rue des Canonniers,
Lille F-59000, France
e-mail: christophe.talin@univ-lille1.fr

S.H. Christensen et al. (eds.), *Engineering Identities, Epistemologies and Values*, Philosophy of Engineering and Technology 21,
DOI 10.1007/978-3-319-16172-3_13

Introduction

The degradation of landscapes by steam-powered industrial technology has emerged as a social concern since the nineteenth century. Soon the first national-scale environmental laws were voted in several countries. But, it took a century for environmental protection to become a global issue. New words needed to be coined, like "Ökologie" by the German zoologist Haeckel in 1866 and "ecosystem" by the English botanist Tansley in 1935; the principles of ecology had to develop and the science of ecology to emerge as a distinct discipline. Ecological thought and environmental concern expanded in the twentieth century and the first global initiative appeared in the 1970s with the UN's first major conference on international environmental issues. Since the 1987 Brundtland Report, a new concept has been proposed and widely accepted to combine in a single expression developmental and environmental issues: "sustainable development". Proposed by experts and defined as a development which "ensure[s] that it meets the needs of the present generation without compromising the ability of future nations to meet their own needs", it was popularized at the 1992 Rio Summit. Since then, it has disseminated rapidly among laypeople and been included in educational programs all over the world. The 36th chapter of the Agenda 21 on "Education, Public Awareness and Training" was 1 of the 4 among 40 chapters to be singled out at the UN Commission for Sustainable Development for special work programs. Today, this goal is pursued by most educational programs in the world, also in engineering education. Since the turn of the twenty-first century, companies, and especially multinational corporations, have been considered as unavoidable sustainable development actors. It is also widely accepted that industry is highly concerned and that the engineering profession can play a key role in delivering sustainability.

Although definitions of the engineer differ from one country to another, there is sufficient commonality to assess that the members of this profession are directly concerned with the challenges of sustainable development. While some surveys have been made to determine what engineering students know about sustainable development (Azapagic et al. 2005), there is a lack of information about the engineers' attitudes once they have left university.

Our research goal intends to fill this void. It is based on an extensive survey conducted online, in 2011, by the French National Council of Scientists and Graduate Engineers (CNISF, called today IESF). Of the 39,000 survey respondents, 27,000 engineers answered to an optional part of the questionnaire, which we have designed, dealing with social, ethical and professional values. The data were analyzed with SPSS.

In this chapter, we focus on the items dealing with environmental issues, and particularly with six statements, which belong to the "revised" New Ecological Paradigm (NEP) (Dunlap et al. 2000). These statements, which constitute a "short" NEP (Bozonnet 2010b) have been included in the fourth wave of the European Values Study (EVS, called previously European Values Survey), in 2008. EVS is a large-scale, cross-national and longitudinal survey the first wave of which was in

1981. The French part of the 2008 survey was conducted by Pierre Brechon (Bréchon and Galland 2010). In the first part of this chapter, we present the outcomes which confirm our first hypothesis on the specificity of engineers' attitude in comparison which their fellow citizens. In the second part, we show evidence in favor of rejecting our second hypothesis on the influence of the graduates' demographic characteristics on environmental attitudes. In the last part, we show that there are strong links between environmental attitudes and the engineers' attitude to other fields of values like political and religious values.

Research Questions and Hypotheses

Since the 1970s, the environmental issue has become a central concern throughout the world, thanks in part to a better understanding of the interconnection between environment, economy and quality of life (Carson 1962). Besides, major technological disasters generated awareness amongst the public of the dangers posed to the natural environment by human activity (Lagadec 1981). In the late 1970s, the first green parties were founded. In countries all over the world, departments and ministries were created dedicated to this cause. This period also saw the birth of a new field of investigations in the social sciences: environmental sociology (Dunlap and Van Liere 1978, Catton and Dunlap 1978, Dunlap and Catton 1979).

Since 1987 and the publication of the Brundtland Report by the World Commission on Environment and Development, a new phrase has become the slogan of our contemporary societies: "sustainable development" (Brundtland 1987). The ideas covered by this new concept were not entirely original, but the expression and the definition proposed in the UN report disseminated widely. Since the Rio Summit in 1992, sustainable development has become a global cause and the phrase "think globally, act locally", the new mantra of the late twentieth century.

Because of its complexity, sustainable development requires to be dealt with by a set of very different actors and not only government and experts. Yet, after the Rio Convention, it took 10 more years (until the Johannesburg Summit, in 2002) for the business world to be recognized as a major player in this field. The UN report has put forward in a new way the responsibility of the business world – alongside that of government – in the implementation of a more sustainable development (ONU 2002).

If the business world is called to be concerned about its environmental and social impacts, the industrial world is even more concerned because technical development is at the roots of many environmental problems. Although definitions of "the engineers" (who they are and what they do) may vary from one country to another, the type of knowledge and activities of engineers, as well as their work environment make them appear as actors "involved" in the environmental issues. They are not necessarily personally sensitive but they cannot, as members of their professional group at least, escape their responsibility. Obviously, engineers are aware of this unique position. This is evidenced by the presence of environmental topics in major engineering conferences and in most training for more than 30 years.

In the United States, the first codes of ethics for engineers have existed since the beginning of the twentieth century. They have long concerned solely internal issues within the profession. The environmental issue first appeared in 1977 in the code of ethics of the American Society of Civil Engineers (ASCE), in a very modest way. New proposals to transform this recommendation into a stronger commitment in 1984 and 1995, met strong resistance from the profession. The 1996 version introduced a reference to sustainable development in canon 1 along with "their" definition of sustainable development (ASCE 1977, 2006).

> Engineers shall hold paramount the safety, health and welfare of the public in the performance of their duty (fundamental canon 1, ASCE code of ethics, 1976)[1]
>
> Engineers should be committed to improving the environment to enhance the quality of life. (provision set forth in paragraph (f) in the guidelines to practice for canon 1)
>
> Engineers shall perform services in such a manner as to husband the world's resources and the natural and cultured environment for the benefit of the present and future generations (canon 8, proposed in 1984 but not included)[2]
>
> Engineers shall perform services that help sustain the world's resources and meet long-term human needs, while protecting the natural and cultural environment (revised canon 8 proposed in 1995, but again not included)
>
> Engineers shall hold paramount the safety, health and welfare of the public and shall strive to comply with the principles of sustainable development in the performance of their duty (fundamental canon 1, ASCE code of ethics 1996)
>
> Sustainable development is the challenge of meeting human needs for natural resources, industrial products, energy, food, transportation, shelter, and effective waste management while conserving and protecting environmental quality and the natural resource base essential for future development. (definition adopted by ASCE in 1996)[3]

Environment has also been mentioned, since 1990, in the code of ethics of the world's largest engineering association by members: the Institute of Electrical and Electronics Engineers (IEEE 1990). It is still present in 2006, when the code was revised, with no change in the first article.

> We, the members of the IEEE, in recognition of the importance of our technologies in affecting the quality of life throughout the world, and in accepting a personal obligation to our profession, its members and the communities we serve, do hereby commit ourselves to the highest ethical and professional conduct and agree: (1) to accept responsibility in making decisions consistent with the safety, health and welfare of the public, and to disclose promptly factors that might endanger the public or the environment

---

[1] Adopted in 1976, le code was effective in 1977.

[2] This canon was proposed by the ASCE's Environmental Impact Analysis Research Council, but not proposed to the Board of Directors because the Professional Activities Committee voted against (ASCE 2006).

[3] In October 2009, the ASCE Board of Direction adopted a new definition: "Sustainable Development is the process of applying natural, human, and economic resources to enhance the safety, welfare, and quality of life for all of society while maintaining the availability of the remaining natural resources".

The first code of ethics of the European Federation of National Engineering Associations (FEANI), in 1992, includes an article dealing with environment (FEANI 1992). The version adopted shortly after by the French National Engineers Association (CNISF, today IESF) was more cautious. The "*Charte d'éthique*" which replaced in 2001 the "*code de déontologie*"[4] evokes environment in a more straightforward way in several articles.

> The engineer takes into account the health and safety of the public and contributes to environment protection in a reasonable manner (*une protection raisonnée de l'environnement*). (CNISF 1996, translation by the authors)

> The engineer is aware and makes the public aware of the impact of technical achievements on the environment. (CNISF 2001, art. 3)

> The engineer acts according to the principles of 'sustainable development'. (CNISF 2001, art. 4, translation by the authors)

From this evolution, we come to our research question: "to what extent have French engineers (not just their official spokespersons) adopted the view of the CNISF/ IESF Charter?" In a hypothetico-deductive approach, we propose to observe the relationships between dependent and independent variables corresponding to specific hypotheses that we seek to test or to invalidate (Popper 1973). So, what do we know about the engineering profession in France and what hypothesis can we formulate?

1. Engineers have a special position in the social and the economic world, also in the relationship between society and its natural environment. Their professional group is considered to bear responsibility for many environmental problems, and sometimes also for solutions: in both cases, engineers are supposed to know and be able to do what a laymen might not know or not be able to do, or to a lesser extent.
2. Previous research suggests that engineers are more optimistic than their fellow citizens about the social impacts of technology. Indeed, a survey "Engineering Science and Society" (ISS) conducted in 1999 showed that 68 % of French graduate engineers considered that technical progress brings more good than harm to humanity (2 % believe that progress brings more harm and 28 % that it brings almost as much harm as good) (Didier 2008). A survey by the Centre for the Study of French Political Life about science conducted at the same period shows that, among the French, more than half of the respondents considered that science brings as much good as bad and 45 % that it does more good than harm (CEVIPOF 2001).
3. The analysis of the codes of ethics promulgated in various countries shows an emergence, among the engineering profession, of a concern for environment, although it is prudent and rather late in comparison with the rest of society.

[4] Both expressions are translated by "code of ethics" in English, but the expression "*code de déontologie*" is usually reserved in France for a professional code which is legally binding.

So our first hypothesis is:

Hypothesis 1 The engineers' environmental attitude differs from those of the French and is marked by greater optimism vis-à-vis the environmental impact of technical development.

For the development of the second hypothesis, we start from the observation that the engineering profession in France has undergone profound changes over the last 20 years:

1. The engineering profession is still largely masculine. It is estimated that nationally, the proportion of women among practicing engineers is 17 %. But things are changing and the profession feminizes. The share of female graduates among engineers who are under 30, is estimated at 26 % (Darsch and Longuet 2011).
2. The flow of graduates increases and tends to accelerate. There are more and more young engineers (under 30). This may generate a generation – or age– effect, particularly since the environment issues are still quite recent.
3. The way to access to the engineering degree has evolved over the past 30 years. In 2010, 85 % of graduates obtained their grade through initial training, with an increase of students coming through the parallel admissions track, enabling university students to enter engineering schools; 11 % became engineers through continuing education and 5 % in apprentice status, which has been proposed since the 1990s only.

Hence our second hypothesis:

Hypothesis 2 as young people have been sensitized to environmental issues since their youth and women are supposed to be more sensitive to environmental issues than other members of our societies (because they are supposed to hold more holistic views), young engineers and female engineers express a greater sensitivity towards the environmental impact of techniques than their colleagues.

Finally, the ISS survey conducted in 1999, with French graduate engineers, highlighted strong links between the political and religious attitudes of respondents and their professional ethics (Didier 2008).

1. Practicing Catholics Engineers (22 % of respondents) appeared more sensitive to social issues but less sensitive to environmental issues, than their colleagues;
2. Left-wing engineers (26 % of respondents) seemed more concerned about the potential negative impacts of technology and they agree more often with the environmental and anti-nuclear movements than their colleagues; they seemed to have more confidence in the capacity of democratic debate to guide the country's technical choices.

So our third hypothesis is:

Hypothesis 3 The values that carry engineers – in the field of morality, religion and politics – affect their environmental attitudes.

To test our research hypotheses we have developed a series of questions that were included in the annual survey of the French National Council of Scientists and

Graduate Engineers conducted in April 2011. More than 39,000 graduate engineers responded to the general survey, and more than 27,000 to the optional module on the values that we designed. This part consisted of 50 variables dealing with opinions and behaviors on ethics, morals, religion and politics. Ten variables were explicitly devoted to environmental attitudes including the 6 among the 15 items which compose the "revised" New Ecological Paradigm. To make comparisons with the attitudes of French in general, we relied on the European Values Study (EVS) conducted in France in 2008.

Specific Traits of the Engineers' Environmental Attitudes

Although our investigation focuses on the contemporary period, it seems useful to recall the evolution of French public opinion about the environment over recent decades. Most investigations dealing with the French and the environment show a high stability in attitudes over the past 15 years. More than eight in ten French say they are very sensitive or rather sensitive to the environment. If we look closer at those who are very sensitive or if we use a finer indicator, such as the "deep sensitivity" (11 % of the French in 2001) we observe an overrepresentation in this category of executives, graduates and households with the highest incomes (Bigot 2002). Social status, education and income appear to be linked with environmental deep sensitivity. How about French engineers who are all graduates of higher education, benefit from rather high social status and come for many of them from well-off families?

The outcome of our research is that the graduate engineers' attitude towards environment differs from that of the other French people and that the variations we observed are highly significant (Table 13.1). But although engineers share the social characteristics of people with deep sensitivity to environmental issues, they don't share their opinion. Overall, engineers appear much more confident in "the ability of the genius of man to maintain our Earth viable", which is consistent with their training and profession. More surprising is that they reject more the idea the "destiny of man is to dominate nature" although they contribute to make this domination possible. Regarding the fragility of nature, which is evoked in items 2 and 4 (two items which are negatively correlated in the French population), the attitude of the engineers is again very different from that of the French. While an overwhelming majority of French (95 %) are concerned about the consequences of human activities and while only a small minority of them (16 %) believes that nature is able to cope with the damage, the engineers' opinion is divided on both issues (51 % agreement for the two items which are negatively correlated). Finally, the engineers' answers also differ from those of the French about the two items concerning the future (the inability to support population growth and the fear of the occurrence of a major ecological crisis): engineers are much less concerned by the occurrence of an environmental catastrophe if any change is made to the current development

Table 13.1 Agreement of the French and engineers with six items on environment

		French 2008	"Cadres" 2008	Eng. 2011
1	We are approaching the limit of the population number the earth can support (Overpopulation)	48	45	67
2	When humans interfere with nature it often produces disastrous consequences (Disaster)	95	93	51
3	Human ingenuity will insure that we do not make the earth unlivable (Ingenuity)	51	57	87
4	The balance of nature is strong enough to cope with the impact of modern industrial nations (Strength)	16	16	51
5	Humans were meant to rule over the rest of nature (Domination)	23	21	8
6	If things continue on their present course, we will soon experience a major ecological catastrophe (Catastrophe)	89	83	14

Table 13.2 Correlation matrix for the European values study using Somer's D (Bozonnet 2010)

French (EVS)	Variables					
Variables	Overpop.	Disaster	Ingenuity	Strength	Domin.	Catastr.
Overpopulation	1	0.20	-0.03	-0.04	0.03	0.16
Disaster	0.15	1	-0.08	-0.23	-0.16	0.34
Ingenuity	-0.02	-0.11	1	0.34	0.24	-0.19
Strength	-0.03	-0.28	0.30	1	0.35	-0.33
Domination	0.02	-0.21	0.23	0.38	1	-0.18
Catastrophe	0.14	0.40	-0.17	-0.32	-0.16	1

(14 % versus 89 %). However, they are much more worried than their fellow citizens by the impacts of population growth (67 % versus 48 %) (Tables 13.2 and 13.3).

So, the engineers' attitude differs from that of the French in general. It is marked by a strong optimism towards technical development and the strength of nature. But it also differs from that of the French "*cadres*" (executives). What is striking at first glance is the proximity of the answers given by the executives in the 2008 European Value Study with the answers given by the French in general, in the same survey. The executives trust a bit more than the other French on the ingenuity of man to solve environmental problems and believe a little less that an ecological disaster will come if nothing changes (with a rate which is very high compared to engineers). When comparing the engineers and executives, engineers appear significantly less pessimistic concerning major ecological disasters, and much more confident in human abilities. Not only the risk of disturbing nature appears less problematic to them (51 %, versus 93 % of executives), but they also believe more in the genius of man to keep Earth livable (87 % versus 57 % of executives). Much more than executives, they think there's more to worry about population growth (67 % versus 44 % of executives). Optimistic regarding techniques, they appear more pessimistic in the other fields.

The correlation matrix reveals important links between "Overpopulation", "Disaster" and "Catastrophe" on the one hand and between "Ingenuity", "Strength"

Table 13.3 Correlation matrix for the engineers survey using Somer's D (in the 2011 survey)

Engineers (IESF)	Variables					
Variables	Overpop.	Disaster	Ingenuity	Strength	Domin.	Catastr.
Overpopulation	1	0,10	0,15	-011	-0,23	-0,17
Disaster	0,11	1	0,25	-0,25	-0,17	-0,14
Ingenuity	0,08	0,12	1	-0,13	-0,25	-0,18
Strength	-0,12	-0,25	-0,27	1	0,32	0,30
Domination	-0,08	-0,05	-0,17	0,11	1	0,14
Catastrophe	-0,10	-0,07	-0,19	0,15	0,21	1

and "Domination" on the other. To agree with these last three items is to show confidence in the future and in humans' ability to "manage" the environment. However to agree to the other three items "Overpopulation", "Disaster" and "Catastrophe" is to demonstrate pessimism or at least anxiety. The negative correlation between the two groups of items means that not only do they constitute different elective universes but also, these worlds appear in opposition

The attitudes of the engineers are characterized by trust and optimism towards technical development. The millenarian discourse about the end of the world due to environmental catastrophes, where the disastrous consequences of human intervention seem decisive, has little effect on them. Their attitude towards environment is in clear dissonance compared to other occupational groups, including executives (*cadres*) which they belong to. It should also be noted that not only their opinions differ greatly from those of their fellow citizens, but the very structure of their environmental attitudes is different: two items that are the most linked among engineers are among those that repel most among French people: "The balance of nature is strong enough to cope with the impact of modern industrial nations" and "If things continue on their present course, we will soon experience a major ecological catastrophe".

A Low Correlation with Demographic Variables

Among the hypotheses that we have formulated, some concern the impact of demographic diversity on environmental attitudes. One can indeed wonder how feminization, rejuvenation and diversification of routes into the profession are likely to generate specific environmental attitudes?

While differences of position with respect to the items of the New Environmental Paradigm between men and women are not very significant for the French population as a whole, it is quite different among executives (Table 13.4). Indeed, for four items on the 6, there is a difference of more than 4 %. Thus, women executives believe less than men in the capacity of human ingenuity to maintain our Earth livable (−13 %). They show less agreement with the idea that "If things continue on their present course, we will soon experience a major ecological catastrophe" (−12 %). Finally they believe less than men that "the balance of nature is strong

Table 13.4 Answers to the NEP according to gender

	French sample EVS 2008			Sample of "Cadres" 2008			Graduate engineers 2011		
Overpopulation	Average	M	F	Average	M	F	Average	M	F
Disaster	48	49	48	45	45	45	67	67	68
Ingenuity	95	94	96	93	93	94	51	51	51
Strength	51	50	52	57	62	49	87	87	88
Domination	16	16	16	16	18	11	51	51	50
Catastrophe	23	24	22	21	23	19	8	9	7

enough to cope with the impact of modern industrial nations" and that "Human were meant to rule over the rest of nature" (−7 % for both items).

Overall, women executives seem to be less confident in the capacity of nature or of the human genius. This may explain that they believe less than men that the human destiny is to master nature and they fear more the possibility of a technological disaster. One explanation may be advanced. Women who work as executives are better educated than other women. They may have a more critical eye over the relationships between humans and nature and take more distance from the dominant model valued by males. These results lead us to believe that the acquisition of a higher social and cultural status allows women to situate themselves in terms of environmental attitudes outside the dominant male model. This explanation is consistent with many studies on the importance of work in the emancipation of women.

However, the next step of our analysis provides more surprises. Indeed, the differences between the attitudes of men and women disappear completely when analyzing the data from the engineers' survey. The difference between male and female engineers varies up to a maximum of 3 % (for the item "disaster") and the average variation is 1.33 (for all the items)[5] which is less than the variation for the entire population (1.83) and much less than that of the *cadres* (6,16). In other words, while belonging to the professional group of *cadres* generates different environmental attitude depending on gender (women appear less confident in human ingenuity and in the strength of nature, more dubious about the mission of human to dominate nature and more aware of the risks of an environmental catastrophe), belonging to the engineering profession annihilates this gender difference. Within the engineering profession – with a high education and techno-scientific expertise, and largely male – the difference of opinion regarding indicators of NEP disappears between male and female. Of course, the elements of explanation are plural and it is not ours to decide. However, we are inclined to believe that the engineers' workplace influence and the predispositions for science that led them to undertake engineering studies are two important explanatory factors of female engineer professional identity – and ethos.

If gender generates little difference among engineers with respect to environmental attitudes, age proves slightly more discriminating. Younger engineers appear

[5] The average is calculated from the absolute differences.

a bit more sensitive to environmental issues. This is also what sociologist Jean-Paul Bozonnet observed by analyzing the effects of age on the responses to NEP in the French part of EVS in 2008 (Bozonnet 2010a). Younger people are more skeptical about the balance of nature (48 % under 30 believe that the balance of nature is strong enough to cope with industrial damage, versus 58 % of those over 60). They agree less than their elders with the statement that the destiny of humans would be to dominate nature (8 % versus 11 %). However, they are less worried about the possibility of a major disaster (13 % under 30 versus 19 % over 60). Overall, even if the amplitude of the variations is low, the engineers under 30 appear more concerned about environment than the engineers who are older than 60. Because they were born at the same time as the concept of sustainable development and its widespread distribution, they were sensitized early to the environmental issues, which have been debated a lot in the public arena. Concerning the social diversification of the engineering population linked to the multiplication of access ways into the profession, we just note that the engineers who graduated from the most prestigious schools (called group "A+"[6]) differ from their congeners. They are less trustful than the other engineers in the ability of human ingenuity to insure that we do not make the earth unlivable, but are more confident in the capacity of nature to regenerate itself. In addition, they agree more often than the other engineers that if nothing changes, a major ecological catastrophe could occur (Table 13.5).

Regarding the influence of the type of educational route, variations are not very significant (except for items "Strength" and "Catastrophe"). We note however that the attitudes of the engineers who entered engineering school through parallel admission (i.e. after a first degree at university rather than after a preparatory class) are opposite of those of graduates from the most selective schools (A+). A link seems to appear between the symbolic hierarchy of the engineering schools and the "best way" to get into engineering education and the graduates' environmental attitudes. The more their study profile approaches the traditional and historical "best way" *(voie royale)* to the diploma (i.e. scientific preparatory class followed by an engineering program in a A+ school), the less they seem concerned about nature, the more they trust in human ingenuity to keep the Earth habitable and paradoxically, the more they are concerned about the occurrence of an environmental catastrophe.

Thus, neither the gender of the respondent, nor their age, appears to be factors that explain their environmental attitudes. We observe, however, some variation related to the type of engineering educational track, even if they are not all significant. Hypothesis 2 is therefore not confirmed.

[6] In the annual list published by the magazine *L'Etudiant*, the group of engineering schools called "A +" is composed mainly of Parisian very prestigious schools. They represent 18 % of the engineering students population.

Table 13.5 Respondents' agreement with the 6 NEP items

	A+[a] (Eng. school	*Prepa* before school	Aver-age	Other Eng. School	Parallel access to Eng.
We are approaching the limit of the population number the earth can support (Overpopulation)	68	68	67	67	67
When humans interfere with nature it often produces disastrous consequences (Disaster)	50	51	51	52	53
Human ingenuity will insure that we do not make the earth unlivable (Ingenuity)	80	86	87	88	88
The balance of nature is strong enough to cope with the impact of modern industrial nations (Strength)	58	53	51	49	47
Human were meant to rule over the rest of nature (Domination)	10	9	8	8	7
If things continue on their present course, we will soon experience a major ecological catastrophe (Catastrophe)	19	15	14	13	12

[a] In this table, "A+" and "Other engineering schools" refers to the question "from what school did you graduate?" where we distinguish the graduates from the most prestigious schools from the others. "*Prepa* before school" and "parallel access" refers to the question "what was your training before entering the engineering school?". We have grouped graduates who went through a traditional or an integrated preparatory class from those who have entered the engineering degree in other ways (i.e. after a first cycle at university, or after a 2 years programs in a technical school). 24 % of the students belong to this second group

The Engineers' Environmental Attitudes and Their Others Values

Is the way people conceive good and evil related with the world of environmental attitudes? In our survey, a question was asked about moral attitude.[7] This variable provides enough evidence of correlations with environmental attitudes. Within the population of engineers, 15 % believe that in moral matters there are clear lines that are valid in all situations ("hardliners"), 62 % think it depends on the circumstances ("conditional") and 23 % are not found in either of the two proposals ("moderate"). The "hardliners" are less alarmist against the risk of overpopulation (63 % versus 68 % of all engineers and 48 % of French) and more likely to agree with the idea that the destiny of Man is to dominate nature (13 % versus 8 % of engineers and 23 % of French). Their view is close to that of the French in general about these two

[7] The question the respondents had to answer, was: *"Here are three statements which people sometimes make when discussing good and evil. Which one comes closest to your own point of view? (1) There are absolutely clear guidelines about what is good and evil. These always apply to everyone, whatever the circumstances. (2) There can never be absolutely clear guidelines about what is good and evil (3) I disagree with both statements."*

Table 13.6 Moral type, religious attitude and the NEP items

	Overpop.	Disaster	Ingenuity	Strength	Domin.	Catastr.
Average	68	52	87	51	8	14
Hardliner	63	45	84	60	13	23
Moderate	71	48	84	49	7	14
Conditional	68	54	88	50	8	12
Religious	64	43	86	59	10	21
Non religious	69	54	88	51	7	11
Atheist	70	58	86	44	8	11
French	**48**	**95**	**51**	**16**	**3**	**89**

items (Table 13.6). However, they have different opinions about the strength of nature (60 % think it is strong enough to compensate for the industrial damage versus 51 % of engineers and 16 % of French) (Table 13.6).

The engineers' religious attitude generates differences of environmental attitudes. Engineers who define themselves as being religious are less concerned than other engineers by the risk of overpopulation. They are also less concerned about the environmental risks that may result from human actions than the "non-religious" and even less than "atheists". They are more likely to believe that nature is strong enough to compensate for the damage caused by the industrialized countries than "non-religious" and even more than "atheists" (59 % versus 50 % and 44 %).[8] They are however, more sensitive to the risks of a major ecological disaster (21 %) than "non-religious" and "atheists" (11 %). It should be noted that these four environmental variables, the correlation with the subjective religious feeling is greater than the previous one about moral attitude and far higher than the demographic criteria.

The choice between freedom and equality is highly correlated with people's symbolic universe.[9] Engineers who prefer "equality" (and represent 45 % of the population) are much more sensitive to the risk of overcrowding that those who value more "freedom" (76 % show concern versus 61 % of the "pro- freedom"). They give slightly more credence to the ingenuity of man to solve environmental problems and have less confidence in the soundness of Nature (45 % versus 57 %). Rejecting the idea that the destiny of man is to dominate nature, they are also less pessimistic than the average about the risk of ecological disasters. They promote an "ecological discourse" based on greater solidarity and human intervention in the process of evolution of the planet. They seem both more concerned over the current

---

[8] This result is consistent with the trends of greater technical optimism among practicing Catholics engineers compared to other engineers in the ISS survey conducted among engineers in northern France (Didier 2008, p. 160; Didier 2009).

[9] The question the respondents had to answer, was:"*I find that both freedom and equality are important. But if I were to choose one or the other: (1) I would consider personal freedom more important, that is, everyone can live in freedom and develop without hindrance (2) I would consider equality more important, that is, that nobody is underprivileged and that social class differences are not so strong; (3) I don't know.*"

Table 13.7 Political attitude and the 6 NEP items

		Overpop.	Disaster	Ingenuity	Strength	Domin.	Catastr.
Average		68	52	87	51	8	14
Equality		76	52	89	45	6	11
Freedom		61	51	85	57	11	17
Engineers should engage	++	74	48	87	54	8	17
	+	69	53	87	52	8	14
	no	47	53	81	48	14	16
Political interest	+, ++	69	50	85	54	9	17
	−,—	66	54	90	48	7	11
French		**48**	**95**	**51**	**16**	**3**	**89**

situation and future but also more confident in the ability of man to face the situation (Table 13.7).

The most politicized engineers are less often than the average concerned about the risks of natural disasters caused by human activity. They are also less confident in the genius of human to protect the environment but believe more in the ability of Nature to compensate for the errors caused by the industrialized countries. Finally, they are much more likely to believe in the possibility of a major ecological disaster (17 % versus 11 % of those who reported "little" or "not interest at all" for politics). Moreover, a large majority of engineers who responded to the survey agree with the statement that "the engineer must commit to a transformation of society" (85 %, including 21 % who say agreed "strongly"). These "pro-commitment" engineers show a strong concern about population growth (74 % versus 47 % of engineers who answered "no" to this question) and to a lesser extent, they have confidence in the ability of nature to absorb damage due to industrial development (54 % versus 48 %). Moreover, they believe a little more than the average engineer in the ingenuity of Man (87 % versus 80 %) and do not agree at all with the statement that the destiny of man would be to dominate nature (8 % versus 14 %).

On a number of aspects, one could highlight the influence of the religious attitudes of respondents on their environmental attitudes. Variations also exist when considering the criteria of political interest and commitment of engineers to transform society. The engineers who are more interested in politics – and those who advocate greater involvement of the profession in the res publica – have a conception of the relationship of men to the environment which differs from other engineers. They believe that the balance of nature is strong enough to withstand industrial damage while worrying about the possibility of a major ecological catastrophe "if thing continue on their present course". Overall, subjective criteria seem most relevant to explain the different environmental attitudes within the engineering profession that demographics. They offer an intensity of correlation two to three times higher than the demographic criteria. From this fact, we can conclude that the ethical stance, the preference of the respondents for liberalism or egalitarianism, as well as their religious and political attitudes are important variables to take into account to understand the environmental attitudes of French engineers. Hence, hypothesis 3 is confirmed.

Conclusion

Our research question finds, at the conclusion of this article, an affirmative answer. All hypotheses lead to conclude that environmental dynamics occur at different levels. On the one hand, engineers, differ about environmental attitudes from both the average French and the executives. On the other hand, engineers are driven by values. The different dynamics at work, not exclusive of each other, have their genesis in a series of factors – more endogenous than exogenous – probably joining them in a complex manner to form a symbolic system capable of structuring intensely the universe of representations, beliefs and behaviors related to the environment. We believe we have helped to show some kind of pluralism in a profession dominated by the scientific paradigm, and often seen as homogeneous or monolithic.

Regarding the explanation of the singularity of the attitudes of engineers, the analyses presented here have eliminated assumptions rather than offered immediate response. The lack of gender variation leads us to say that the singularity of the engineers' attitude cannot be explained by the strong masculinity of socio-professional group (83 % men). Another outcome is that the feminization of the profession is unlikely to produce a transformation of the environmental attitudes of the engineers, because their attitudes do not seem to differ from those of their colleagues on that topic. The strong correlations between moral attitudes, religious and political engineers and their environmental attitudes made us update a relative pluralism within the profession. They do not allow us to advance causal explanations of environmental attitudes. Finally, a draft analysis of the respondents' attitude according to their type of engineering education – although not at the heart of this work – opens up new avenues of research. It may contribute to better understand the profession's environmental attitudes.

References

ASCE (American Society of Civil Engineers). (1977). Code of ethics.

ASCE (American Society of Civil Engineers). (2006). *Development of sustainability provisions in ASCE code of ethics*. http://www.asce.org/Publications/ASCE-News/2008/11_November/A-Question-of-Ethics/

Azapagic, A., Slobodan, P., & Shallcross, D. (2005). How much do engineering students know about sustainable development? The findings of an international survey and possible implications for the engineering curriculum. *European Journal of Engineering Education, 30*(1), 1–19.

Bigot, R. (2002). *L'opinion et les comportements des Français en matière d'environnement*. Paris: Centre de recherche pour l'étude et l'observation des conditions de vie.

Bozonnet, J.-P. (2010a). L'écocentrisme, un grand récit protestataire, mais faiblement engagé. In P. Bréchon & O. Galland (Eds.), *L'individualisation des valeurs* (pp. 119–140). Paris: Armand Colin.

Bozonnet, J.-P. (2010b). Short new ecological paradigm in 30 European countries: First results. Gothenburg: Sociology on the move. XVII ISA World Congress of Sociology.

Bréchon, P., & Galland, O. (Eds.). (2010). *L'individualisation des valeurs*. Paris: Armand Colin.

Brundtland, G. H. (1987). *Notre avenir à tous, Rapport de la Commission mondiale sur l'environnement et le développement de l'ONU.* Oslo: United Nation.

Carson, R. (1962). *Silent spring*. Boston: Houghton Mifflin.

Catton, W. R., Jr., & Dunlap, R. E. (1978). Environmental sociology: A new paradigm. *The American Sociologist, 13*, 41–49.

CEVIPOF (Centre d'Étude de la Vie Politique Française). (2001). *Images de la science*. Paris: Cevipof.

CNISF (Conseil National des Ingénieurs et Scientifiques de France). (1996). Code de déontologie. Paris: CNISF.

CNISF (Conseil National des Ingénieurs et Scientifiques de France). (2001). Charte d'éthique de l'ingénieur. Paris: CNISF.

Darsch, C., & Longuet, L. (Eds.). (2011). *Observatoire des ingénieurs. Enquête 2011*. Paris: Ingénieurs et scientifiques de France.

Didier, C. (2008). *Les ingénieurs et l'éthique*. Paris: Lavoisier.

Didier, C. (2009). Religious and political values and the engineering ethos. In S. H. Christensen, M. Meganck, & B. Delahousse (Eds.), *Engineering in context* (pp. 417–434). Aarhus: Academia.

Dunlap, R. E., & Catton, W. (1979). Environmental sociology. *Annual Review of Sociology, 5*, 243–273.

Dunlap, R. E., & Van Liere, K. D. (1978). The "new environmental paradigm": A proposed measuring instrument and preliminary results. *Journal of Environmental Education, 9*(1), 10–19.

Dunlap, R. E., Van Liere, K. D., Mertig, A. G., & Jones, R. E. (2000). Measuring endorsement of the new ecological paradigm: A revised NEP scale. *Journal of Social Issues, 56*(3), 425–442.

FEANI (Fédération Européenne des Associations Nationales d'Ingénieurs). (1992). Code de conduite. Brussels: FEANI.

IEEE (Institute of Electrical and Electronics Engineers). (1990). Code of ethics. http://ieee-ies.org/resources/media/about/history/ieee_codeofethics.pdf

Lagadec, P. (1981). *La civilisation du risque*. Paris: Seuil.

Organisation des Nations Unies. (2002). *Rapport du Sommet mondial pour le développement durable*. Johannesburg: UNO.

Popper, K. (1973). *La logique de la découverte scientifique*. Paris: Payot.

Christelle Didier B.S. in Electrochemistry Engineering, M.A. in Education, Ph.D. in Sociology from Ecole des Hautes Etudes en Sciences Sociales (EHESS), Paris. From 1993 to 2013, Assistant Professor, Lille University, France, Ethics Department. Assistant Professor, Charles de Gaulle University of Lille, Education Department. Member of CIREL (EA 4354). Co-author of *Ethique industrielle* (DeBoeck, Brussels 1998), author of *Penser l'éthique des ingénieurs* (PUF, Paris, 2008) and *Les ingénieurs et l'éthique. Pour un regard sociologique* (Hermes 2008). She has published many articles on ethics and social responsibility in the engineering profession and education, and on the engineering profession's values (from interviews and extensive surveys). Research areas: engineering ethics and values, including historical, cultural and gender perspective, sustainable development and corporate social responsibility, social responsibility.

Kristoff Talin M.A. in Sociology and Ph.D. in Political Science. He is Researcher At the French National Center for Scientific Research (CNRS), co-director of the Research Center in Sociology and Economy of Lille, France (CNRS Laboratory, UMR 8019). He has published many articles, and two books, on individual values in a comparative perspective. He is a specialist of quantitative analysis. Research areas: engineering values, including political and social perspective.

Chapter 14
Engineering as a Technological Way of World-Making

Sylvain Lavelle

Abstract In *Ways of Worldmaking*, Goodman examined the various ways of making worlds by comparing the activity of science with that of art. It is however regrettable that he did not regard the activity of engineering, thus viewed as a kind of technical art, as a genuine *way of worldmaking*. Engineering not only deals with the design and the production of technical artefacts but is also concerned with their use in as far as it tends to 'make a world' of a certain kind. One of the main problems in the philosophy of engineering is precisely to determine the nature of the relations between design and production on the one hand and use on the other. Thus the examination of the engineering way of worldmaking leads to focus on the modalities of human action (necessity, obligation, possibility, permission, etc.) in a world designed and produced by and through technology. In this respect, the constitution of some technological networks and frameworks calls for looking at the web of modalities of human agency as entailed by the engineering process.

Keywords World-making • Technical artefacts • Constitution • Modalities

Introduction

Winston Churchill once said: "We shape our dwellings thereafter our dwellings shape us" (Winston Churchill, House of Commons, October the 28th, 1948). He thus suggested that we, humans or subjects, are the actors by which things or objects are being designed and built. But then, the objects are designed and used by humans in such a way that they come to shape their daily way of living and behaving, if not their way of feeling and thinking.

One could enlarge Churchill's statement and consider the way the objects we design, produce and use are the same around the world and shape human conduct

S. Lavelle (✉)
Centre for Ethics, Technology and Society, Department of Humanities,
ICAM Paris-Sénart Engineering School, 2 Allée des Savoirs,
77127 Lieusaint-Sénart, France
e-mail: sylvain.lavelle@icam.fr

S.H. Christensen et al. (eds.), *Engineering Identities, Epistemologies and Values*, Philosophy of Engineering and Technology 21,
DOI 10.1007/978-3-319-16172-3_14

and thought in the same way. This raises the question of making the world identical in all its parts through the expansion of material and social standards applying to a set of technical artefacts. The latter at a certain level of systemic integration are assumed to be related to one another in a network but also to function as a framework on a global and local scale. The idea of a standard world that materially and socially speaking would be more or less the same in all its parts is frightening for some, or reassuring for the others. It echoes Friedman's myth of the "Flat World" and leads to ask the question of the standardisation of objects over the whole planet then leading to a single standard world (Friedman 2007).

Technology in general usually designates an activity that encompasses design and production of potentially useful artefacts, while engineering in particular designates an activity that is one part of the technology process, though an essential one. The use of technical artefacts in human business is the common aim of the technology process, while the transformation of the natural and human world can be viewed as its common end. It can be said that technology especially in the sense and in the form of engineering equals making a world or, to use Nelson Goodman's expression, is a *way of worldmaking*. Unfortunately, the philosopher who authored the expression "worldmaking" confined it to science and art in the aesthetical sense ("fine art") and he did not expand it to art in the technical sense ("useful art"). It is then valuable to examine to what extent technology in general and engineering in particular can be taken to be a way of worldmaking in Goodman's sense, or possibly in another sense.

The "making of a world" refers to the way technical artefacts as designed by engineers are arranged and connected and then form a network and a framework for humans. It is hardly disputable that engineering as a dynamic process of worldmaking concerns not only the design and the production of technical artefacts, but also their use. In addition, engineering as a creative and productive process is twofold and combines a descriptive-factual part together with a prescriptive-normative part. The classical opposition in philosophy between the "Is" ("what things are") and the "Ought" ("what things ought to be") is also relevant for the engineering way of worldmaking (Lavelle 2006). Technology as engineering indeed can be viewed as far as its ordinary process is concerned as an activity that is both science-based and society-oriented. Hence a set of questions: (1) What is it to "make a world"? (2) In what sense can engineering be considered as a "way of worldmaking"? (3) Can the notion of technological worldmaking be expressed in terms of "constitution" and "modalities" (necessity, possibility, etc.)?

Ways of Worldmaking

The "World" is quite vague a word and is commonly used in order to designate the set of things and beings that populate our environment. The question of the world definition and delineation (what it is, where it begins, and where it ends) makes it quite obvious that a world is not something very definite and delineated. A world can be my home, my garden, my family, my school, my city, my territory, my district, my country, and possibly all of that to be viewed as a whole. A world can be natural, material, mental, social; it can also be small (a *microcosm*), big (a *macrososm*), or something in between (a *mesocosm*), depending on the scale you choose. In addition, a world is not only something that is thought over, it is also something that is made up, hence the notion of worldmaking.

What a World Is

The World as a whole is no doubt an *Idea* in the critical sense that Kant gives to this word, and certainly not a *Concept* that could be related to a set of empirical data (Kant 1787, 2008). Kant used to define the world as the set of all phenomena and in the transcendental sense as the absolute totality of the whole set of existing things. In this respect, the World is something that can be thought about, but not strictly speaking something that can be fully known, for it would exceed the capacities of human cognition and experience. Imagine if someone were to ask you to list the set of elements that exist in the World: a full life would not suffice for this Herculean task.

"The world is the totality of facts, not of things", the "philosopher-engineer" Wittgenstein said (Wittgenstein 1998); yet the world of the engineers is a world of things, not facts (McCarthy 2010). A world is not merely something that is thought about by human beings: a world is basically something that is *made* up on the basis of some criteria that helps determine more or less precisely the bonds of a world. One can hardly imagine what a world would be without any human determination of its boundaries, aspects, elements, conditions, etc., which means that a world is always a "world-for-someone". Otherwise, if it is not a 'world-for-someone', the world is just a piece of space and time, in other words, some location as related to some duration. Nevertheless, it is not easy to clarify what "making a world" actually means when it comes to technology taken as a technical art, as compared to science and to art in the aesthetical sense.

Let us imagine that an old man, John, once an aviator on a flight tanker, creates a wonderful garden at home so nicely arranged that he spends almost all of his time in there. When he used to be an aviator on a flight tanker, he spent most of his time outside his home, and the world for him was not only his plane and his boat, his sleeping and dining rooms, his navy colleagues, but also the sky, the sea and to a lesser extent, the earth. More precisely, the world was the piece of sky and the piece of sea that he would daily fly through or over and experience in its various aspects (shape, colour, temperature,…). After he retired, the world actually meant something like his home and his garden more than his plane, his boat, or the sky and the sea in some exotic regions. Of course, since he experienced on a daily basis what flying through the sky and over the sea is about, he has kept in his memory some remainders of that time that help him figure out what and how the world is. So he is quite aware that the world is not bound to his home and his garden, and if he would not experience it, at least, he would be taught and informed about it since his very childhood. But the fact is, once retired, he has no longer experienced flying through the sky and over the sea, and what now he does experience about them are the pictures and the comments he can watch and listen to on his television. This is sometimes a very good substitute to a certain personal experience of a world, but even if the TV programme is well done, it does not provide a direct active experience of it.

Basically, what makes one think that *a* world is *the* world (for instance, a single planet) is the common version that is given when using a map and the fact that this version is confirmed by some other versions (geography books, Internet maps, TV programmes, etc.). But *the* world is actually always *one* world, or more precisely some pieces of the world, even if one has several versions possible of his or her world at hand.

Wholes, Parts and Relations

One can propose several possible definitions of the world and insist either on the conceptual aspects or on the experiential aspects of it. The philosophical view on the world encompasses the phenomenological notion as well as the logical notion of it, alike the *lifeworld* in Husserl, or the *world of things* in Carnap (Ryckman 2007). Both insights and approaches, though supposedly opposite, refer in different proportions to the material-natural or the social-cultural immediate and ordinary environment to which humans relate and in which they live.

A world in the *logical* definition can be *in extension* the indefinitely open series of existing things and *in intension* the general shape, the overall aspect of reality. However the logical concept of "world" is in a way too formal if one considers the world as it is experienced by human beings. One can then shift to a *phenomenological* definition: a world is a set of intentional relations (beliefs, desires, expectations, affects, etc.) of humans to things and beings that constitute their experience as familiar or unfamiliar. A world can be viewed as a set of regular possibilities for cognition, volition and action enabling someone to experience certain things and events and, more broadly, to experience a certain way of life (Ihde 1990).

However, a world is not a mere set of elements, but mostly refers to the specific *connections* and *arrangements* in terms of relations, structures and situations that one can identify among those elements. For instance, *World 1* can be different from *World 2* for they have different elements (*a b c* for *World 1* and *a b d* for *World 2*), or because they have different relations between the same elements (*a b c* for *World 1* and *a c b* for *World 3*):

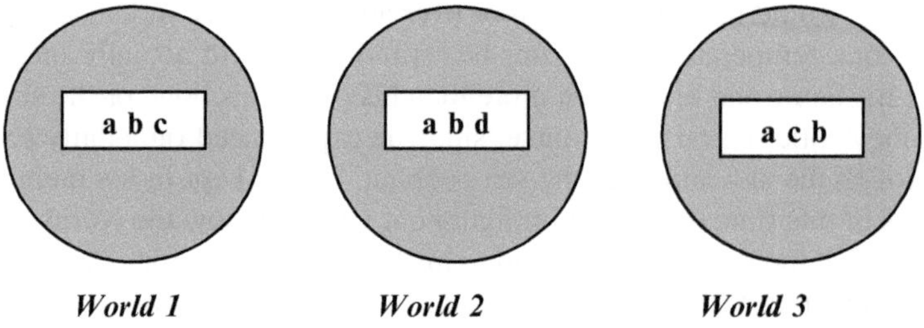

Worlds, elements and relations

The World as a part is possibly easier to delineate than the world as a whole. For instance, if you compare the North and the South of the Earth, you will find different worlds on the basis of their components:

- World 1 (Europe): oaks, apples, wheat, foxes, dogs, cats, cars, boats, planes, computers,...
- World 2 (Africa): baobabs, bananas, sorgho, lions, elephants, hyenas, cars, boats, planes, computers,...

Of course, you would find lions and elephants in the numerous zoos of Europe, and conversely, you would find dogs and cats as pets in Africa. But the important point is the population of things or beings in each area and the relationship to those things and beings as part of our world. If you take the example of bananas, you will find them quite freely in Europe, although this fruit does not grow there, and this is due to the technical organisation of collecting, freighting and distributing products. In this respect, technology has changed our world to the point that we can buy bananas, a fruit from outside Europe, every day and everywhere inside Europe, almost as if it were a local fruit. However, it is one thing to say that, in both worlds, you can find computers, it is another thing to say that, in one world, 80 % of the population uses a computer on a daily basis, whereas, in the other world, 10 % of the population uses a computer on a daily basis. In both cases, it is the same world component (a computer), but it is not the same world, for the relationship to the world component is not the same: it is an ordinary tool in one world, but it is a marginal machine in the other world.

Moreover, the notion of world does not refer to a static reality, but rather to a *dynamic process* of change in which new things or new beings come into play. Thus, one can compare World 2 in its Version A and World 2 in its Version B:

- World 2 Version A (Africa): baobabs, bananas, sorgho, lions, elephants, hyenas,...
- World 2 Version B (Africa): baobabs, bananas, sorgho, lions, elephants, hyenas + *cars, boats, planes, computers,...*

There are indeed several possible versions of the world, as suggested by Goodman, and that depends upon the selection we operate regarding the world's elements and their relations.

Making a World

It is the merit of Goodman who authored the expression *way of worldmaking* to set up a perspective on how humans make a world, be this world that of science or that of art. The question of the unity or plurality of the world according to Goodman cannot be examined without considering in advance the criteria that give birth to a unique structure. For it is from these criteria that versions of the world are developed by everyone in a more or less aware manner.

In James' pluralism, the world is made of several *parts* (James, *A pluralistic Universe* 1909, 1996); in Goodman's pluralism, the world is made of several *versions*, based upon a variety of *frames of reference* (*Ways of Worldmaking*). For instance:

1. "Under frame of reference A, the sun always moves".
2. "Under frame of reference B, the sun never moves"

As Goodman (1978, pp. 2–3) suggests:

> Frames of reference…seem to belong less to what is described than to the system of description: and each of the two statements relates what is described to such a system. If I ask about the world, you can offer to tell me how it is under one or more frames of reference; but if I insist that you tell me how it is apart from all frames, what can you say? We are confined to ways of describing whatever is described. Our universe, so to speak, consists of these ways rather than of a world or of worlds.

For Goodman, there is no doubt that there is a plurality of versions of the world for there is a plurality of ways of structuring aspects of the world whose meanings are interpreted and valued in different ways for different individuals. Versions of the world that Goodman speaks about are like some unique perspectives from a framework that is specific to an individual. Goodman does not claim that it is impossible to produce any convergence between versions of the world by different individuals. But he argues that it is unlikely that these versions do not differ in at least one aspect, be it a minor one.

For Goodman, there is no neutral world prior to the human activity and language and modes of organization of our existence are not found in the world, but built to make a world (Goodman 1978, p. 20):

> The fact that there are several different versions of the world is hardly debatable. The question seems virtually empty know how there are world-in-themselves if any…We might… take the real world to be that of some one of the alternative right versions (or groups of them bound together by some principle of reductibility or translatability) and regard all others as versions of that same world differing from the standard version in accountable way. The physicist takes his world as the real one…the phenomenalist regards the perceptual world as fundamental…For the man-in-the-street, most versions depart in some ways from the familiar serviceable world he has jerry-built from fragments of scientific and artistic tradition and from his own struggle for survival. This world, indeed, is the one most often taken as real; for reality in a world, like realism in a picture, is largely a matter of habit. Ironically, then, our passion for one world is satisfied, at different times and for different purposes, in many different ways.

Thus, according to Goodman, versions of the world are symbolic systems which may have different forms and be expressed in words, sounds, images, dances and all sorts of symbols. Worlds or versions of the world are made from symbols for man is an animal whose language is flexible, who makes the world with words and who composes reality through language.

The making of a world can be achieved through several operations:

- *Composition* and *decomposition*: we gather and cut into parts by which we divide the existing worlds into sub-worlds, and we analyze their features and build new relationships until we can combine parts of it in a new way.

- *Weighting*: we weigh the differences between the worlds that can depend only on the greater or lesser emphasis, so that the relevant parts in the world cannot be in another world, and vice versa.
- *Ordering*: we order and group objects before incorporating them, and these ways are built to order the world, they depend on objects and contexts that we consider.
- *Deletion* and *supplementation*: we often remove old material, add new material and we collect and add the parts we need.
- *Deformation*: we reform the world or destroy its original form then the re-formations can be viewed as corrections or corruptions.

One of the critiques of the version-based philosophy of worldmaking pointed out that Goodman oscillated between several meanings of the term "world" of which he never attempted to give a definition (Scheffler 1980). One could add to this critique a world is not only a version (a *mental-linguistic system*) but also a set of concrete connections and arrangements of things (a *material-social system*) that can be designed and produced by a technical work.

Engineering as a Way of Worldmaking

It might be that technology as an engineering process is a specific way a worldmaking in that it not only provides some intellectual or artistic *versions* of the world, but also some *fashions*, or material and social shapes. In that sense, "worldmaking" in engineering refers less to a linguistic approach like in Goodman's than to an instrumental capacity of fixing and changing the general and particular material and social shape of the world.

Minds, Matters and Acts

The version of a world as made by engineering is not just what you bear in mind alike some scientific or artistic pictures or images (*imago mundi*). It is actually more about what mind puts into *matter* through the mediation of some human acts in order to give things their structure, their function and to some extent, their significance. In this respect, as suggested by Natasha MacCarthy in her comment on Wittgenstein's *Tractatus logico-philosophicus*, the world of technology is "a world of things, not facts": "Engineering is a practical pursuit, ultimately focused on the real world, not the idealized conditions explored in the lab or the armchair. Its very nature and purpose requires that engineering deals with complexity, contingency and context" (McCarthy 2010).

For instance, if one takes the example of information and communication technologies, on can distinguish different frames or versions, as suggested by Goodman, although the technological equipment is the same:

- In *Frame A* ("information society and global village"), the use of computers as connected and arranged in so making an information network and framework for the users is viewed as making a *world of communication.*
- In *Frame B* ("control society and global war"), the use of computers as connected and arranged in so making an information network and framework for the users is viewed as making a *world of alienation.*

Examples of frames of references

Frame A
"information society and global village"

computer systems + information networks/frameworks = a world of communication

Frame B
"control society and global war"

computer systems + information networks/frameworks = a world of alienation

In both worlds, the technological equipment is the same, but the frame on the basis of which the actors give some significance to them in terms of "worldmaking" is really different (Lavelle 2007).

The problem in this version-based approach to worldmaking as suggested by Goodman is its *idealistic tropism* that tends to neglect the realities of technical artefacts to be viewed as a set of concrete material things (Kroes 2012). A world is not just a scientific or an artistic version, it is also a material and social *fashion* not reducible to a mere worldview. Both material and social shaping of the world though technology and engineering produces some consequences on the *actions* of human beings in terms of gain and loss of capacities, in terms of knowledge, action, will, experience and significance (Latour 2005).

For example, one can take the example of what using ICT's as designed and produced with the help of engineers actually entails in terms of gain and loss of capacities for the users:

Technological shaping of human capacities in the case of ICT's

	Knowledge	Action	Will	Experience	Significance
Gain	Wide access to information	Interactive behaviour	Curiosity for the world	Open inquiry and discussion	Power and ubiquity
	Diversity of sources	Reflex of web surfing	Will to participate	Involvement in networks	Sense of the present
Loss	Lack of validation	Media-based activity	Problems with effort/desire	Reduction of reality	Anonymity and loneliness
	Patchwork culture	Less handwriting	Need to be connected	Decline of direct contacts	Loss of the past and the future

As Langdon Winner (Winner, 1986) suggests, technologies are ways of materially and then socially ordering our world and our actions in space and time:

> The things we call 'technologies' are ways of building order in our world. Many technical devices and systems important in everyday life contain possibilities for many different ways of ordering human activity. Consciously or unconsciously, deliberately or inadvertently, societies choose structures for technologies that influence how people are going to work, communicate, travel, consume, and so forth over a very long time. In the processes by which structuring decisions are made, different people are situated differently and possess unequal degrees of power as well as unequal levels of awareness. By far the greatest latitude of choice exists the very first time a particular instrument, system, or technique is introduced. Because choices tend to become strongly fixed in material equipment, economic investment, and social habit, the original flexibility vanishes for all practical purposes once the initial commitments are made. In that sense technological innovations are similar to legislative acts or political foundings that establish a framework for public order that will endure over many generations. For that reason the same careful attention one would give to the rules, roles, and relationships of politics must also be given to such things as the building of highways, the creation of television networks, and the tailoring of seemingly insignificant features on new machines. The issues that divide or unite people in society are settled not only in the institutions and practices of politics proper, but also, and less obviously, in tangible arrangements of steel and concrete, wires and semiconductors, nuts and bolts.

This view on technology and the kind of bounds they impose upon humans could also be well illustrated by a concrete situation in another space-time scale. You are an engineer working in a leading world industry: can you refuse to use a computer? "Can" here means several things: (a) Do you have the material possibility to do it? (b) Do you have the social possibility to do it? (c) Do you even have the mental possibility to do it? Technological worldmaking characteristically entails this kind of "modal" situation in which human actions are materially and socially necessary, possible or impossible (Lavelle 2009).

Technology, Engineering and Worldmaking

An overview of the technological process will hold together several stages ranging from design and production to the use of technical artefacts. However, the problem at stake is not so much the ever relative difference between each category as the difference inside each category. Thus one can figure out that the category of designers may count some design people in the technical but also in the aesthetical sense, while the category of producers would include engineers and operators. As to the users, it could be relevant to make a difference between sub-categories such as: (a) *primary users* of technical artefacts who use them as a network but also as a framework for the action of (b) *secondary users* of technical artefacts who benefit from these technical artefacts, but who also undergo the organisation of their world and their lives on the basis of this network/framework as designed and produced by the primary users with help of designers and producers.

For instance, if you need as a secondary user to travel by plane to a country in which biometric passports and detection tests are compulsory at the airport, then you will have to walk through several control systems put together by primary users of this technology:

Ways of worldmaking in air travelling

Airport road

Airport room

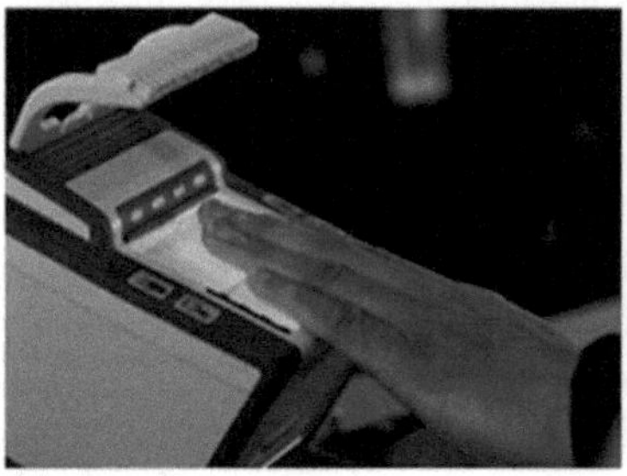

Biometric passport

Security check

Outside the plane

Inside the plane

If one accepts the distinction between several types of users, namely the primary and the secondary users, then one should emphasize the way the use of some technical artefacts as elaborated by designers and producers by primary users actually shape the use of secondary users. Basically, the idea of a technological process as forming a chain of actions is on the one hand that engineers design and produce technical artefacts, while on the other hand primary users, thanks to technical artefacts, design and produce a *world* for the secondary users. In other words, you can design and produce a technical artefact, but depending on the way some users use it, you design and produce a world for the other users.

Types of Technological Worldmaking

One can distinguish between several types of worldmaking in the technological process and emphasize the specificity of constitutive and modal way of worldmaking in the shaping of ordinary life practices and theories:

(TP): a *Technological Process* is a set of operations of design and production of some technical artefacts that shape their use in a certain material and social context.

(TPW): a *Technological Process of Worldmaking* is a set of operations of design and production of some technical artefacts that shape their use in a certain material and social context *by making a world* for the users.

(TPCW): a *Technological Process of Constitutive Worldmaking* is a set of operations of design and production of some technical artefacts that shape their use in a certain material and social context by *making a world that* constitutes *the ordinary life practices and theorie*s of the users.

(TPMW): a *Technological Process of Modal Worldmaking* is a set of operations of design and production of some technical artefacts that shape their use in a certain material and social context by *making a world that* modalises *the ordinary life practices and theories* of the users.

A technology as designed, produced and used respectively by engineers or producers and by users or consumers involves a process of ordinary practice-shaping and theory-shaping. It is a *practice-shaping* process in that the techniques as used within a socio-technical network – that also functions as a socio-technical framework – shape the modalities of ordinary human *action* and then bring about a set of practical conditions that the users must adapt. It is also a *theory-shaping* process in that the techniques also shape the modalities of ordinary human *thought* and carry about a set of "theoretical" options, some worldviews that the users are urged or at least incentivised to adopt. In fact, the production of ordinary theories is as much a factor of adaptation of humans to a technical system as the production of ordinary practices. For technology as a process of *modalisation* expressed in terms of necessity or obligation, possibility and impossibility, etc. not only shapes matters and acts, but also minds.

A Modal Constitution of Things and Humans

The notion of *Constitution* is used both in logical and phenomenological philosophy (from Carnap to Husserl) to designate the linguistic or mental foundation or characterisation of the world objects and of the objects' structures and relations. It is also a useful approach as regards the kind of objects that are humanly shaped and that one usually calls artefacts – especially technical artefacts, if one can distinguish them from aesthetical artefacts. I would like here to follow a sort of "third way" in between the logical and the phenomenological approaches in taking together the conceptual and the experiential analysis. In other words, I would like to suggest through the notion of *modal constitution* of things and humans – and further of human practices and theories – that the structure of relations between objects such as socio-technical artefacts can be identified as a network and a framework from a conceptual side as well as an experiential side.

Constitution of Everyday Life

The notion of Constitution that I refer to borrows from Rudder Baker's *Metaphysics of Everyday Life*, even if I rather insist on the notion of *modal constitution* which quite differs from her view (Rudder Baker 2007, p. 32):

> Constitution is a very general relation, ubiquitous in the world. It is a relation that may hold between granite slabs and war memorials, between pieces of metal and traffic signs, between DNA molecules and genes, between pieces of paper and dollar bills – things of basically different kinds that are spatially coincident. The fundamental idea of constitution is this: when a thing of one primary kind is in certain circumstances, a thing of another kind – a new thing, with new causal power – comes to exist. When an octagonal piece of metal is in circumstances of being painted red with marks of the shape S-T-O-P, and is in an environment that has certain conventions and laws, a new thing – a traffic sign – comes into existence. A traffic sign is a different kind of thing, with different causal powers, from a scrap piece of metal that you find in your garage. Yet the traffic sign does not exist separately from the constituting piece of metal. Constitution is a relation of unity – unity without identity.

The notion of Constitution also applies to technical artefacts as opposed to natural things or objects and as defined by Rudder Baker, who focuses on the technical artefacts, rather than the aesthetical (*idem*, p. 51):

> Technical artefacts (are) the material products of our endeavours to attain practical goals. Such artefacts are object intentionally made *to serve a given purpose*. Artefacts with practical functions are everywhere. We sleep in beds; we are awakened by clocks; we eat with knives and forks; we drive cars; we write with computers (or with pencils); we manufacture nails. Without artefacts, there would be no recognizable human life…Typically artefacts are constituted by aggregate of things…Artefacts have proper functions that they are (intentionally) designed and produced to perform (whether they perform it their proper functions or not). Artefacts have *intended* functions, which are obviously normative…We cannot understand the world we live in without presupposing normativity.

It can be said as regards the assumption of normativity that the nature of an artefact lies in its proper function – what it was designed to do, the purpose of which it was produced –, that is, its intended function. And the proper function of an artefact is determined by the intentions of its designer and/or producer.

If one refers to the general definition of "constitution", if x constitutes y at t, and y's primary kind is G, then x is in "G-favourable circumstances" at t. In the case of boats, there are two kinds of "G-favourable circumstances": (1) the circumstances in which a boat may come into existence; (2) the circumstances in which an existing boat continues to exist. For (1), the circumstances are the following: (a) the aggregate must be in the presence of one or more persons who know how to build a boat from the items in the aggregate, and who either intend to build a boat from the items in the aggregate or whose activity is directed by someone who intends to have a boat built from the items in the aggregate (b) the items in the aggregate must be manipulated by such persons (either manually or by machine) in ways that execute their productive intentions or of those directing the persons; (c) the result of the manipulation must satisfy the productive intentions of the persons.

Idea of a Modal Constitution

One could mention many critics to this idea of Constitution as elaborated by Rudder Baker, especially concerning the notion of "favourable circumstances". I would just like to show that the idea of a *constitution of things and humans* can be expressed in modal terms and emphasize the modal conditions of human agency as related to artefacts. In fact, what counts as a world for an individual or a group is also the *web of modalities* (or modal web) that shapes his or her action in terms of necessity, contingency, possibility and impossibility ("theoretical-epistemic" terms), or in terms of obligation, liberty, permission and prohibition ("practical-deontic" terms). For instance, if I can use a car to make my daily 15 miles journey to work instead of using a bicycle, then after a few years, I will certainly take the use of a bicycle to be something impossible and the use of a car to be something necessary.

This means that we qualify the means and ends in modal terms (necessity, possibility, etc.), what we may call the *modalisation* of the world's elements and their relations. The modalisation entails that the world is not just a set of elements, but also *a set of modal relations between those elements* that shapes the modalities of human action and even of human thought in structural and situational terms as interpreted by modal judgment.

The idea of a *Modal Constitution* is the following:

(i) A constitution can be expressed in *modal terms* (necessity, obligation, contingency, liberty, possibility, impossibility, permissibility, impermissibility) regarding the scope of potential and actual human actions and thoughts as allowed or restricted by a network and a framework of socio-technical artefacts.

(ii) A modal constitution reflects the way a technology as a *network* and a *framework* of socio-technical artefacts not only equals the production of human techniques, but also entails the production of human practices and even of human theories.

One can define the classical way of articulating structures, functions and use of an artefact by using the notion of constitution in the following sense:

1. *Constitution Rule*: An artefact *A* is designed, produced and/or used for achieving a goal *G* through an action *X* in a context *C*.

If one takes into consideration the *modalities* of action, then the constitution rule can be expressed in the following way:

2. *Modal Constitution Rule*: An artefact *A* is designed, produced and/or used for modally achieving a goal *G* through an action *X* in a context *C*.

The notion of modal constitution also suggests that some elements are related to some other elements and then form a *web of modal conditions*. For example, X can travel by plane to country *C*, but provided he owns a biometric passport, otherwise he would be able to travel only to countries *non C*.

The basic model for the modal web referring to the relations between modalities can be expressed in the following way:

3. *Modal Web Constitution Rule*: An artefact *A* is designed, produced and/or used for modally achieving a goal *G* through an action *X* that modally implies an action *Y* in a context *C*.

This formal model can be translated into a less formal view that gives some kind of visual representation of the modal dynamics of human agency as long as it is concerned with actions taking place within a technological network and framework. Thus one can think of a circle that represents a modal *scope of possibilities* for a given human action the size of which varies according to the elements and the elements' relations that constitute the evolving technological network and framework.

Worldmaking technologies and modal scope of possibilities

Constitution 1.......................Constitution 2..............Constitution 3

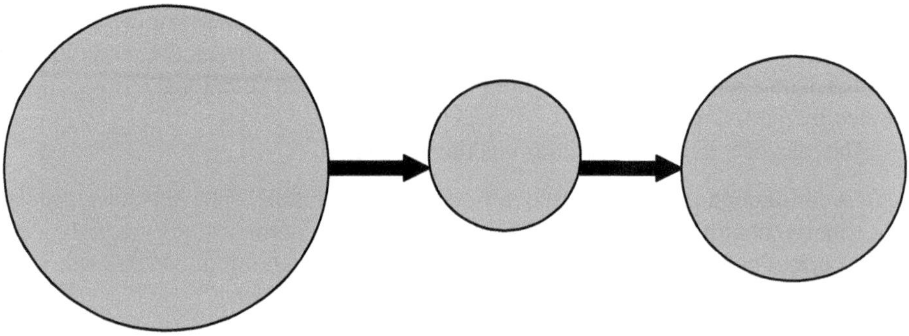

For Constitution 1, the scope of possibilities is maximal, while for Constitution 2, due to some changes in the technological components of the network/framework, the scope of possibilities is minimal. This is the case, for instance, when a law makes a biometric passport compulsory to travel to country *C* and a user of a plane cannot sail across the ocean several weeks and so has to fly to go there.

Modal Analysis and Synthesis

One can suggest that the modal constitution of an artefact (or rather a set of artefacts) is set up in a given context in the shape of a network that also functions as a framework for the users. One can then make a modal analysis as well as a modal synthesis of the modal web of actions in order to reveal the set of implications of the artefact's material and social network/framework.

Let us take the example of Ivan Denissovitch, the famous hero of Solzhenitsyn's novel (Solzhenitsyn 1962, 2000), and let us suppose that Ivan managed to escape the Soviet Goulag and then managed to join the "Free World":

> Ivan Denissovitch is very happy to escape the Soviet Goulag and to join the Free World that he views as a kind of promised land. After arriving in the Free World, Ivan is trying to get directly to a plane in an airport, but he is biped when walking around a detection system. He is harassed by the police asking for his biometric passport, whereas he can just exhibit his paper-made and hand-written documents. Then, Ivan cannot find a telephone functioning with coins, or ask the people around; they are all running and escaping, talking alone and holding their hand on their right ear. He wants to hire a room in a low-cost hotel, but he is requested by an answering machine to send a confirmation by the Internet. When arriving at the hotel with a taxi, which displays road information seemingly coming from space, he intends to discuss with someone in order to explain his poor situation. But he finds nobody in the hotel: everything is automatic...provided you have a credit card, and so a bank account. Hopefully, Ivan Denissovitch remembers the Goulag, and can enjoy the freedom of sleeping outside, without money, in the Free World.

If we have a look at the table below, it appears that Ivan Denissovitch uses technical means that belong to Worlds 1, 2 or 3, but not to World 4:

Functions	World 1	World 2	World 3	World 4
Locomotion L	Walk	Boat	Train, Car	Plane
Communication C	Voice	Telegraph	Telephone	Mob. phone
Identification I	Name	Identity card	Passport	Bio. passp.
Reservation R	Mail	Telephone	Fax	Internet
Payment P	Coin	Note	Cheque	Credit card

Worlds and Technologies

We can now express this specific modal constitution as related to a specific network/framework and to a specific worldmaking process in using a modal analysis and a modal synthesis. In the presentation of the modal analysis and the modal synthesis,

I will not use some logical symbols as required in modal logic, such as $\Diamond$ for "Possibility" or $\Box$ for "Necessity". I will use the terms as expressed in ordinary language, but I will keep mainly to the terms as used in *epistemic* and *deontic* logic (Necessity/Obligation, Possibility/Permissibility, Impossibility/Impermissibility, Contingency/Liberty).

(I) **Modal analysis**

In the *modal analysis*, one makes an analysis of each modality in order to explicit the range of modal options at stake for one type of action:

Locomotion (L)
Necessity/Obligation: X has to take the plane to travel to country C.
Possibility/Permissibility: X can take the car, the train or the boat if he/she needs not to travel to country C.
Impossibility/Impermissibility: X cannot walk if he/she travels across the sea or the ocean to country C.
Contingency/Liberty: X can take the plane or the boat to travel across the sea or the ocean to country C.

Communication (C)
Necessity/Obligation: X has to use a mobile phone to be able to call from anywhere.
Possibility/Permissibility: X can use a telephone if he or/she can access it in some public or private locations.
Impossibility/Impermissibility: X cannot use a telegraph.
Contingency/Liberty: X can use a mobile phone or a telephone if he/she does not need to be able to call from anywhere.

Identification (I)
Necessity/Obligation: X has to use a biometric passport to travel to country C.
Possibility/Permissibility: X can use a classical passport or an identity card to travel to some countries non C.
Impossibility/Impermissibility: X cannot use a mere name as such.
Contingency/Liberty: X can use a passport or a biometric passport in some countries non C.

Reservation (R)
Necessity/Obligation: X has to use an Internet reservation for hotel H.
Possibility/Permissibility: X can use a fax or telephone for hotel non H.
Impossibility/Impermissibility: X cannot use a paper mail reservation for hotel H.
Contingency/Liberty: X can use an Internet, a fax or a telephone reservation.

Payment (P)
Necessity/Obligation: X must use a credit card for payment.
Possibility/Permissibility: X can use notes for the payment of hotel H.

Impossibility/Impermissibility: X cannot use cheques for the payment of hotel H.
Contingency/Liberty: X can use credit card or a cheque for the payment of hotel H.

(II) **Modal synthesis**

In the *modal synthesis*, one makes a synthesis of all the actions implied for each modality:

Necessity N (or Obligation O)
Locomotion: X has to take the plane to travel to country C.
Communication: X has to use a mobile phone to call from anywhere.
Identification: X has to use a biometric passport to travel to country C.
Reservation: X has to use an Internet reservation for hotel H.
Payment: X must use a credit card for payment for hotel H.

Possibility P (or Permissibility P*)
Locomotion: X can take the car, the train or the boat if he/she needs not to travel to country C.
Communication: X can use a telephone if he/she can access it in some public or private locations.
Identification: X can use a classical passport or an identity card for some countries non C.
Reservation: X can use a fax or telephone for hotel non H.
Payment: X can use notes for the payment of hotel H.

Impossibility I (or Impermissibility I*)
Locomotion: X cannot walk if he/she travels across the sea or the ocean to country C.
Communication: X cannot use a telegraph.
Identification: X cannot merely use his/her name.
Reservation: X cannot use a paper mail reservation for hotel H.
Payment: X cannot use cheques.

Contingency C (or Liberty L)
Locomotion: X can take the plane or the boat to travel across the sea or the ocean to country C.
Communication: X can use a mobile phone or a telephone if he/she needs not to call from anywhere.
Identification: X can use a passport or a biometric passport in some countries non C.
Reservation: X can use an Internet, a fax or a telephone reservation.
Payment: X can use credit card or a cheque.

The modal analysis and synthesis explicit the web of modalities as conceptualised and as experienced by some human beings in terms of their actions and as regards their modality in the context of a worldmaking technological network and framework.

Conclusion

Technology and engineering are certainly alike art and science one of the several possible ways of worldmaking in that they shape our material and social environment as well as our daily lives at home, at work, in transportation or on vacation. The technological making of a world through engineering can be viewed as a socio-technical process that counts several stages, namely those of design, production and use of technical artefacts. It is important to understand better the extent to which a set of technical artefacts are arranged and connected so that they come to constitute a network as well as a framework for the actors or the agents who use them. One can say that, at a certain level of integration of technical artefacts, what is at the very beginning a mere combination of artificial things becomes at the end a genuine artificial system that bounds the actions of the users for better or worse.

The idea of constitution suggests that the distribution of technical artefacts in space and time to form a network/framework is not a matter of chance but the result of a series of plans and scripts. One can support the idea of a *contextual* constitution of things and humans and argue that for an object to be a technical artefact it depends upon some criteria to be satisfied and upon some "favourable circumstances" to be met. This view is not wrong, basically, although it appears insufficient mainly for reasons of contextual indeterminacy, but also, so to speak, for reasons of *modal indeterminacy*. The modal approach to constitution makes it explicit that the relations between the elements of a network that otherwise functions as a framework is not only situation-sensitive but also structure-productive.

The *modal constitution* of things and humans requires a "model of modalities", an analysis and a synthesis of the modalities of human actions in a material and social context shaped by a web of connections and arrangements. These modalities refer to the classical modal terms (necessity/obligation, possibility/permissibility, impossibility/impermissibility, contingency/liberty). But they are not just some formal concepts, they are also meant to reflect some informal experiences of human beings who produce some situated judgments as lived and expressed in modal terms. Moreover, this view is not a static picture, but refers to a dynamic process in which the scope of possibilities of the users can vary depending upon the variation of the elements and their relations that make a world.

Contrary to what Goodman suggested, a world is not a mere scientific or artistic version: it is also a complex material and social organisation of things that is *made* by humans and that *makes* their lives.

References

Friedman, T. L. (2007). *The world is flat: A brief history of the twenty-first century*. New York: Picador.

Goodman, N. (1978). *Ways of worldmaking*. Indianapolis: Hackett.

Ihde, D. (1990). *Technology and the lifeworld*. Bloomington: Indiana University Press.

James, W. (1909, 1996). *A pluralistic universe*. Lincoln: University of Nebraska Press.
Kant, E. (1787, 2008). *Critique of pure reason*. New York: Cosimo.
Kroes, P. (2012). *Technical artefacts: Creations of mind and matter*. New York: Springer.
Latour, B. (2005). *Reassembling the social*. Oxford: Oxford University Press.
Lavelle, S. (2006). *Science, technologie et éthique*. Paris: Ellipses.
Lavelle, S. (2007). Technology and engineering in context. In S. H. Christensen, B. Delahousse, & M. Meganck (Eds.), *Engineering in context*. Aarhus: Academica.
Lavelle, S. (2009). Politiques des artefacts Ce que les choses font et ne font pas. *Cités, 39*(3), 39–51.
McCarthy, N. (2010). A world of things, not facts. In I. Van de Poel & D. Goldberg (Eds.), *Philosophy and engineering: An emerging agenda* (pp. 265–273). New York: Springer.
Rudder Baker, L. (2007). *Metaphysics of everyday life*. Cambridge University Press: Cambridge, UK.
Ryckman, T. (2007). Carnap and Husserl. In R. Creath (Ed.), *The Cambridge companion to Carnap*. New York: Cambridge University Press.
Scheffler, I. (1980). The wonderful worlds of Goodman. *Synthese, 45*(2), 201–209.
Solzhenitsyn, A. (1962, 2000). *One day in the life of Ivan Denisovich*. Penguin Modern Classics: London, UK.
Winner, L. (1986). Do artifacts have politics? In L. Winner, *The whale and the reactor: A search for limits in an age of high technology* (pp. 19–39). Chicago: University of Chicago Press.
Wittgenstein, L. (1998). *Tractatus logico-philosophicus*. Mineola: Dover.

Sylvain Lavelle Ph.D. in Philosophy, University of Paris-Sorbonne, 2000. Degree in Political Sciences from University of Paris and Bordeaux, and in Natural Sciences from University of Paris and Aix-Marseille. After 2 years as an assistant at Sorbonne 1998–2000, he has taught philosophy, epistemology and ethics in the department of humanities of an engineering school, ICAM Lille. He is currently working at ICAM-Paris and is Director of the Center for Ethics, Technology and Society (CETS) in Lille and an associate researcher at the Ecole des Hautes Etudes en Sciences Sociales (EHESS-GSPR). Sylvain Lavelle has been involved over the past years in several research projects funded by the European Commission (TRUSTNET, COWAM, EGAIS) and the French National Re-search Agency (PARTHAGE). He is an author and/or an editor of several books and of many articles, chapters and research documents, including *Science, Technologie et Éthique* (Ellipses, Paris, 2006), *Ethical Governance of Emerging Technologies Development* (IGI Global, New York, 2013) and *La Société en Action*, (Herman, Paris, 2013).

Chapter 15
The Nuclear Pipeline: Integrating Nuclear Power and Climate Change

Jen Schneider, Abraham S.D. Tidwell, and Savannah Avgerinos Fitzwater

Abstract This chapter focuses on nuclear scientists and engineers, and the effectiveness of small-scale interventions that could be made to prepare them to consider novel kinds of climate disruptions and how such considerations could affect plant design and operations. Events at Fukushima in 2011 prompted renewed attention to nuclear safety. Soon after, scientists recorded record-breaking global temperatures, particularly during the summer of 2012. Perhaps as a result of these two events, academics and the media have begun asking whether nuclear power plants are robust to natural events beyond the range of available historical data (beyond design basis), including climate-related events such as increasing drought and rising cooling-water temperatures. Science policy scholars, scientists, and engineers outside nuclear science and engineering have begun to pose such questions and model possible effects. This study demonstrates there is almost no public discourse and very little professional discourse within the nuclear science and engineering community on this topic. We posit that this is largely because of the insular culture and professionalization standards of nuclear science and engineering, which could limit the effectiveness of curricular interventions made in engineering education.

Keywords Nuclear engineering • Climate change • Nuclear power • Global warming • Engineering education

---

J. Schneider (✉)
Department of Public Policy and Administration, Environmental Research Building, Office 5135, Boise State University, 1910 University Drive, Boise, ID 83725, USA
e-mail: jenschneider@boisestate.edu

A.S.D. Tidwell
Consortium for Science, Policy and Outcomes, Arizona State UniversityTempe, P.O. Box 875603, Tucson, AZ 85287-5603, USA
e-mail: Abraham.Tidwell@asu.edu

S.A. Fitzwater
Liberal Arts and International Studies, Colorado School of Mines, 301 Stratton Hall, 1005 14th Street, Golden, CO 80401, USA
e-mail: savannah.fitzwater@gmail.com

S.H. Christensen et al. (eds.), *Engineering Identities, Epistemologies and Values*, Philosophy of Engineering and Technology 21,
DOI 10.1007/978-3-319-16172-3_15

Introduction

The research in this chapter emerged from a larger project that asked the question: can engineering education be reformed to better prepare engineering students to incorporate climate variability, as a result of climate change, in their engineering design work? The focus of this work has been on infrastructure resilience in particular: if increasingly severe flooding will adversely affect engineered structures such as dams and levees, for example, should not engineers be thinking differently about design bases, assumptions, and their interactions with other experts, policymakers, and the public?

Initial work from the project took the form of ethnographic interviews with faculty members in engineering and the sciences at our university, the Colorado School of Mines (CSM). Findings from that study suggested that there were tremendous challenges facing efforts to reform engineering education in order to better integrate climate change, not the least of which were complex faculty attitudes toward the topic itself, but also institutional disincentives and barriers (Lucena et al. 2011). In fact, there was a significant absence of climate-change related curriculum in the university as a whole. This initial work may give us some context for understanding why nuclear science and engineering educators are ignoring climate change as it relates to plant vulnerability: it simply is not yet part of institutional cultures.

Furthermore, in 2011, there was a dearth of scholarly approaches to thinking about the resiliency of nuclear power plants in response to climate change. The authors of this chapter, along with other students involved in the initial stages of this research project, were only able to find one or two articles in peer-reviewed sources dealing with the subject (e.g., Kopytko and Perkins 2011). The majority of other sources on the topic, such as websites and blog entries, treated the topic in a fairly non-academic fashion. The Fukushima Dai-ichi crisis happened in the spring of 2011, and pundits and nuclear experts alike were renewing their focus on nuclear plant resilience, but little of that seemed to be translating into scholarly assessments of U.S. plants and climate change.

The year 2012, however, was to teach us differently. Climate scientists reported that they were recording the highest global temperatures on record. Droughts became increasingly severe in already-dry parts of the globe; rising sea levels and increasingly intense floods struck other parts. The summer was intensely hot in Western Europe and the Eastern United States, areas with high concentrations of nuclear power plants. Blogs, press releases, news stories, and articles in the scientific press (such as *Science Magazine* and *Nature Climate Change*) began reporting that increasing numbers of nuclear power plants were being shut down because their cooling waters were too hot, either for intake or for discharge. Clearly, there was something to this issue, despite first appearances.

This study emerged from these initial experiences, and the "failure" of our initial research into this topic. Science policy scholars, scientists, and engineers outside of nuclear science and engineering have begun to think critically about nuclear plant responses to climate disruptions. This chapter demonstrates, however, that there is almost no public discourse and very little professional discourse within the nuclear science and engineering (NSE) community on this topic. We posit that this is largely because of the insular culture and professionalization standards of nuclear science and engineering, which – along with material obstacles, such as access to specifics of plant design and security restrictions – limit the effectiveness of curricular interventions made in engineering education at the undergraduate and graduate levels.

Methodology

One emphasis of this chapter is on performing a literature review of relevant policy and nuclear science and engineering journals on the topic of nuclear power and climate change. The Center for Science, Technology, and Policy Research (CSTPR) has identified a list of influential journals in science and environmental policy. We used this as a starting point for conducting a search on nuclear power and climate change, focusing our work on those journals from the list that look specifically at environmental policy. We also chose to search the journals *Nature* and *Science Magazine* because of their stature and wide readerships, particularly in relation to climate science and policy commentary. We limited our initial search of the journal *Nature* to "Research" articles: future work will need to expand to include the hundreds of commentaries and news pieces related to our topic. The publications we searched are listed below.

1. *Nature*
2. *Science Magazine*
3. *Climatic Change*
4. *Environmental Research Letters*
5. *Global Environmental Change*
6. *Weather, Climate, and Society*

We conducted a general search for any original research articles containing the phrases "nuclear power" or "nuclear energy" in the text. From these articles, the search was further refined to focus on those articles that refer to nuclear power and energy in the context of climate change. We also chose to review two types of publications in nuclear science and engineering: the five English-only journals with the highest rated impact factors according to the Web of Science, and more operations-oriented publications produced by the American Nuclear Society (ANS), a professional organization of scientists, engineers, and other professionals devoted to the peaceful applications of nuclear science and technology. ANS has 10,500 members in 46 countries, and its publications are widely read by practicing nuclear scientists, engineers, and students, (http://www.new.ans.org/about/history/).

What became immediately clear is that highly theoretical research published by the top journals in nuclear science does not address the primary concerns of power plant operators. These top journals were searched and, yielding no results, eventually excluded. On the other hand, it seemed possible that ANS publications might address climate change as affecting nuclear power plant production in some way. Because the focus of the paper is on electricity generation, we focused our search to articles that addressed the nexus of climate change and commercial-scale nuclear power plant production or small modular reactor (SMR) design. The following ANS publications were also searched using the keywords "climate change," "global warming," "climate disruption," and "climate variability":

1. *Fusion Science and Technology*
2. *Nuclear Science and Engineering*
3. *Nuclear Technology*
4. *Nuclear News* (magazine)

Finally, we also searched the websites of the two most prominent nuclear industry organizations, the World Nuclear Association (WNA) and the Nuclear Energy Institute (NEI) for articles addressing nuclear plant resilience to climate change. Both organizations have a strong web presence, and a core mission of communication with the public. Both could be described as industry front or trade groups.

We read abstracts for all articles that resulted from these searches, and then conducted a keyword search for "nuclear" within the article to determine its relevance to our topics and hypotheses and to look for repeating patterns, approaches, and attitudes toward nuclear. Articles that were irrelevant to our topic, or which only addressed climate change and nuclear power tangentially, were excluded. Policy modeling papers that explored nuclear power more thoroughly, particularly in relation to emissions targets, were included.

Articles were coded by summarizing each relevant article and then we organized these notes by thematic unit in order to determine relevant categories (Frey et al. 2000). For example, as we searched the policy journals, we looked for articles that addressed how nuclear power was understood as a response to climate change. In some of these cases, nuclear power took on the role of being a key decarbonization technology. In others, the articles reflected policy scholars' concerns about public resistance to nuclear power plant builds. Therefore, we added a category titled "public acceptance of nuclear power as a response to climate change." We added categories in this way as they emerged from the coding and then organized articles into categories in order to understand general patterns in the literature.

Nuclear Power as Climate Policy

In their 2004 *Science Magazine* article "Stabilization Wedges: Solving the Climate Problem for the Next 50 Years with Current Technologies," Stephen Pacala and Robert Socolow famously argue that the challenges posed by climate change could be solved in the "first half of this century simply by scaling up what we already know how to do (Pacala and Socolow 2004). Pacala and Socolow demonstrate how this could be done by dividing carbon emissions into a series of seven "wedges," with the wedges representing avoided carbon emissions resulting from technological and social improvements in energy efficiency or development. Accomplishing the reduction implied by the wedges would limit atmospheric CO_2 to 500 parts per million (ppm) – an amount the authors argue "would prevent most damaging climate change."

Pacala and Socolow identify at least 15 strategies for accomplishing the wedged reductions that involve scaling up existing energy production or consumption technologies. In retrospect, we now see that some wedges are more feasible than others. For example, the authors argue that society could accomplish one wedge of CO_2 reductions by replacing 1,400 GW of low-efficiency coal-fired power plants with gas plants. Unlike some of the other strategies – the ones requiring significant performance from CO_2 Capture and Storage (CCS) facilities, for example – this particular strategy might prove to be economically and politically feasible depending on the development of shale gas, which in turn faces substantial public opposition.

Pacala and Socolow's article was historically significant in part because it galvanized discussions about concrete climate policy action, and because its central assumptions are representative of persistent climate policy debates. For example, their work forces us to ask whether we can accomplish climate goals using technologies we have, or whether we must launch a "Manhattan Project" for new energy research and development. This is arguably an unresolved and ongoing climate policy debate. The Pacala and Socolow proposal also focuses on mitigation strategies, as opposed to adaptation strategies, another contentious area of debate. Ongoing research regarding the feasibility of their ideas continues (e.g., see Pielke 2010; Socolow 2011).

Pacala and Socolow's article is interesting as context: it is typical of how those concerned with policy think about nuclear power, which is as a mitigation response to climate change. Nuclear fission is a stand-alone strategy on Pacala and Socolow's list; the authors argue that doubling our current nuclear capacity to 700 GW would accomplish the reduction of one wedge of CO_2 emissions by 2054. This optimistic outlook for nuclear seems both unrealistic and unreasonable now; however, this article was written before the meltdowns at the Fukushima Daiichi reactors in 2011 and the consequent reduction in German and Japanese electricity production from nuclear power. The article was also published before the 2008 global recession, making it still possible for US and European utilities or states and developing-world economies to imagine investing billions in building large-scale nuclear reactors. Such building continues in China, of course, but at a somewhat less aggressive pace than before (Yiyu 2012).

Despite these concerns, the promise of nuclear power as a response to climate change persists. It is appealing because it promises to provide massive amounts of baseload power, if we continue with large-scale plant building, while emitting a reasonably small amount of CO_2 into the atmosphere, particularly when compared with coal or gas plants. It does not face the same storage problems that renewable technologies such as wind and solar face, and it is easily transmitted over existing power distribution networks. Furthermore, many of the licensing hurdles that nuclear plants have faced since the 1980s promise to be reduced by an improved and streamlined regulatory regime, which might improve building efficiencies and therefore reduce cost overruns. In addition, SMRs are receiving increasing attention as a possible response to the problems posed by huge power plants; they may be more economical and more easily maintained and licensed than large plants, and answer a philosophical resistance to large-scale energy production (e.g., Winner 1986). Arguments such as we have listed here in favor of the nuclear solution are frequently made by those arguing for a "moonshot" approach to energy development and climate policy (e.g., Bryce 2010).

A wide swath of thinkers and scholars think of nuclear power as a climate mitigation strategy. Advocates of the "nuclear renaissance" – including some formerly notable anti-nuclear environmentalists such as Patrick Moore (Madrigal 2007) and Gwyneth Cravens (2007) – tout nuclear as the truly environmental response to climate change, and a relatively safe technology in comparison with fossil fuel electricity production. Similarly, policy scholars also see nuclear power as a key decarbonization strategy, particularly in cases where stringent emissions targets have been set. There are a few minor exceptions and corollaries to this generalization, but it is notable that, in general, scholars assume nuclear is a feasible and/or necessary climate solution (Bosetti et al. 2009; Bush and Harvey 1997; Buttel et al. 1990; den Elzen et al. 2008; Green 2000; de Lucena et al. 2010; Luderer et al. 2012; Mander et al. 2007; Mohnen et al. 1991; Myhrvold and Caldeira 2012; Riahi et al. 2011; Rosenberg and Scott 1994; Schultz et al. 2003; Urban et al. 2009). The industry groups – WNA and NEI – also beat this drum exceptionally loudly. WNA and NEI mention climate change frequently on their websites, but only in the context of justifying a nuclear renaissance.

In critical policy terms, however, one could argue that the future of nuclear is anything but bright, and the possibility of a nuclear renaissance looks bleak. There is no doubt that, if we did not have social, economic, or ethical concerns, nuclear power would be an excellent technological response to climate change, but such a world in which power plants operate separate from their socio-political context does not exist. Nuclear opponents are, of course, stridently opposed to such expansion of nuclear power. Furthermore, there is some concern or suspicion among a number of policy scholars as to how countries will meet seemingly unrealistic emissions targets and/or avoid catastrophic climate change when prevailing trends against nuclear are unfavorable, and the immense number of reactors that would be required to successfully meet such targets are economically infeasible (Collier and Löfstedt 1997; Davis et al. 2010; Pielke 2009a, b; Webber 2007; Yue and Sun 2003). Other scholars suggest that while nuclear power might be a promising decarbonization technology, political and social forces may prevent its implementation. Public views on these two issues – nuclear power and climate change – are incredibly complex (Bostrom et al. 2012; Carr-Cornish et al. 2011; Pidgeon et al. 2008; Rayner 1993; Truelove and Greenberg 2012; van Vuuren et al. 2007).

Yet only a few policy scholars openly critique nuclear power as a response to climate change, citing concerns over its technocratic nature, which has implications for social justice and community engagement, or its inability to respond quickly and with flexibility to changing climate policies (McEvoy and Wilder 2012; Trancik 2006). Scholars are more likely to take a social constructionist view of nuclear power, examining when and how it is invoked or used in the context of climate change (Rogers-Hayden et al. 2011).

Does Climate Change Pose a Risk to Nuclear Power?

The nuclear industry, generally speaking, invokes climate change primarily as a justification for supporting a nuclear renaissance in the United States and abroad. Popular books on the subject, including Gwyneth Cravens' *Power to Save the World* (2007), Herbst and Hopley's *Nuclear Energy Now* (2007), and Charles Ferguson's *Nuclear Energy: What Everyone Needs to Know* (2011) make this argument, and industry websites such as those of the WNA and NEI frequently speak about nuclear power's value in mitigating climate change. Referring to climate change is a central component of contemporary nuclear renaissance discourse. The study of how nuclear power plants will respond to climate change itself, however, is still clearly in its infancy. Engineers, scientists, regulators, and policymakers who are involved in designing and maintaining the existing environment across multiple sectors are struggling with how to retrofit, plan around, or otherwise prepare for climate disruptions, which are still uncertain and difficult to predict at local scales. Nuclear plant design, construction, and maintenance are no different in terms of facing similar challenges. Yet we can begin to identify some vulnerabilities that nuclear plants in particular face as a result of increasing global temperatures and disruptions, and which are being addressed in the policy literature:

Ability to Discharge Cooling Water This is the most prominent concern expressed in the literature (Aaheim et al. 2012; Golombek et al. 2012; Rübbelke and Vögele 2012; Van Vliet et al. 2012; Vine 2012). During the 1970s, nuclear power plants came under intense public and environmental scrutiny due to their contributions to the "thermal pollution" of rivers and lakes used to discharge cooling water. As a result, governments created regulations that required plants not to exceed certain temperatures when discharging their cooling water. However, as global temperatures rise, the temperature of rivers and lakes may also rise. Power plants may not be able to discharge warm cooling waters into rivers and lakes or risk exceeding temperature levels. Consequently, some regulators are actually raising the allowable temperatures for discharge to accommodate the plants (Godoy 2006).

Access to Cooling Water There is some concern among experts outside of NSE that, as droughts get more severe and access to water more contested, nuclear plants may have to undergo extensive retrofitting measures (such as having intake pipes lowered), endure closures (temporary or otherwise), or politically maneuver for water rights in highly contested policy terrains (Rübbelke and Vögele 2012; Stillwell et al. 2011; Van Vliet et al. 2012). There has been some public concern about how rapidly changing ecosystems may affect oceanic cooling waters. For example, some have predicted a "jellyfish apocalypse": As ocean waters warm, scientists expect species such as jellyfish to thrive. Jellyfish are also easily sucked into intake pipes, fouling up cooling water mechanisms for power plants (Attrill et al. 2007; see Eng 2012). However, in general the policy literature there is mostly concern about drought and water availability.

Siting Problems Some scholars are also concerned about how shifting climate patterns or melting permafrost will affect the buildings of the plants themselves, or change the way plant builders have historically chosen sites, i.e., near bodies of water that have historically been replenished by significant snowfall (Bulygina et al. 2011; Nelson et al. 2001).

The significance of nuclear plant shutdowns or efficiency losses resulting from climate disruptions is presently hard to quantify, and even harder to predict for the future. Yet, some research suggests that the frequency and/or severity of nuclear plant efficiency losses or shutdowns are increasing as the climate warms because of the reasons described above. Van Vliet et al. (2012), suggest that the Southern and Eastern US will be most vulnerable, along with the Southwestern and Southeastern parts of Europe (p. 2). These authors argue that substantial losses in efficiency in Europe occurred in 2003, 2006, and 2009, and in the US in 2006–2007 (p. 1). Recent news reports also tell of efficiency losses or shutdowns in US plants, occurring in 2011 and 2012, at plants that had not seen such shutdowns before.

The elements identified above form the framework of the paradox in understanding the relationship between nuclear power and climate change. As Rübbelke and Vögele put it in a 2011 article in *Climatic Change*:

> Nuclear energy is frequently regarded as a vehicle to reduce CO2 emissions and thus to combat global warming. Yet, there is also a reverse interrelation: the nuclear power sector is negatively affected by climate change, since cooling processes of power plants are likely to be impaired by climate-change related extreme weather events like droughts and heat waves.

For reasons discussed below, the nuclear industry does not seem to be paying much attention to this paradox.

Industry Responses

Those in the nuclear industry are clearly aware of these issues, but according to our research, frame them as falling under typical operations and maintenance. One example is illustrative. The NEI features a blog called "NEI Nuclear Notes." It was in this blog that we found the *only* nuclear science, engineering, or industry responses to nuclear power plant resilience and climate change. On the blog, the author responds to concerns about rising temperatures of cooling waters with the following points ("The Truth" 2006):

1. These issues are not particular to nuclear plants but to all thermoelectric plants, which "account for over 80 % [sic] of all electricity generated on the planet;"
2. Usually, nuclear plants don't have to shut down altogether, like wind power generation did during the 2006 heat waves (because the wind wasn't blowing);

3. And nuclear plants can be designed to "minimize water usage" and therefore not suffer from droughts.

These responses are not wholly satisfying: they do not address our central paradox, which is that the industry has represented itself as a meaningful solution to climate change, yet has not adequately or meaningfully addressed its vulnerabilities to significant climate disruption. Furthermore, while plants *may* be designed to minimize water usage, that does not mean they *can* currently operate in that way, especially in areas not historically accustomed to drought-like conditions, but which may encounter them as the global climate continues to change. Instead, existing nuclear plants have primarily been designed using historical environmental and climate data rather than predictive data.

In a related post on *NEI Notes*, the author writes that nuclear is being held to an unfair standard compared with coal-fired or gas-fired plants; these plants are also vulnerable to supply disruptions, particularly when overwhelmed with peak loads, such as during a heat wave, or even during an exceptionally harsh winter. The author also notes that environmental regulations for thermal pollution are probably "over-conservative and not based on today's best available science," though no link to this science is provided (Countering More Propaganda 2006). Again, this answer seems unsatisfying, perhaps because the nuclear industry has made so much of efficiency gains and high capacities, and of their ability to mitigate climate change. Furthermore, risks of nuclear accidents increase with extensive wear and tear of infrastructure, and the scale of a potential nuclear accident may far exceed that of a gas-fired or coal accident. If a nuclear plant must be shut down for a long period of time due to water-related issues, then questions about when to begin the very expensive decommissioning process must be raised.

Finally, in a third blog post, *NEI Notes* provides a meaningful answer to our paradox when it briefly suggests that the industry is attempting to respond to the thermal pollution problem, in some cases by building "small cooling towers to pre-cool discharge water." Still, the author does not explore this in depth, and soon returns to the problems with wind power's intermittency and coal and natural gas's inability to address meaningfully the climate problem (Revisiting Nuclear 2012).

Though neither scholarly nor complete, at the very least these blog posts acknowledge the problems raised by van Vliet et al., and attempt to respond. By contrast, in our study of the scholarly literature and the literature produced by the ANS we found *not one instance* of scholars, scientists, or engineers addressing in research articles or public discourse the potential challenges posed to nuclear plants by climate change.

Discussion and Conclusion

The research above reveals a possible discourse gap between those concerned about nuclear power plant resilience and those who design and operate nuclear power plants but who do not seem to be discussing resilience in terms of climate change at all. Here, we attempt here to describe why this discourse gap might be occurring, proposing these possible explanations in a spirit of humility and inquiry. It is possible, for example, that our methodology is not sound; we acknowledge that those who operate and maintain nuclear power plants may be having these discussions, as might industry insiders, behind closed doors. The Nuclear Regulatory Commission (NRC), the regulatory body within the U.S. responsible for the safe use of radioactive materials for civilian use, may be accounting for climate disruptions out of the public eye, and may be doing so using terms other than "global warming" or "climate change." When we presented our research to energy policy experts and to the NSE students and faculty at CSM, however, none raised the possibility that we missed important publications or venues where these discussions were happening. To the contrary, their impressions matched our own: these discussions are *not* occurring in nuclear science and engineering. The possible reasons for *why* they are not occurring are many. They are outlined here.

Climate Skepticism First, research emerging from this research project confirmed what many of us who had worked as engineering educators for years already suspected: engineers, generally speaking, may be more resistant to "believing in" or taking seriously climate change and climate change research. These findings are discussed in detail in Lucena et al. (2011).

Furthermore, it seems important to acknowledge that not *all* scientists support the consensus on anthropogenic climate change; some scientific disciplines, including Physics, may be more likely to be skeptical of this consensus. Most NSE programs have some foundation in Physics or Mechanical Engineering programs. Myanna Lahsen (2008), Laura Nader (1981), and others have convincingly argued that physical scientists and engineering, as disciplines, encourage skepticism as a cultural value. In fact, a number of prominent physicists – particularly those who made their careers during the postwar era – have also emerged as strong doubters of climate change theory. It seems possible, if not likely, that nuclear scientists and engineers might absorb this culture from professors and mentors, and that such beliefs might translate into an unwillingness to engage in discussions about nuclear power plant resilience to climate change.

Disciplinary Factors Another possible explanation grapples with the nature of the development of the nuclear engineering discipline and curriculum itself. According to Sean F. Johnston, the peculiar nature of the development of nuclear engineering as a discipline made nuclear engineers subservient to the state and the technology that brought them into existence. Consequently, nuclear engineers had little opportunity to establish social and economic relationships common to other professions; develop a coherent and cross-cutting curriculum; and, in the case of the United Kingdom, play a major part in the industry they were created to serve. Summarizing the current status of the nuclear engineering discipline, Johnston (2012) states:

> If the field of nuclear engineering were considered in the framework of development psychology, the neutron's children [nuclear engineers] might be perceived as suffering from arrested development, peculiar idiosyncrasies and worldview, insecure self-image, weak communication skills, and poor socialization with their peers. The gradual estrangement of

their governments and the traumatic experiences of Three Mile Island, Chernobyl, and Fukushima furthered shaped their identity.

Without the opportunity to properly develop as a discipline, nuclear engineering, as the segment of NSE that deals most directly with nuclear plant operations, possibly never developed the educational capacities necessary to grapple with the challenges posed by climate change.

Climate Change as a Wicked Problem A third explanation for this discourse gap, in addition to the cultural ones outlined above, is the nature of climate change itself. As many scholars have noted, climate change is particularly difficult to communicate and plan for because it is a diffuse and ubiquitous threat; it evolves relatively slowly compared with the rapid pace of human news and policy cycles; and its effects will be uneven and unpredictable at the local scale. This makes it a so-called "wicked" problem. These characteristics could lead to any number of responses from scientists and engineers. One response could be fatalism: if climate change is going to be as bad as some scientists predict, then nuclear power plant efficiency will be low on the totem pole of priorities. Another response might be denial; the scope and the severity of the climate problem is so great, it is perhaps easier to discount it as the delusional ramblings of greedy climate scientists. Or, perhaps, a third response – and one that we feel is quite likely – is that nuclear scientists and engineers might see climate change as just another engineering problem. They believe they know how to engineer for severe weather, earthquakes, even terrorists crashing jetliners into plants. Increasing drought or rising sea levels can be dealt with as well. This also appears to be the position the NRC took when recently responding to public concern about potential rising sea levels at a plant applying for re-licensing. The correspondence from the NRC to a concerned letter-writer in response to these sea level concerns was as follows (Nuclear Regulatory Commission 2012):

> The NRC has multiple processes to evaluate the adequacy of current plant operations and licensing bases. Should the NRC become aware at any time of information calling into question the continued safe operations of any nuclear power plant … the NRC will take the appropriate actions as part of the agency's ongoing safety oversight, regardless of whether those plants have sought or are seeking a renewed license.

This response further supports the argument that scientists and engineers within the nuclear field believe that impacts from climate change can be tackled like any other engineering problem. Furthermore, the NRC does not appear to be concerned about the impacts from climate change until they pose specific threats, which does not indicate much concern for long term planning.

There is also a tendency in NSE to argue that nuclear *has* addressed concerns such as drought, cooling water temperatures, and safety, primarily through the development of Generation-IV designs which do not require water for coolant, or which can be developed and deployed on a local scale in the form of small modular reactors (SMRs). In our experiences, face-to-face discussions about plant resilience almost always end up focused on these future, not-yet-implemented technologies. However, such future technologies are yet to be deployed on a utility-scale, potentially face numerous social, economic, and political challenges, and do nothing to address the vulnerabilities of the more than 400 plants that already exist worldwide. This rhetorical shift to future technologies also serves to emphasize the supremacy of technical discourse and the technological fix, ground upon which scientists and engineers often feel most comfortable (see Weinberg 1994).

Political Economy A fourth explanation for the discourse gap has to do with politics and economics. On a practical level, the costs of adaptation to climate change might be tremendous. Retrofitting plants for risks that *may* occur but that are not accounted for in a plant's existing design basis is unpalatable from an industry perspective. Fluctuations in the economy exacerbate financial concerns, particularly in the wake of the 2008 financial crisis. Furthermore, following the Fukushima crisis and the push to denuclearize electricity production in many parts of the world, the industry has seemingly more pressing concerns. In the words of one colleague from NSE, "We [the industry] just have other priorities right now." Such a sentiment is congruous with political approaches to risk in the United States in general, where we tend to be reactive when crisis hits, rather than proactive in trying to head off the crisis to begin with.

Similarly, while climate change might provide a positive justification for building more nuclear power plants (the "nuclear renaissance" reasoning) there are strong disincentives to explore the downsides climate change might pose. From a public relations perspective, it is difficult to argue that climate change is both good for the industry *and* bad for it, and the industry is not likely to point out its potential weaknesses in this regard. Furthermore, many utilities own not only nuclear power plants but also coal-fired and gas-fired power plants, which are a primary source of the very greenhouse gases that are accelerating climate change to begin with. The industry and its public relations organizations must be careful in how they address this paradox.

The NSE Pipeline A fifth explanation might unite all of those presented above, and, from our point of view, is most compelling. This explanation is largely an instrumental one, and suggests that, quite simply, there is just no space in NSE to meaningfully take up climate change and plant resiliency as a research question. When Jen presented this research to a group of NSE students at CSM, one asked, "Why don't you start an academic journal about this?" She replied, "Why don't *you*?" The discussion that ensued suggested that such a thing was atypical of the culture of NSE. There are no panels at the annual ANS conference on such topics; no publication venues; no funding opportunities; no classes; no publications. There do not even seem to be casual side conversations on the topic.

If this is true, then there are significant ramifications for those interested in engineering reform that advocates introducing climate change more meaningfully into these engineering students' curricula. If not meaningfully addressed by their professors, colleagues, future employers, journal editors, or conference organizers, climate change becomes just another concern of the liberal arts professor, and a dodgy one at that. If the entire professional pipeline of NSE, from diploma to retirement, is built to shut out such concerns, what hope do micro-interventions or modest reforms have? Such classroom interventions will only be successful if integrated into "pipeline" efforts that target practicing scientists and engineers, professors, employers, and policymakers, in addition to undergraduate and graduate students.

Which brings us to our last possibility, which is the possibility that climate change really does not pose much of a threat to nuclear power plants, especially when compared to other concerns we might have. We acknowledge that this may be the case. Yet, we would make the case to nuclear scientists and engineers that their voices are needed at the table of this particular discussion in order to determine if inaction is the best course. We would encourage them to propose a panel at ANS, to write an editorial on the subject, to post blog entries, and to push their colleagues and professors to address the question. Perhaps the answer is an easy one, perhaps not. In any event, the question of whether nuclear power plants are resilient to climate change is being asked by the public, the media, and experts outside of NSE; pretending that the question has not been posed is not a good strategy and, at worst, could backfire on the industry the way so many issues of public concern have in the past.

Acknowledgments The writing of this paper would not have been possible without funding from the National Science Foundation. We would also like to thank the members of the Climate Change Engineering Partnership for their support and feedback as the ideas for this paper evolved. The authors received valuable feedback from audience members at the Western Energy Policy Research Conference in Boise, Idaho (2012), the International Conference on Culture, Politics, and Climate Change in Boulder, Colorado (2012), and the Nuclear Science and Engineering research seminar (2013).

References

Aaheim, A., Amundsen, H., Dokken, T., & Wei, T. (2012). Impacts and adaptation to climate change in European economies. *Global Environmental Change, 22*(4), 959–968.

Attrill, M. J., Wright, J., & Edwards, M. (2007). Climate-related increases in jellyfish frequency suggest a more gelatinous future for the North Sea. *Limonology and Oceanography, 52*(1), 480–485.

Bosetti, V., Carraro, C., Sgobbi, A., & Tavoni, M. (2009). Delayed action and uncertain stabilisation targets how much will the delay cost? *Climatic Change, 96*(3), 299–312.

Bostrom, A., O'Connor, R. E., Böhm, G., Hanss, D., Bodi, O., Ekström, F., Halder, P., Jeschke, S., Mack, B., Mei, Q., Rosentrater, L., Sandve, A., & Sælensminde, I. (2012). Causal thinking and support for climate change policies: International survey findings. *Global Environmental Change, 22*(1), 210–222.

Bryce, R. (2010). *Power hungry: The myths of 'Green' energy and the real fuels of the future*. New York: Perseus Books.

Bulygina, O. N., Groisman, P. Y., Razuvaev, V., & Korshunova, N. N. (2011). Changes in snow cover characteristics over Northern Eurasia since 1966. *Environmental Research Letters, 6*(4), 045204.

Bush, E. J., & Harvey, L. D. D. (1997). Joint implementation and the ultimate objective of the United Nations framework convention on climate change. *Global Environmental Change, 7*(3), 265–285.

Buttel, F. H., Hawkins, A. P., & Power, A. G. (1990). From limits to growth to global change: Constraints and contradictions in the evolution of environmental science and ideology. *Global Environmental Change, 1*(1), 57–66.

Carr-Cornish, S., Ashworth, P., Gardner, J., & Fraser, S. (2011). Exploring the orientations which characterize the likely public acceptance of low emission energy technologies. *Climate Change, 107*(3–4), 549–565.

Collier, U., & Löffstedt, R. E. (1997). Think globally, act locally: Local climate change and energy policies in Sweden and the UK. *Global Environmental Change, 7*(1), 25–40.

Countering More Propaganda. (2006). *NEI nuclear notes*. August 8. Available at http://neinuclear-notes.blogspot.com/2006/08/countering-more-propaganda.html. Accessed 16 Aug 2012.

Cravens, G. (2007). *Power to save the world: The truth about nuclear energy*. New York: Random House.

Davis, S. J., Caldeira, K., & Matthews, H. D. (2010). Future CO_2 emissions and climate change from existing energy infrastructure. *Science, 329*(5997), 1330–1333.

de Lucena, A. F. P., Schaeffer, R., & Szklo, A. S. (2010). Least-cost adaptation options for global climate change impacts on the Brazilian electric power system. *Global Environmental Change, 20*(2), 342–350.

den Elzen, M. G. J., Lucas, P. L., & van Vuuren, D. P. (2008). Regional abatement action and costs under allocation schemes for emission allowances for achieving low CO_2-equivalent concentrations. *Climate Change, 90*(3), 243–268.

Eng, J. (2012, April 27). Diablo Canyon nuclear plant in California knocked offline by jellyfish-like creature called salp. *NBC News*. Available at http://usnews.nbcnews.com/_news/2012/04/27/11432974-diablo-canyon-nuclear-plant-in-california-knocked-offline-by-jellyfish-like-creature-called-salp?lite. Accessed 17 Aug 2012.

Ferguson, C. D. (2011). *Nuclear energy: What everyone needs to know*. New York: Oxford University Press.

Frey, L. R., Botan, C. H., & Kreps, G. L. (2000). *Investigating communication: An introduction to research methods*. Boston: Allyn and Bacon.

Godoy, H. (2006). European heat wave shows limits of nuclear energy. *Common dreams*, July 28. http://www.commondreams.org/headlines06/0728-06.htm. Accessed 16 Aug 2012.

Golombek, R., Kittelsen, S., & Haddeland, I. (2012). Climate change: Impacts on electricity markets in Western Europe. *Climatic Change, 113*(2), 357–370.

Green, C. (2000). Potential scale-related problems in estimating the costs of CO_2 mitigation policies. *Climate Change, 44*(3), 331–349.

Herbst, A. M., & Hopley, G. W. (2007). *Nuclear energy now: Why the time has come for the world's most misunderstood energy source*. Hoboken: Wiley.

Johnston, S. F. (2012). *The neutron's children: Nuclear engineers and the shaping of identity*. Oxford: Oxford University Press.

Kopytko, N., & Perkins, J. (2011). Climate change, nuclear power, and the adaptation–mitigation dilemma. *Energy Policy, 39*(1), 318–333.

Lahsen, M. (2008). Experiences of modernity in the greenhouse: A cultural analysis of a physicist "Trio" supporting the backlash against global warming. *Global Environmental Change, 18*(1), 204–219.

Lucena, J., Delborne, J., Johnson, K., Leydens, J., Munakata-Marr, J., & Schneider, J. (2011). Integration of climate change in the analysis and design of engineered systems: Barriers and opportunities for engineering education. *Proceedings of the ASME 2011 international mechanical engineering congress & exposition*, Denver, 11–17 Nov 2011.

Luderer, G., Bosetti, V., Jakob, M., Leimbach, M., Steckel, J. C., Waisman H., & Edenhofer, O. (2012). The economics of decarbonizing the energy system – Results and insights from the RECIPE model intercomparison. *Climatic Change, 114*(1), 9–37.

Madrigal, A. (2007, November 19). Co-founder of Greenpeace envisions a nuclear future. *Wired Magazine*. http://www.wired.com/science/planetearth/news/2007/11/moore_qa. Accessed 17 Aug 2012.

Mander, S., Bows, A., Anderson, K., Shackley, S., Agnolucci, P., & Ekins, P. (2007). Uncertainty and the Tyndall decarbonisation scenarios. *Global Environmental Change, 17*(1), 25–36.

McEvoy, J., & Wilder, M. (2012). Discourse and desalination: Potential impacts of proposed climate change adaptation interventions in the Arizona–Sonora border region. *Global Environmental Change, 22*(2), 353–363.

Mohnen, V. A., Goldstein, W., & Wang, W.-C. (1991). The conflict over global warming: The application of scientific research to policy choices. *Global Environmental Change, 1*(2), 109–123.
Myhrvold, N. P., & Caldeira, K. (2012). Greenhouse gases, climate change and the transition from coal to low-carbon electricity. *Environmental Research Letters, 7*(1), 014019.
Nader, L. (1981). Barriers to thinking new about energy. *Physics Today, 34*(2), 99–104.
Nelson, F. E., Anisimov, O. A., & Shiklomanov, N. I. (2001). Subsidence risk from thawing permafrost. *Nature, 410*(6831), 889–890.
Nuclear Regulatory Commission. (2012, October 17). Letter to Bruce Skud of No More Fukushimas. http://pbadupws.nrc.gov/docs/ML1227/ML12279A288.pdf. Accessed 21 Jan 2013.
Pacala, S., & Socolow, R. (2004). Stabilization wedges: Solving the climate problem for the next 50 years with current technologies. *Science, 305*(5686), 968–972.
Pidgeon, N. F., Lorenzoni, I., & Poortinga, W. (2008). Climate change or nuclear power – no thanks! A quantitative study of public perceptions and risk framing in Britain. *Global Environmental Change, 18*(1), 69–85.
Pielke, R., Jr. (2009a). The British climate change act: A critical evaluation and proposed alternative approach. *Environmental Research Letters, 4*(2), 024010.
Pielke, R., Jr. (2009b). Mamizu climate policy: An evaluation of Japanese carbon emissions reduction targets. *Environmental Research Letters, 4*(4), 044001.
Pielke, R., Jr. (2010). *The climate fix: What scientists and politicians won't tell you about global warming*. New York: Basic Books.
Rayner, S. (1993). Prospects for CO2 emissions reduction policy in the USA. *Global Environmental Change, 3*(1), 12–31.
Revisiting Nuclear Energy and Cooling Water. (2012, June 6). *NEI Nuclear Notes*. http://neinuclearnotes.blogspot.com/2012/06/revisiting-nuclear-energy-and-cooling.html. Accessed 16 Aug 2012.
Riahi, K., Rao, S., Krey, V., Cho, C., Chirkov, V., Fischer, G., Kindermann, G., Nakicenovic, N., & Rafaj, P. (2011). RCP 8.5 – A scenario of comparatively high greenhouse gas emissions. *Climate Change, 7*(1), 33–57.
Rogers-Hayden, T., Hatton, F., & Lorenzoni, I. (2011). 'Energy security' and 'Climate change': Constructing UK energy discursive realities. *Global Environmental Change, 21*(1), 134–142.
Rosenberg, N. J., & Scott, M. J. (1994). Implications of policies to prevent climate change for future food security. *Global Environmental Change, 4*(1), 49–62.
Rübbelke, D., & Vögele, S. (2012). Short-term distributional consequences of climate change impacts on the power sector: Who gains and who loses? *Climatic Change, 116*, 191–206.
Socolow, R. (2011, September 27). Wedges reaffirmed. *Bulletin of the Atomic Scientists*. http://thebulletin.org/web-edition/features/wedges-reaffirmed. Accessed 13 Aug 2012.
Schultz, M. G., Diehl, T., Brasseur, G. P., & Zittel, W. (2003). Air pollution and climate-forcing impacts of a global hydrogen economy. *Science, 302*(5645), 624–627.
Stillwell, A. S., Clayton, M. E., & Webber, M. E. (2011). Technical analysis of a river basin-based model of advanced power plant cooling technologies for mitigating water management challenges. *Environmental Research Letters, 6*(3), 034015.
The Truth about Nuclear Power and Increased Water Temperatures. (2006, August 17). *NEI Nuclear Notes*. http://neinuclearnotes.blogspot.com/2006/08/truth-about-nuclear-power-and.html. Accessed 16 Aug 2012.
Trancik, J. E. (2006). Scale and innovation in the energy sector: A focus on photovoltaics and nuclear fission. *Environmental Research Letters, 1*(1), 014009.
Truelove, H., & Greenberg, M. (2012). Who has become more open to nuclear power because of climate change? *Climatic Change, 116*, 1–21.
Urban, F., Benders, R. M. J., & Moll, H. C. (2009). Renewable and low-carbon energies as mitigation options of climate change for China. *Climate Change, 94*(1–2), 169–188.
van Vliet, M. T. H., Yearsley, J. R., Ludwig, F., Vogele, S., Lettenmaier, D. P., & Kabat, P. (2012). Vulnerability of US and European electricity supply to climate change. *Nature Climate Change, 2*(9), 676–681.

van Vuuren, D., den Elzen, M., Lucas, P., Eickhout, B., Strengers, B., van Ruijven, B., Wonink, S., & van Houdt, R. (2007). Stabilizing greenhouse gas concentrations at low levels: An assessment of reduction strategies and costs. *Climatic Change, 81*(2), 119–159.
Vine, E. (2012). Adaptation of California's electricity sector to climate change. *Climatic Change, 111*(1), 75–99.
Webber, M. E. (2007). The water intensity of the transitional hydrogen economy. *Environmental Research Letters, 2*(3), 034007.
Weinberg, A. M. (1994). *The first nuclear era: The life and times of a technological fixer*. Woodbury: AIP Press.
Winner, L. (1986). *The whale and the reactor: A search for limits in an age of high technology*. Chicago: University of Chicago Press.
Yiyu, L. (2012, February 1). Nuclear approvals to be resumed at slower rate. *China Daily*. http://www.chinadaily.com.cn/cndy/2012-02/01/content_14514229.htm. Accessed 17 Feb 2012.
Yue, C.-D., & Sun, C.-H. (2003). Climate protection and newly industrialized countries: Dilemmas and opportunities in Taiwan. *Global Environmental Change, 13*(1), 31–42.

Jen Schneider Ph.D. in Cultural Studies from Claremont Graduate University. Associate Professor, Public Policy and Administration, Boise State University, USA. Co-author with Juan Lucena and Jon Leydens of *Engineering and Sustainable Community Development*. Past projects have addressed the role communication plays in environmental crises such as climate change and engineering for development work. Current projects address the role of rhetoric, media, and communication in energy controversies in the United States, with a special focus on fossil fuel production and consumption. Jen is completing a co-authored book titled *Under Pressure: Coal Industry Rhetoric in the Age of Neoliberalism*.

Abraham S.D. Tidwell B.S. Chemical Engineering, Virginia Tech. M.S. in the Master of International Political Economy of Resources (MIPER) program at the Colorado School of Mines. Ph.D. student at Arizona State University. At the Colorado School of Mines, Abraham worked on a multitude of topics including public lands management and policy, uranium milling policy in Colorado, and communication studies of nuclear energy controversies. Abraham's master's thesis work examined the controversy surrounding a proposed uranium mill in western Colorado – the first of its kind in 30 years – and how this controversy adds to our understanding of the intersection of nuclear technology and society in the twenty-first century American West, a space he has termed *The New Nuclear West*. In addition, Abraham co-authored an entry in *American Political Culture: An Encyclopedia* on nuclear power and U.S. politics with Jen Schneider and Savannah Avgerinos Fitzwater.

Savannah Avgerinos Fitzwater B.S. in Nuclear Engineering from Missouri S&T and currently pursuing an M.S. in Nuclear Engineering and Masters in International Political Economy of Resources at Colorado School of Mines. Research interests include the politics of nuclear weapons, the education of nuclear engineers and the associated long term impacts, and the intersection of nuclear weapons and national identity. Co-authored an entry focusing on nuclear power and U.S. politics for *American Political Culture: An Encyclopedia* with Jen Schneider and Abraham S.D. Tidwell.

Chapter 16
Societal Implications of the Smart Grid: Challenges for Engineering

Joseph Herkert and Timothy Kostyk

Abstract The smart grid, which would combine advanced information and communication technologies with a new generation of electric power production, transmission, and distribution technologies, has been highly touted as a solution to modernizing the U.S. electric grid while simultaneously addressing other policy goals such as improving energy efficiency and expanding the use of renewable energy resources. As with any large scale socio-technical system, however, the smart grid raises a number of societal issues that are interwoven with its technical capabilities. This chapter discusses three such issues – privacy, security, and equity – and argues for addressing them concurrent with the development of the smart grid, as well as educational reforms that will better position engineers to recognize and address such issues.

Keywords Smart grid • Privacy • Security • Equity • Education

Introduction

The world's electricity systems face a number of challenges, including ageing infrastructure, continued growth in demand, the integration of increasing numbers of variable renewable energy sources and electric vehicles, the need to improve the security of supply and the need to lower carbon emissions (IEA 2011). In the United States and some other countries, the electrical power grid is deteriorating. In the U.S. the annual number of large power outages has been increasing since the late

J. Herkert (✉)
College of Letters and Sciences, Consortium for Science, Policy & Outcomes, Arizona State University, 250D Santa Catalina Hall, 7271 E. Sonoran Arroyo, Mesa, AZ 85212, USA
e-mail: joseph.herkert@asu.edu

T. Kostyk
College of Letters and Sciences, Bachelor of Arts in Interdisciplinary Studies (*BIS*) and Organizational Leadership (OGL) programs, Arizona State University, 411 North Central, Mail Code 1901 Phoenix, AZ 85004-0696, USA
e-mail: Timothy.Kostyk@asu.edu

S.H. Christensen et al. (eds.), *Engineering Identities, Epistemologies and Values*, Philosophy of Engineering and Technology 21,
DOI 10.1007/978-3-319-16172-3_16

1990s (Amin and Schewe 2007). For two consecutive days in July of 2012, India experienced blackouts that took down large portions of the country's power grid. The second outage was the largest in history, leaving more than 600 million people, nearly a tenth of the world's population, without electricity (Romero 2012). The numbers, duration, and impact of power failures have severe implications for an energy-intensive way of life, economic stability, and even national security.

One proposed engineering response is widely known as the "smart grid." The increasing occurrences of outages and instances of cyber intrusions between 2000 and 2008 were considered so threatening to U.S. economic viability and security that the federal government, as part of the American Recovery and Reinvestment Act of 2009, earmarked more than $3.3 billion in smart grid technology development grants and an additional $615 million for smart grid storage, monitoring, and technology viability as an initial investment in building the smart grid. In addition, utilities have begun to mount demonstration projects and government and professional societies have begun the development of smart grid standards. Worldwide, investment in smart grid technologies totaled nearly $14 billion in 2012, topped by the U.S. at $4.3 billion and China at $3.2 billion. Major investments also occurred in the rest of Asia and the European Union (Rogers 2013).

The smart grid will be comprised of three fundamental structural elements: replacement of aging core physical infrastructure items including transmission lines and switching equipment with more efficient and reliable newer technologies; two-way distributed and loosely coupled supply and demand connectivity to the grid, which allows consumers to supply electricity through technologies such as photovoltaic cells and wind power; and, most importantly, highly optimized two-way information and communication technology (ICT) systems architectures and networks that control the grid through process- and rule-based programs to match power demand with supply in order to improve efficient use of energy resources. The conceptual model at the core of the smart grid is based upon a framework developed by the U.S. National Institute of Standards and Technology (NIST) that is composed of seven distinct domains – Markets, Operations, Service Provider, Bulk Generation, Transmission, Distribution and the Customer – and the resulting relationships among the domains (see Fig. 16.1).

One aspect of the NIST model is especially noteworthy; the domain model is based on a services based architecture (known as "actor-application") where each domain can literally exist anywhere. A home or business can possess generation capabilities transmitted to a distribution point within a building or plant, maintained by a control panel on a computer with excess power sold to a neighbor or across the country by markets controlled by internet based companies. For example, as noted in a recent *New York Times* article, "Google won federal approval in February to buy and sell electricity on American electricity markets." It also plans to offer "tools for measuring the electricity consumption of home appliances through partnerships with companies like General Electric" (Bhanoo 2010). In the future the intelligence to control these services is predicted to be "cloud" based internet applications much like the banking system of today (NIST 2012).

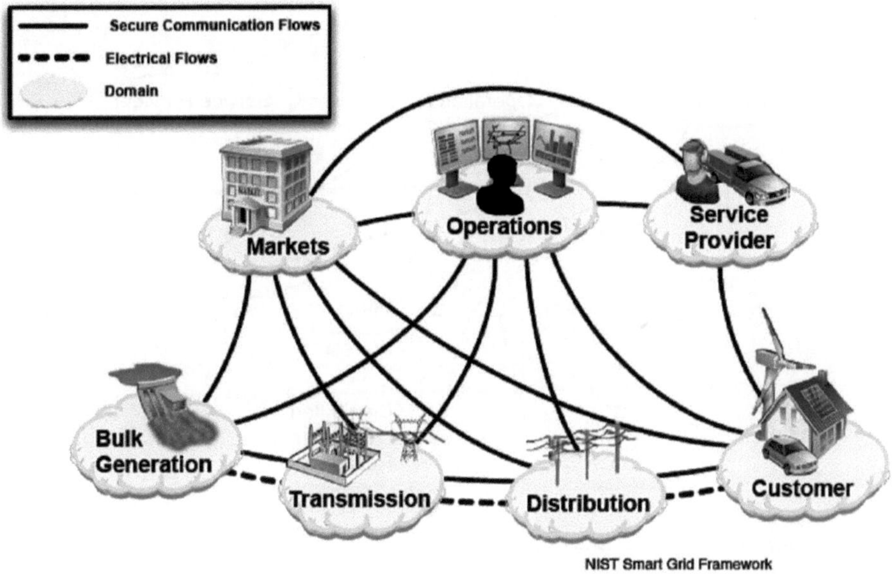

Fig. 16.1 NIST smart grid framework (NIST 2012, p. 42)

The European Union (EU) plan for Smart Grid development has taken the NIST model to the next level of complexity and flexibility. The EU model, developed by the Smart Grid Coordination Group of three standards organizations, Comité Européen de Normalisation (CEN), Comité Européen de Normalisation Électrotechnique (CENELEC), and European Telecommunications Standards Institute (ETSI), extends the NIST model by incorporating two additional actors: Distributed Energy Resources and Microgrid technologies architecture (see Fig. 16.2). Together these additional actors allow for a Smart Grid that is more modular in design with the ability to integrate power sources which can be isolated from the main grid into smaller grids. This extended NIST model allows more resilience and security by allowing the smaller grids the ability to disconnect from the large grid in the case of security breaches or disruptions or damage to other parts of the physical grid. A side benefit from a segmented grid built on Microgrid technologies is the ability to introduce privacy based data restrictions within individual Microgrids. (CEN-CENELEC-ETSI 2012).

Benefits of the Smart Grid

The fundamental differences between the existing grid and the smart grid are the ICT and distributed connectivity capabilities where the solid lines in Figs. 16.1 and 16.2 represent data networks which can exist via the internet or cloud and where the components can exist within the same building or across the country. It has been estimated that a smart grid could save U.S. utilities and their customers as much as

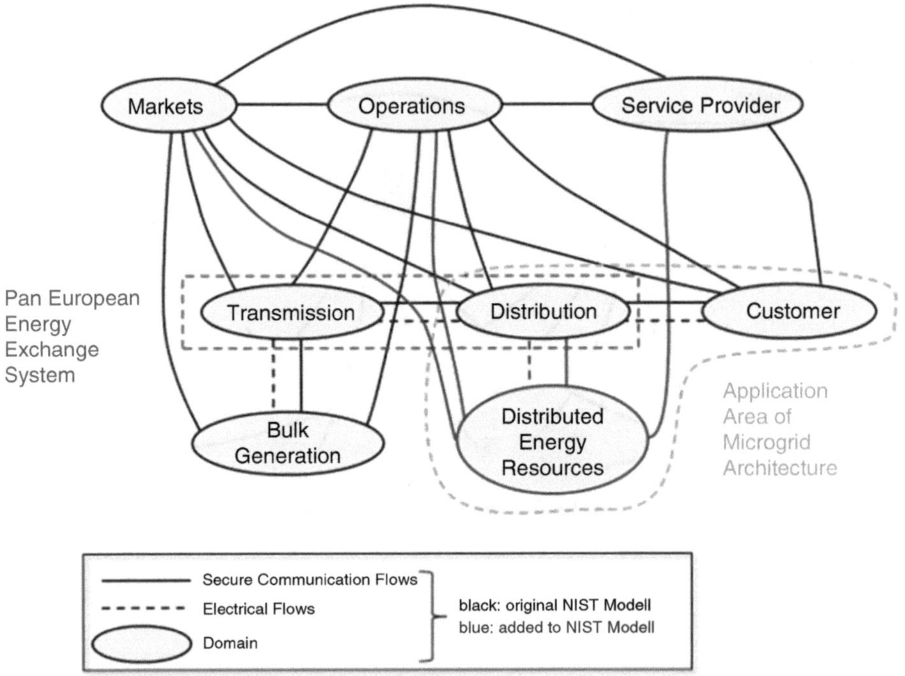

Fig. 16.2 EU extension of NIST framework (CEN-CENELEC_ETSI 2012, p. 21)

$20.4 billion annually by 2030 (Zeller 2010); however, the potential benefits of the smart grid extend well beyond the energy cost savings.

Amin (2004) has noted: "All economic and societal progress depends on a reliable and efficient energy infrastructure; for instance, banking and finance depend on the robustness of electric power, cable, and wireless telecommunications. Transportation systems including military and commercial aircraft and land and sea vessels depend on communication and energy networks. The linkages between electric power grid, telecommunications, and couplings of electric generation with oil, water, and gas pipelines are ever increasing and continue to be a lynchpin of energy supply networks." According to the International Energy Agency (IEA 2011) the Smart Grid has six key characteristics that will contribute to a stronger energy infrastructure and thus provide enhanced economic benefits (See Table 16.1).

Many of these characteristics rely upon the smart grid's ICT backbone. For example, an important aspect of Characteristic 2 (see Table 16.1) is the ability, utilizing smart grid ICT technologies, to spread the risk of price shocks in conventional fuels for power generation to other forms of power generation using substitute or renewable sources. Prior to the development of smart grid technologies it was difficult or even impossible to integrate these power sources on a large scale into the traditional transmission system. Additionally, smart grid technologies, specifically emerging Microgrid technologies (Farhangi 2010), provide the consumer the ability to interface with multiple power sources, thereby allowing individuals and businesses the opportunity to seamlessly replace or augment more expensive power

Table 16.1 Smart grid characteristics (IEA 2011)

Characteristic	Description
1. Enables informed participation by customers	Consumers help balance supply and demand, and ensure reliability by modifying the way they use and purchase electricity. These modifications come as a result of consumers having choices that motivate different purchasing patterns and behavior. These choices involve new technologies, new information about their electricity use, and new forms of electricity pricing and incentives
2. Accommodates all generation and storage options	A smart grid accommodates not only large, centralized power plants, but also the growing array of customer-sited distributed energy resources. Integration of these resources – including renewables, small-scale combined heat and power, and energy storage – will increase rapidly all along the value chain, from suppliers to marketers to customers
3. Enables new products, services and markets	Correctly designed and operated markets efficiently create an opportunity for consumers to choose among competing services. Some of the independent grid variables that must be explicitly managed are energy, capacity, location, time, rate of change and quality. Markets can play a major role in the management of these variables. Regulators, owners/operators and consumers need the flexibility to modify the rules of business to suit operating and market conditions
4. Provides the power quality for the range of needs	Not all commercial enterprises, and certainly not all residential customers, need the same quality of power. A smart grid supplies varying grades (and prices) of power. The cost of premium power-quality features can be included in the electrical service contract. Advanced control methods monitor essential components, enabling rapid diagnosis and solutions to events that impact power quality, such as lightning, switching surges, line faults and harmonic sources
5. Optimizes asset utilization and operating efficiency	A smart grid applies the latest technologies to optimize the use of its assets. For example, optimized capacity can be attainable with dynamic ratings, which allow assets to be used at greater loads by continuously sensing and rating their capacities. Maintenance efficiency can be optimized with condition-based maintenance, which signals the need for equipment maintenance at precisely the right time. System-control devices can be adjusted to reduce losses and eliminate congestion. Operating efficiency increases when selecting the least-cost energy-delivery system available through these types of system-control devices
6. Provides resiliency to disturbances, attacks and natural disasters	Resiliency refers to the ability of a system to react to unexpected events by isolating problematic elements while the rest of the system is restored to normal operation. These self-healing actions result in reduced interruption of service to consumers and help service providers better manage the delivery infrastructure

with cheaper power or even their own generated power, which if produced in excess of personal demand could be sold on the open market.

Characteristic 4 (see Table 16.2) refers primarily to centralized power and the pricing of electricity based upon the power quality needs of various customers. What the table fails to illuminate is that power quality can be augmented by consumers of power through devices or even by segmenting sections of their power

Table 16.2 Potential privacy consequences of the smart grid (EPIC, n.d.)

1. Identity theft
2. Determine personal behavior patterns
3. Determine specific appliances used
4. Perform real-time surveillance
5. Reveal activities through residual data
6. Targeted home invasions (latch key children, elderly, etc.)
7. Provide accidental invasions
8. Activity censorship
9. Decisions and actions based upon inaccurate data
10. Profiling
11. Unwanted publicity and embarrassment
12. Tracking behavior of renters/leasers
13. Behavior tracking (possible combination with personal behavior patterns)
14. Public aggregated searches revealing individual behavior

infrastructure or by merely purchasing power from multiple suppliers which guarantee levels of power quality through service level agreements (SLA) (Gustavsson and Ståhl 2010). This is a common ICT practice in many companies for all aspects of the technical infrastructures and the ICT infrastructure which supports them. This blending of responsibility and source of power based upon quality not only allows for competitive pricing among suppliers but also allows the consumer the means to control costs through conditioning their own power when protecting valuable pieces of equipment which are vital economic assets of companies and individuals alike.

Characteristic 5 (see Table 16.2) points to the Smart Grid's ability to manage the grid in a holistic manner (Nampuraja 2011) using an ICT approach known as IT Service Management (ITSM). ITSM is a process-based practice that aligns the delivery of services with the needs of the customer. An important aspect of ITSM is its ability to manage assets of entire systems through a parallel and simultaneous system of Service Support (SS) and Service Delivery (SD). Both SS and SD delicately balance asset management and maintenance while insuring virtually uninterrupted service at agreed upon levels, once again based upon the SLA model. (See ITIL, n.d. for a more detailed discussion of ITSM.)

The traditional grid which we live with today will transform to an "end state" grid which ultimately may look different than currently planned models. As noted by the U.S. Department of Energy (DOE):

> The Smart Grid will consist of millions of pieces and parts – controls, computers, power lines, and new technologies and equipment. It will take some time for all the technologies to be perfected, equipment installed, and systems tested before it comes fully on line. And

> it won't happen all at once – the Smart Grid is evolving, piece by piece, over the next decade or so. Once mature, the Smart Grid will likely bring the same kind of transformation that the Internet has already brought to the way we live, work, play, and learn (DOE n.d.).

While the innovative features of the smart grid hold great potential for improved energy efficiency through better management of consumer demand and improved stewardship of energy resources including greater utilization of renewable generation, they also pose a number of social and ethical challenges including: protecting the privacy of consumer usage information; securing the grid from attacks by foreign nations, terrorists, and malevolent hackers; and ensuring social justice in determining the price of electric power service. As with many new technologies the engineers engaged in developing the smart grid often overlook such issues or only turn to considering them once the technical standards and specifications have been settled. Failure to address these issues in a timely manner, however, may result in delays in establishing the smart grid and undermine its potential. Engineers and others involved in developing the smart grid need to examine ways to address organizational, social, and ethical dimensions that distributed generation and more extensive efforts to influence consumer usage patterns will raise. The cost of doing so would amount to an insignificant fraction of the projected necessary investments.

Preparing engineers to recognize and address such issues presents a significant challenge for engineering education. While several models of curriculum change to incorporate smart grid concepts have been proposed (e.g., Reed and Stanchina 2010; Sluss 2011), most focus solely on the technical aspects of the smart grid to the neglect of privacy, security, equity, and other social and ethical issues.

Privacy

As is the case for many other modern ICT applications such as the Internet and geographical positioning system (GPS), ensuring consumer privacy will be a challenge for the smart grid. Up until now our personal energy usage had been recorded by simple consumption metrics such as kilowatt hours measured using a conventional meter attached to a home or business. In the initial transition to a smart grid, utilities have begun to install "smart meters" that can provide feedback to the utility and customers (as often as every 15 s) on such factors as time of use of electricity; 65 million residential smart meters are expected to be in service in the U.S. by 2020 (Zeller 2010). Since many appliances have a unique "load signature," smart meter data can be analyzed to determine the types of appliances and other equipment consumers are using (Bleicher 2010). In the future, as more demand side technologies are developed, the smart grid could have the capability to monitor and control the usage of every plugged-in electrical device, which would allow the electric utility to turn the device off during times of peak demand to balance load across the grid. For the privilege of acquiring data and controlling consumer electrical devices utility companies may charge a reduced rate. Alternatively, rate structures that vary by time of day or fuel source (coal vs. wind, for example) may be instituted in order to influence consumer energy usage behaviors.

As we move from theory to design, the emerging smart grid will become a vast ICT network populated with a diverse set of data acquisition devices capable of tracking the source, ownership, performance, and behavioral characteristics of each connected component. The smart grid technologies with the potential to be privacy invasive include "smart" power meters, energy monitoring and control software programs, and monitoring chips built into devices that consume electricity.

In addition to control and monitoring functions, however, the smart grid will have the ability to collect, aggregate, and store individual consumer usage data such as the temporal pattern of electricity usage and the number, type, and usage of electrical appliances and electronic devices. Analysis of this data could reveal such information as home occupation patterns, the number of occupants, and the manufacturer and usage of individual devices – valuable to utility planners but additionally to marketing agencies, insurance companies (property, health, and life) and, potentially, criminals (for example, outsiders may be able to tell when a home is occupied, determine the type of security system, and learn other sensitive information). As Table 16.2 indicates, the list of potential privacy implications of the smart grid is extensive.

Collection and storage of data are only part of the issue. Ultimately, the privacy implications of the smart grid rest upon who *owns* consumer data (Cardenas and Safavi-Naini 2012). Much like the data acquired by supermarket bar-code scanners and loyalty cards, data on specific devices in homes and consumers' patterns of energy use will become a prized resource. Electric utilities or third-party vendors may sell personal data to other organizations to defray costs or simply to increase profits.

Data available through the smart grid will not necessarily be limited to electrical usage data. For example, as noted in an article in *Computerworld*: "GE is even building a smart refrigerator that will be able to read the bar codes of food containers. It'll be able to keep track of what's been bought, what recipes can be made from the food it contains and what should be on next week's grocery list (Cline 2009)."

The PowerMeter application developed by Google is an example of how third-party vendors may become involved in the management of smart grid data. An Internet-based application, PowerMeter received real-time information from utility smart meters and energy management devices and provided customers with access to their home electricity consumption on their personal iGoogle home page. According to McDaniel and McLaughlin (2009): "Although Google has yet to announce the final privacy policy for this service, early versions leave the door open to the company using this information for commercial purposes, such as marketing individual or aggregate usage statistics to third parties." Though Google discontinued the service in 2011, it is only one of many data-hungry organizations racing to develop smart grid monitoring equipment and data systems.

Of course, like supermarket loyalty cards, utility customers may be willing to give up some of their personal data if they think it is being used benignly and if they are getting something in return (such as reduced prices or rates). Up to now, however, utilities have not had to deal with consumer energy usage data on this scale; they may be unwilling to incur the added expense of protecting consumer data from illegitimate uses or reassuring consumers that this data is protected adequately. The implications have not escaped the privacy watchdogs or even high-ranking U.S. federal government officials. Indeed, former Commerce Secretary Gary Locke warned that privacy concerns might be the "Achilles' heel" of the smart grid. Achieving public acceptance of the smart grid may prove difficult if privacy concerns are not addressed in a proactive manner.

One reflection of consumer concern over the smart grid and privacy is that public controversies, utility commission investigations, and legal cases have already begun to emerge in several places including Nevada, Colorado, Maryland, Illinois and Texas in the United States (Mufson 2011; Cardenas and Safavi-Naini 2012) as well as in other countries including the Netherlands, Australia, and the United Kingdom (Global Smart Grid Federation 2012). Perhaps the best known smart grid case, which involved a class-action lawsuit, is Bakersfield, California where customers of Pacific Gas and Electric Company (PG&E) claimed that their utility bills rose significantly after installation of smart meters (Chediak 2009). In an Illinois court case with privacy implications, compulsory smart meters are being contested on the grounds of violation of 4th Amendment protections against privacy invasion and illegal search (Munkittrick 2012).

There are no federal laws on the books in the U. S. specifically regarding the smart grid (Munkittrick 2012) and existing privacy laws have limited application (McDaniel and McLaughlin 2009). There has, however, been no lack of federal government studies on the smart grid and privacy issues, including reports by the National Institute of Standards and Technology (NIST) (2010), the Department of Energy (DoE) (2010), and the Congressional Research Service (CRS) (Murrill et al. 2012). According to legal blogger David Munkittrick (2012), the reports recommended the following guidelines for smart grid development:

- Appoint personnel responsible for data security and privacy.
- Regularly audit privacy procedures.
- Establish procedures for law enforcement data requests.
- Provide notice to consumers in advance of collection and use of energy use data.
- Aggregate and anonymize data in a way that personal information or activities cannot be determined.
- Keep personal information only as long as necessary to accomplish the purpose for which it was collected.
- Allow individuals access to their personal energy data to correct inaccuracies.

NIST has also established a privacy working group under the framework of its Cyber Security Working Group. Per Cardenas and Safavi-Naini (2012): "The goal of the Privacy group is to identify and clearly describe privacy concerns with energy usage data and to propose ways to mitigate these concerns. In addition, the group strives to clarify privacy expectations, practices, and rights with regard to the Smart Grid."

Utility regulators in the U. S. have been sensitive to the smart grid privacy issue for more than a decade. In 2000, the National Association of Regulatory Utility Commissioners (NARUC) passed a "Resolution Urging the Adoption of General Privacy Principles For State Commission Use in Considering the Privacy Implications of the Use of Utility Customer Information" including provisions relating to the importance of privacy interests, customer determination of the degree of privacy extended to them, required informed consent by consumers for use of non-service or non-billing related data, and provision of data to third parties pursuant only to utility commission approval (NARUC 2000). More recently, NARUC (2011) passed a "Resolution on Smart Grid Principles" which includes sections on protections for vulnerable consumer groups, access to data by consumers and third parties (subject to informed consent of the consumers), and the importance of maintaining consumer privacy.

In 2011, The California Public Utility Commission (CPUC) became the first state commission to promulgate regulations on the privacy and security of consumer usage data that the Center for Democracy and Technology describes as "…a remarkable achievement that merits the attention of not only utility commissions in other states but also of stakeholders in other sectors, for it shows that a comprehensive privacy and data security framework can be crafted that supports both technology innovation and consumer protection" (Dempsey 2011).

The issue of privacy and the emerging Smart Grid is becoming noticed worldwide. In a recent directive the European Commission Data Protection Working Party alerted the public to the potentials of Smart Grid data acquisitions: "The Europe-wide rollout of 'smart metering systems' enables massive collection of personal information from European households, thus far unprecedented in the level of detail and comprehensive coverage: smart metering may enable tracking what members of a household do within the privacy of their own homes and thus building detailed profiles of all individuals based on their domestic activities" (European Commission Article 29 Data Protection Working Party 2013, p. 4).

A number of approaches to smart grid design aimed at protecting consumer privacy have also been proposed including anonymizing sensitive consumer data by distinguishing high-frequency metering data from low frequency data or by data aggregation; power routing to prevent individual appliance data from being detected at the meter; and optimizing sampling frequency to balance data needs and privacy concerns (Cardenas and Safavi-Naini 2012). A more comprehensive approach, involving organizational as well as technical innovation, that was adopted by San Diego Gas and Electric (SDG&E) in 2012, is to apply the "Privacy by Design (PbD)" principles developed by Ann Cavoukian, Information and Privacy Commissioner of Ontario, Canada:

> Privacy by Design (PbD) principles may be integrated right from the start as utilities begin their Smart Grid implementations, thus helping to make sure that customer information is protected. Embracing a positive-sum model whereby privacy, security and energy conservation may be achieved in unison is key to ensuring consumer confidence in electricity providers as Smart Grid projects are initiated. In addition, customer satisfaction with and trust

> of Smart Grid initiatives is an integral factor in the success of energy conservation and other goals of Smart Grid efforts. (Cavoukian and Winn 2012, p. 5).

PbD is based on seven Foundational Principles (Cavoukian and Winn 2012, p. 6):

1. Proactive not Reactive; Preventative not Remedial
2. Privacy as the Default Setting
3. Privacy Embedded into Design
4. Full Functionality – Positive-Sum, not Zero-Sum
5. End-to-End Security – Full Lifecycle Protection
6. Visibility and Transparency – Keep it Open
7. Respect for User Privacy – Keep it User-Centric

These principles were adopted for the smart grid context in a collaboration between SDG&E and Ann Cavoukian's group (see Table 16.3).

In addition to providing a roadmap for utilities, PbD provides an excellent framework for educating engineers on the importance of privacy. As Meldal et al. (2008) have shown, privacy and related concerns can and should be incorporated in both general education and engineering curricula:

> With the ever-increasing embedding of interconnected computing platforms at the core of our lives and of society, the successful trust systems issues education of the population in general and of the engineering professionals in particular becomes a matter of critical societal concern.
>
> Educational institutions benefit from taking an holistic approach to teaching security- and trust-related topics. The very ubiquity of the challenge can be made a vehicle for education, allowing for a pervasive injection of the concepts (and underlying technological and political challenges) of the interplay of security, trust, privacy and technology throughout the core as well as the discipline-specific curriculum components. (Meldal et al. 2008, p. 8)

With so much awareness of the importance of privacy on the part of nations, federal agencies, regulators, and smart grid technology designers, one would think

Table 16.3 Smart grid privacy principles (Cavoukian and Winn 2012, p. 13)

1. Smart Grid systems should feature privacy principles in their overall project governance frame work and proactively embed privacy requirements into their designs, in order to prevent privacy-invasive events from occurring
2. Smart Grid systems must ensure that privacy is the default – the "no action required" mode of protecting one's privacy – its presence being assured
3. Smart Grid systems must make privacy a core functionality in the design and architecture of Smart Grid systems and practices – an essential design feature
4. Smart Grid systems must avoid any unnecessary trade-offs between privacy and legitimate objectives of Smart Grid projects
5. Smart Grid systems must embed privacy end-to-end, throughout the entire life cycle of any personal information collected
6. Smart Grid systems must be visible and transparent to consumers – engaging in accountable business practices – to ensure that new Smart Grid systems operate according to stated objectives
7. Smart Grid systems must be designed with respect for consumer privacy, as a core foundational requirement

that privacy concerns will be comprehensively addressed. As seen in other areas of emerging technology, however, legal and ethical responses often lag far behind such issues (Marchant et al. 2011). Ultimately, the problem won't be solved until consumers are convinced their privacy is being preserved. As noted by the Electronic Privacy Information Center (EPIC, n.d.): "The key to privacy protection is to have the user maintain control over the collection, use, reuse, and sharing of personal information including their use of electricity."

Security

Unsurprisingly, many security aspects of the smart grid look like those of the Internet. Although the Internet has not been designated as the primary source of ICT communications, the smart grid will more than likely mature into a system that will utilize the Internet as its backbone. To secure both the informational and power-carrying capacity of the smart grid two important features must be addressed: the physical security of power and ICT networks and equipment and the security of huge databases and computers that analyze the data. The smart grid of the future will integrate both these networks creating the ability for either one to cause disruption to the other. Examples abound where highly automated systems have been brought to a halt or damaged by failures or security breaches in their ICT backbones (e.g., failures in automated securities trading, cyber warfare damage to Iran's centrifuges for nuclear fuel enrichment, and malevolent hacking resulting in infiltration and shutdown of corporate and government Web sites).

As noted by Kosut et al. (2011), "Future smart grids will likely to be more tightly integrated with the cyber infrastructure for sensing, control, scheduling, dispatch, and billing. Already the current power grid relies on computer and communication networks to manage generation and facilitate communications between users and suppliers. While such integration is essential for a future 'smart' grid, it also makes the power grid more vulnerable to cyber-attacks by adversaries around the globe."

Security breaches in the smart grid could lead to brownouts or even blackouts, and could cause serious, long-term damage to power generation, transmission, and distribution equipment. With the integration of power and ICT networks, power delivery components and even everyday power devices (such as appliances) will become nodes on the Internet. In the future, cyber-attacks such as denial-of-service or virus attacks could cause outages in the smart grid and limit electricity supplies, including critical services such as infrastructure and public safety. These attacks could originate anywhere in the world and could start as easily as introducing false data regarding energy usage across many nodes. What do these concerns mean for the development of security mechanisms, policies, and practices to secure the smart grid? There will be pressure to introduce a wider range of surveillance technologies; such technologies are already at the forefront of many heated debates regarding the intrusion of local, state, and federal governments, and also corporations, into the daily lives of individuals. Security and surveillance systems bring their own data needs, which promise to further erode personal freedoms, including privacy.

One of the most important security concerns of the smart grid is the viability of nation states to protect themselves from having their infrastructures crippled during time of war or as a lead up to hostilities. As the electrical grids of almost all modern societies have become the nerve center for economic, military and vital social systems the attack on these systems could lead to the collapse of entire societies.

As noted by Metke and Ekl (2010), "This vulnerability was considered such a potential risk that the U.S. government identified it as core element of legislation following the New York terrorist attacks of 2011. The need for critical infrastructure protection was first mandated by the Patriot Act of 2001 (Section 1016, the Critical Infrastructure Act of 2001). In 2003, Homeland Security Presidential Directive (HSPD) 7 established the national policy requiring federal departments and agencies to identify and prioritize United States Critical Infrastructure and Key Resources (CIKR) and to protect them from terrorist attacks."

In a 2009 article in the *Wall Street Journal* it was reported that "Cyberspies have penetrated the U.S. electrical grid and left behind software programs that could be used to disrupt the system, according to current and former national-security officials. The spies came from China, Russia and other countries, these officials said, and were believed to be on a mission to navigate the U.S. electrical system and its controls. The intruders haven't sought to damage the power grid or other key infrastructure, but officials warned they could try during a crisis or war" (Gorman 2009). Essentially such attacks could bring a country to its knees even before a single shot was fired. For example, recently a cold war has existed between Iran and the U.S. where a manifestation of hostilities has come in the form of the Stuxnet virus, a sophisticated computer virus deployed during the waning days of the Bush administration in an effort to thwart uranium enrichment in the Iranian government's nuclear program. As Sanger (2012) notes, "It appears to be the first time the United States has repeatedly used cyberweapons to cripple another country's infrastructure, achieving, with computer code, what until then could be accomplished only by bombing a country or sending in agents to plant explosives." Sanger also points out that: "President Obama has repeatedly told his aides that there are risks to using – and particularly to overusing – the weapon. In fact, no country's infrastructure is more dependent on computer systems, and thus more vulnerable to attack, than that of the United States. It is only a matter of time, most experts believe, before it becomes the target of the same kind of weapon that the Americans have used, secretly, against Iran."

The vulnerability of key systems and controls that are in widespread use in the developing smart grid is already apparent. As a recent news article noted: "A widely used system for controlling electricity, heating and other systems inside buildings remains vulnerable to attacks over the Internet, despite warnings from U.S. officials.... Poor security in industrial control systems, including those that run manufacturing facilities and power plants, has become an intense focus for security researchers and hackers alike since 2010 when the Stuxnet virus surfaced." (Menn 2013). The worrisome issue here is that we have identified widespread vulnerabilities in smart grid technologies at the same time as governments and hacker groups have developed internet viruses specifically designed to exploit those vulnerabilities.

Since the smart grid is predominately an intricate web of ICT networks a fully developed smart grid could in fact be the possible entry point of a devastating cyberattack. This kind of futuristic war could wreak havoc upon every aspect of a society since literally any and all electrical devices plugged into the smart grid could be comprised and corrupted.

If the security vulnerabilities discussed above can be identified and managed effectively, the smart grid promises to provide significant economic and social benefits. Indeed, balancing the potential economic benefits with privacy and security concerns will be a key challenge in the development of the smart grid. As NARAC (2011) notes in its "Resolution on Smart Grid Principles:"

> As a condition of approving smart grid investments, State commissions should hold utilities responsible for ensuring that smart grid technologies are deployed in a manner consistent with reasonable and effective cyber and physical security best practices. Smart grid systems should be designed to mitigate risks and enhance the resiliency of the power grid and preserve the accuracy, integrity, and privacy of data. State commissions should...recognize that cyber security requires coordination, adaptability and resiliency that goes beyond standards compliance.... Further, State commissions may want to assure that utilities have recovery plans in the event of a successful cyber or physical threat.

Engineering educators have begun to include security-related topics in smart grid courses and curricula (e.g., Schulz 2011; Shireen et al. 2013), though most treatments are limited to "security" as a technical concept; its social and ethical implications are far less recognized. Approaches advocated by Meldal et al. (2008) (discussed above) which locate topics such as security, privacy and trust in a broader socio-technical context are critically needed in engineering education.

Pricing and Equity

Though not as obvious as privacy and security issues, the smart grid also poses potential problems for equitable pricing of electric power service. The nature of these impacts will depend on whether consumer energy usage is left under utility control or consumers are allowed to make their own usage decisions under variable pricing schemes. The former case would limit consumer autonomy. Some utilities, for example, have expressed an interest in controlling customers' thermostats and other appliances (Levinson 2010). Variable pricing, on the other hand, would place an energy management burden on all residential consumers. Those with lower educational levels, limited Internet access or computer skills, medical or cognitive impairments, or those who simply lack time, resources, or motivation to manage their usage patterns could be at a disadvantage. Both cases will require innovative ratemaking and oversight by public utility commissions and greater coordination and standardization within and among retail service areas.

Though smart meter experiments are just in the beginning stages, there have already been regulatory and legal controversies over such issues as required prepaid service plans for low-income consumers (Ailworth 2009) and alleged price gouging under mandatory switches to smart meters. As noted earlier, a highly publicized controversy over higher bills occurred in Bakersfield, California (Chediak 2009), but protests and law suits have occurred elsewhere including Texas. In both the California and Texas cases, independent studies confirmed the accuracy of the smart meters, but utilities have been cautioned to approach the installation of smart meters with a greater concern for consumer needs and attitudes (Zeller 2010). The UK, for example, has developed a draft consumer engagement strategy (Global Smart Grid Federation 2012).

Issues regarding pricing and equity are far from black and white concerns over the cost of energy. The ability of a smart grid to closely monitor and manage the flow of electrical energy has a dramatic impact on almost every other socio- technical system which together have much influence over everyday lives. In a recent whitepaper, a trusted advisor to the Indian government describe the interrelationship of the Smart Grid to other technological systems:

> A smart grid could also interface with other utilities (gas, water, etc.). New services such as home monitoring, healthcare monitoring, etc. could be unleashed, which could provide new revenue streams to utilities as well as enhance consumer convenience. A power utility with its own network could become an Internet Service Provider, either directly or through a partnership or subsidiary. However, such changes are not only resisted (because of the creation of winners and losers) but also because there is vast uncertainty in how these will evolve. (Tongia 2009, p. 7)

Most of the discussion of equity and the smart grid has focused on the issue of dynamic pricing, i.e., variable electric rates that track the actual costs of providing services (in time-of use blocks or as frequently as "real-time'), with many economists and engineers favoring dynamic pricing on the grounds of economic efficiency. As noted by Faraqui (2010), "The pragmatic school of thought argues that rates should reflect time-variation in costs if the societal benefits from so doing exceed the societal costs. Typically, the societal benefits are associated with avoided capacity and energy costs and the societal costs are associated with implementing [smart metering]." According to the IEA (Heffner 2011), the arguments in favor of dynamic pricing include:

- *Traditional flat rates are not economically efficient and hide cross-subsidies*
- *Contrary to conventional wisdom, low-income customers can and will respond to dynamic price signals*

Faraqui (2010) is particularly adamant on the inadequacy of flat rates: "The opponents of dynamic pricing use the unfairness argument to present their case. But the presumption of unfairness in dynamic pricing rests on an assumption of fairness in today's tariffs. A flat rate that charges the same price around the clock essentially creates a cross-subsidy between consumers that have flatter-than-average load profiles and those that have peakier-than-average load profiles. This cross-subsidy is invisible to most consumers but over a period of time, it can run into the billions of dollars."

For their own part, the critics of dynamic pricing argue that it would not benefit the majority of users (Makovich 2011) and indeed could disadvantage small users and help lead to utility control of consumer loads (Levinson 2010). Because of such concerns the movement toward dynamic pricing has slowed in many jurisdictions and it remains unclear as to what extent and how fast it will be implemented.

As Felder (2011) notes, however, there are other factors in the implementation of the smart grid that raise equity concerns, most notably the compulsory installation of smart meters (and subsequent rate increase). Felder also questions the equity of the standard cost-benefit technique applied by supporters of the smart grid: "It would be a mistake to accept implicitly the assumption that a social cost-benefit analysis is the only equity framework and therefore to assume that if smart grid passes such a test, it should be adopted for both efficiency and equity reasons. Proponents of smart grid may, in effect, be making such an assumption by offering a social cost-benefit analysis as the only criterion for evaluation" (p. 95). Other equity issues highlighted by Felder include the distribution of risk and benefits between the utility and its customers (especially in light of the asymmetry of information) and the distribution of benefits between low-income and higher-income customers.

Ultimately, Felder argues, laws and regulations are needed to ensure equity is appropriately considered in rate-making proceedings:

> Although considerations of efficiency are important, they are not dispositive. Regulatory rulemaking commonly appeals to other values such as providing consumers information so they are better informed about decisions that affect them and they are better able to respond. Ratemaking policy also considers environmental issues, monetary and other support for low-income households, and assigning costs to those that cause them. Each of these considerations suggest individually and collectively that larger customers who consume more electricity than smaller customers should pay more for smart grid, that additional costs imposed on low-income consumers should be offset, at least partially, and that the elements of smart grid that directly and materially improve their lives should be prioritized over those elements that do not. (p. 98)

When applied to engineering education, Felder's argument is similar to recent calls for a focus on social justice in engineering education (Lucena 2013). As Leydens notes (2013): "A more socially just engineering profession will necessitate multiple changes to its pipeline – engineering education. If social justice education is to extend across and within the content of the engineering curriculum, it will need to inform and reform multiple educational components: foundational, design and engineering science – as well as humanities and social science – curricula."

Conclusion

Achieving the smart grid potential while tending to privacy, security, and equity concerns should begin with the realization that the smart grid is a complex sociotechnical system that requires solutions that go beyond the engineering of the grid. Solutions must include thoughtful deliberation by federal and state regulatory agencies, flexible utility responses in addressing consumer concerns and, most importantly, an engineering culture that recognizes and addresses the societal implications of the smart grid upstream in the R&D process and as standards are being developed.

For example, while the National Institute of Standards (NIST) highlighted privacy concerns in a recent report (NIST 2010), the U.S. federal government has yet to enact any smart grid privacy legislation or regulations. By contrast, the California Public Utilities Commission's (CPUC) 2011 decision on protecting privacy and security of consumer data is a landmark ruling that should provide a strong template for other state commissions (CPUC 2011).

One solution for addressing customer concerns regarding the smart grid is to provide opt-out options, such as Pacific Gas and Electric's proposal to permit customers worried about the environment, health, and safety effects of smart meter wireless radio signals to request that the signals be shut off (albeit with a charge for conventional meter reading) (Barringer 2011). More generally, Felder (2010) argues that consumer choice is "the prime benefit that smart grid technologies can provide" (p. 98).

As in the case of the human genome project and nanotechnology, where the U.S. federal funding agencies earmarked a percentage of research funds to examine such issues (Mills and Fleddermann 2005), there is an urgent need to examine the societal implications of the smart grid concurrent with its development. Failure to do so will further threaten civil liberties and social justice in the information age and is likely to pose substantial barriers to public acceptance of the smart grid.

Educating engineers who are prepared to meet the challenges posed by the societal implications of emerging technologies such as the smart grid should be a keystone of efforts to reform engineering curricula for the twenty-first century. Incremental changes such as the linkage of privacy, security and trust advocated by Meldal et al. (2012) are necessary but not sufficient. Ultimately, to prepare engineers for developing a "smart and just grid" (Welsch et al. 2013) will require a revolutionary change in engineering education that places social justice concerns at its core.

Acknowledgments This chapter draws in part on an earlier article (Kostyk and Herkert 2012).

References

Ailworth, E. (2009). Plan for prepaid electricity rejected. *Boston.com*. Retrieved from http://www.boston.com/business/articles/2009/07/23/mass_rejects_utilitys_prepayment_plan_for_low_income_customers/

Amin, M. (2004). Balancing market priorities with security issues. *IEEE Power and Energy Magazine, 2*(4), 30–38.

Amin, M., & Schewe, P. F. (2007). Preventing blackouts. *Scientific American Magazine, 296*(5), 60–67.

Barringer, F. (2011). Pacific gas offers solution to turn off smart meters. *New York Times*. Retrieved from http://www.nytimes.com/2011/03/25/business/energy-environment/25meter.html

Bhanoo, S. N. (2010). Google's energy foray: What's up? *New York Times, Green Blog*. Retrieved from http://green.blogs.nytimes.com/2010/05/05/googles-energy-foray-whats-up/
Bleicher, A. (2010). Privacy on the smart grid. *IEEE Spectrum, online*. Retrieved from http://spectrum.ieee.org/energy/the-smarter-grid/privacy-on-the-smart-grid
Cardenas, A. A., & Safavi-Naini, R. (2012). Security and privacy in the smart grid. In S. K. Das, K. Kant, & N. Zhang (Eds.), *Handbook on securing cyber-physical critical infrastructure* (pp. 637–654). Elsevier/Morgan Kaufman: Burlington, Massachusetts.
Cavoukian, A., & Winn, C. (2012). *Applying privacy by design best practices to SDG&E's smart pricing program*. Information & Privacy Commission, Ontario. Retrieved from http://www.ipc.on.ca/images/Resources/pbd-sdge.pdf
CEN-CENELEC-ETSI Smart Grid Coordination Group. (2012). *CEN-CENELEC-ETSI smart grid coordination group smart grid reference architecture.* Retrieved from http://ec.europa.eu/energy/gas_electricity/smartgrids/doc/xpert_group1_reference_architecture.pdf
Chediak, M. (2009). PG&E faces revolt over smart grid. *BusinessWeek: Technology*. Retrieved from http://www.businessweek.com/technology/content/dec2009/tc20091230_147434.htm
Cline, J. (2009). Will the smart grid protect consumer privacy? Computerworld. Retrieved from http://www.computerworld.com/s/article/9141002/Will_the_smart_grid_protect_consumer_privacy_
CPUC. (2011). Decision adopting rules to protect the privacy and security of the electricity usage data of the customers of Pacific gas and electric company, Southern California Edison Company, and San Diego gas & electric. Decision 11-07-056. California Public Utilities Commission. Retrieved from http://docs.cpuc.ca.gov/PublishedDocs/WORD_PDF/FINAL_DECISION/140369.PDF
Dempsey, J. (2011). California adopts smart grid privacy rule. Center for Democracy & Technology. Retrieved from https://cdt.org/blog/california-adopts-smart-grid-privacy-rule/
DOE. (n.d.). *Building and testing the smart grid.* SmartGrid.Gov, U.S. Department of Energy. Retrieved from http://www.smartgrid.gov/the_smart_grid#smart_grid
DOE. (2010). *Data access and privacy issues related to smart grid technologies.* Department of Energy. Retrieved from http://energy.gov/sites/prod/files/gcprod/documents/Broadband_Report_Data_Privacy_10_5.pdf
EPIC. (n.d.). The smart grid and privacy. EPIC – Electronic Privacy Information Center. Retrieved from http://epic.org/privacy/smartgrid/smartgrid.html
European Commission Article 29 Data Protection Working Party. (2013). *Opinion 04/2013 of the data protection impact assessment template for smart grid and smart metering systems ('DPIA Template'), Prepared by expert group 2 of the commission's smart grid task force.* Retrieved from http://ec.europa.eu/justice/data-protection/article-29/documentation/opinion-recommendation/files/2013/wp205_en.pdf
Farhangi, H. (2010). The path of the smart grid. *IEEE Power and Energy Magazine, 8*(1), 18–28.
Faruqui, A. (2010). The ethics of dynamic pricing. *Electricity Journal, 23*(6), 13–27.
Felder, F. A. (2011). The equity implications of smart grid: Questioning the size and distribution of smart grid costs and benefits. In F. P. Sioshansi (Ed.), *Smart grid: Integrating renewable, distributed & efficient energy* (pp. 85–100). Waltham: Academic.
Heffner, G. (2011). Smart grid – smart customer policy needs. In: *Workshop report for the IEA energy efficiency working party,* April, Paris.
Global Smart Grid Federation. (2012). Global smart grid federation report. Retrieved from http://www.globalsmartgridfederation.org/documents/May31GSGF_report_digital_single.pdf
Gorman, S. (2009). Electricity grid in US penetrated by spies. *Wall Street Journal*. Retrieved from http://online.wsj.com/article/SB123914805204099085.html
Gustavsson, R., & Ståhl, B. (2010). The empowered user – the critical interface to critical infrastructures. In: *5th international conference on critical infrastructure (CRIS),* IEEE, 2010, pp. 1–3
IEA. (2011). Technology roadmap: Smart grids. International Energy Agency. Retrieved from http://www.iea.org/publications/freepublications/publication/name,3972,en.html
ITIL. (n.d.). ITIL official website. Retrieved from http://www.itil-officialsite.com/

Kostyk, T., & Herkert, J. (2012). Societal implications of the emerging smart grid. *Communications of the ACM, 55*(11), 34–36.

Kosut, O., Jia, L., Thomas, R. J., & Tong, L. (2011). Malicious data attacks on the smart grid. *IEEE Transactions on Smart Grid, 2*(4), 645–658.

Levinson, M. (2010). Is the smart grid really a smart idea? *Issues in Science and Technology, 27*(1), 39.

Leydens, J. A. (2013). Integrating social justice into engineering education from the margins: Guidelines for addressing sources of faculty resistance to social justice education. In J. Lucena (Ed.), *Engineering education for social justice* (pp. 179–200). Netherlands: Springer.

Lucena, J. (Ed.). (2013). *Engineering education for social justice*. Netherlands: Springer.

Makovich, L. J. (2011). The smart grid: Separating perception from reality. *Issues in Science and Technology, 27*(3), 61–70.

Marchant, G. E., Allenby, B. R., & Herkert, J. R. (Eds.). (2011). *The growing gap between emerging technologies and legal-ethical oversight: The pacing problem.* Dordrecht, Germany: Springer.

McDaniel, P., & McLaughlin, S. (2009). Security and privacy challenges in the smart grid. *IEEE Security & Privacy, 7*(3), 75–77.

Meldal, S., Gates, K., Smith, R., Su, X. (2008). Security, safety and privacy – pervasive themes for engineering education. In Geza Varady (ed.) *ICEE 2008*, iNEER.

Menn, J. (2013). Researchers warn of cyber flaws in Honeywell control systems. Retrieved from http://www.reuters.com/article/2013/02/05/cybersecurity-controls-idUSL1N0B5LG320130205

Metke, A. R., & Ekl, R. L. (2010). Security technology for smart grid networks. *IEEE Transactions on Smart Grid, 1*(1), 99–107.

Mills, K., & Fleddermann, C. (2005). Getting the best from nanotechnology: Approaching social and ethical implications openly and proactively. *IEEE Technology and Society Magazine, 24*(4), 18–26.

Mufson, S. (2011). Growing field of "Smart Grid" technology faces opposition over pricing, Privacy. *Washington Post*. Retrieved from http://articles.washingtonpost.com/2011-11-11/business/35282623_1_smart-grid-smart-meters-renewable-energy-sources

Munkittrick, D. (2012). Smart grid technology implicates new privacy concerns. Privacy Law Blog. Retrieved from http://privacylaw.proskauer.com/2012/03/articles/data-privacy-laws/smart-grid-technology-implicates-new-privacy-concerns/

Murrill, B. J., Liu, E. C., & Thompson, R. M., II. (2012). Smart meter data: Privacy and cybersecurity. Washington, DC: Congressional Research Service.

Nampuraja, E. (2011). A unified management system for smart grid. In *Proceedings of the IEEE ISGT 2011 India*, pp. 328–333.

NARUC. (2000). *Resolution urging the adoption of general privacy principles for state commission use in considering the privacy implications of the use of utility customer information.* Retrieved from http://www.naruc.org/Resolutions/privacy_principles.pdf

NARUC. (2011). Resolution on smart grid principles. Retrieved from http://www.naruc.org/Resolutions/Resolution on Smart Grid Principles.pdf

NIST. (2010). *Guidelines for smart grid cyber security: Vol. 2, privacy and the smart grid.* U.S. Department of Commerce, Technology Administration, National Institute of Standards and Technology. NISTIR 7628.

NIST. (2012). *NIST framework and roadmap for smart grid interoperability standards, release 2.0.* U.S. Department of Commerce, National Institute of Standards and Technology. NIST Special Publication 1108R2.

Reed, G. F., & Stanchina, W. E. (2010). Smart grid education models for modern electric power system engineering curriculum. *IEEE power and energy society general meeting, IEEE* (pp. 1–5). IEEE: Piscataway, New Jersey.

Rogers, R. (2013). *Smart grid and energy storage installations rising*. Vital signs, Worldwatch Institute. Retrieved from http://vitalsigns.worldwatch.org/vs-trend/smart-grid-and-energy-storage-installations-rising

Romero, J. J. (2012). Blackouts illuminate India's power problems. *IEEE Spectrum, online*. Retrieved from http://spectrum.ieee.org/energy/the-smarter-grid/blackouts-illuminate-indias-power-problems

Sanger, D. E. (2012). Obama order sped up wave of cyberattacks against Iran. *New York Times*. Retrieved from http://www.nytimes.com/2012/06/01/world/middleeast/obama-ordered-wave-of-cyberattacks-against-iran.html

Schulz, N. N. (2011). Integrating smart grid technologies into an electrical and computer engineering curriculum. In*: Innovative smart grid technologies Asia (ISGT), 2011 IEEE PES* (pp. 1–4). IEEE: Piscataway, New Jersey.

Shireen, W., Kotti, R., & Villanueva, J. A. (2013). *ASEE 2013 Annual Conference*. Retrieved from http://www.asee.org/public/conferences/20/papers/7132/view

Sluss, J. J. (2011). Engineering education activities in electric energy systems. *Computer, 44*(4), 97–98.

Tongia, R. (2009). Smart grids white paper. Center for Study of Science, Technology and Policy. Bangalore. Retrieved from http://www.cstep.in/docs/Smart_Grid_Whitepaper_CSTEP.pdf

Welsch, M., Bazilian, M., Howells, M., Divan, D., Elzinga, D., Strbac, G., Fones, L., Keane, A., Gielen, D., Balijepalli, M., Brew-Hammond, A., & Yumkella, K. (2013). Smart and just grids for sub-Saharan Africa: Exploring options. *Renewable and Sustainable Energy Reviews, 20*, 336–352.

Zeller, Jr., T. (2010). "Smart" electric meters draw complaints of inaccuracy. *New York Times*. Retrieved from http://www.nytimes.com/2010/11/13/business/13meter.html

Joseph Herkert B.S. in Electrical Engineering from Southern Methodist University. D.Sc. in Engineering & Policy from Washington University in St. Louis. Lincoln Associate Professor of Ethics and Technology, College of Letters & Sciences and the Consortium for Science, Policy & Outcomes, Arizona State University, USA. He is Co-Editor of *The Growing Gap Between Emerging Technologies and Legal-Ethical Oversight: the Pacing Problem* (Springer, 2011), Editor of *Social, Ethical and Policy Implications of Engineering: Selected Readings* (Wiley/IEEE Press, 2000) and has published numerous articles on engineering ethics and societal implications of technology in engineering, law, social science, and applied ethics journals. He previously served as Editor of IEEE *Technology & Society* and an Associate Editor of *Engineering Studies*. He is a Distinguished Life Member of the Executive Board of the National Institute for Engineering Ethics, a former Chair of the Liberal Education/Engineering and Society Division of the American Society for Engineering Education, and a former President of the IEEE Society on Social Implications of Technology.

Timothy Kostyk B.S. in Engineering, University of Louisville, MBA, Bellarmine University. Doctoral student in Arizona State University's Human and Social Dimensions of Science and Technology Program, a researcher and teacher at Arizona State University, where he: teaches classes on the emerging Smart Grid, is a contributor to the Energy Ethics and Policy ongoing seminar, and a member of the Consortium for Science, Policy, and Outcomes. For the past 3 years this field of study has concentrated on the examination of the world wide effort to redesign, rebuild and remodel the existing electrical grid into what is known as the Smart Grid. The dimension of this study incorporates the human, social, and technological aspects of design, particularly the ethical impact of engineered designs including those related to issues concerning privacy, equity and security. For the past 3 years Tim has extensively studied the ongoing efforts of multiple international and national governmental and professional engineering organizations as they together formulate the vision and develop the plans and standards for the development of the Smart Grid.

Chapter 17
From Engineering Ethics to Engineering Politics

Carl Mitcham and Wang Nan

Abstract Prior to the 1950s, engineering ethics emerged solely within the engineering profession itself. However, after World War II, the first efforts of general philosophical reflection on technology among engineers in Germany gave rise to early engagements between engineers and philosophers on this theme. Then in the 1970s in the United States a second engagement took place between engineers and philosophers, this time stimulated more by philosophers than by engineers. With the separate contributions in the two countries, engineering ethics began to become an interest in many other countries, especially influenced by the American approach, which is much less deeply philosophical than that which emerged in Germany. Outside Germany and the United States, the three places where discussions of engineering ethics most involve philosophers are Denmark, the Netherlands, and China. Finally, as far as the future of this topic is concerned, engineering ethics would benefit from expanding appreciation of the political aspects of engineering, echoing philosophical connections between ethics and political philosophy found in Aristotle.

Keywords Engineering ethics • Engineers • Philosophers • Engineering politics

Introduction

Is engineering ethics part of engineering or part of ethics and therefore of philosophy? Ideally it might be both. Yet prior to the 1950s it would have been difficult to find any discussion of engineering ethics that had much philosophical depth. It was just engineers trying to develop moral guidelines for the profession rather than

C. Mitcham (✉)
Colorado School of Mines, Golden, Colorado, USA

Renmin University, Beijing, China
e-mail: cmitcham@mines.edu

Wang Nan
College of Humanities and Social Sciences, University of Chinese Academy of Sciences, No 19A Yuquan Road, Beijing 100049, China
e-mail: wangnan@ucas.ac.cn

S.H. Christensen et al. (eds.), *Engineering Identities, Epistemologies and Values*, Philosophy of Engineering and Technology 21,
DOI 10.1007/978-3-319-16172-3_17

independent analysis and criticism of moral assumptions common in the profession. During the first half of the twentieth century engineering ethics had about as much to do with philosophy as the work of a shade-tree mechanic has to do with automotive engineering. Then beginning in the 1950s and again in the 1970s, as a result of discussions that took place initially in Germany and independently in the United States, respectively, engineering ethics began to acquire new seriousness in engineering and became a presence in the field of applied ethics.

During the second half of the twentieth century, engineering ethics became increasingly engaged with professionalized philosophy and expanded in a trajectory that moves from discussions based initially in two particular countries through discussions in many countries. The expansion has raised questions related to globalization and suggests the need for a movement from ethics to politics. The present chapter offers a summary sketch of these developments, with some reference to the contexts in which they have taken place. We make a point not just of describing different developments in engineering ethics but of trying also to identify the problems that gave rise to distinct engineering ethics discussions. In conclusion we suggest the emergence of new problems and try to point toward future developments.

Pre-philosophical Origins

In English the terms "engineering ethics" and "ethics in engineering" tend to be interchangeable. The later nevertheless more clearly declares that the subject matter concerns ethical questions in engineering. This means that different conceptions of what engineering is will have implications for engineering ethics.

For purposes of the present discussion, we provisionally adopt Michael Davis's definition of engineering in social and historical terms (Chap. 4). For Davis, like all professions,

> engineering is self-defining (in something other than the classical sense of definition). There is a core, more or less fixed by history at any given time, which decides what is engineering and what is not. This historical core is not a concept but an organization of living practitioners who – by discipline, occupation, and profession – are undoubtedly engineers.

As constituted by discipline, by occupation, and by profession, engineering has undergone continuous emergence from the late 1600s to the present.

One aspect of this emergence of the engineering profession has been the development of engineering ethics. Initially ethics was submerged in occupation. As Davis notes, engineers were first denominated as such in France in 1676 with creation of the military *corps du génie* (engineering corps). The late 1700s and early 1800s witnessed formation in England of the original professional engineering societies as non-military organizations. The Institution of Civil Engineers (ICE, officially founded in 1818 in London, but with roots that go back to the informal Society of Civil Engineers founded by John Smeaton in 1771) had no explicit code of ethics. At the same time, Thomas Tredgold's definition of engineering (formulated for the

ICE in 1828) as "the art of directing the great sources of power in Nature for the use and convenience of man" implicitly associates engineering with the moral theory of David Hume, for whom use and convenience are basic moral categories of human benefit (Mitcham and Muñoz 2010).

Yet it was not until the late 1800s and early 1900s in the United States that engineers began to discuss engineering ethics as such. The first appearance of the term "engineering ethics" in the title of an independent publication comes from a Carnegie Library of Pittsburgh *Monthly Bulletin* (1917) bibliography on the subject. In his preface to the 17-page collection of modestly annotated references, the author notes how it was prepared in response to "numerous requests [which had] come to the Technology Department for material on Engineering Ethics" and explicitly identifies engineering ethics with "ethics for engineers." Ethics for engineers, however, is not the same as ethics in engineering; it subordinates ethics to what engineers do and aims to help them function more effectively as engineers. Ethics in engineering, by contrast, can sometimes make engineering more difficult.

In the references themselves, the first use of "engineering ethics" in a title occurs in an 1893 news story about a discussion in the American Society of Civil Engineers (ASCE, founded in 1852 and the oldest professional engineering society in the United States) concerning the desirability of appointing a committee to explore drafting an ethics code. The resulting discussion in the later 1890s and first decade of the 1900s, together with related discussions in other engineering societies, led eventually to imitation of the professional associations of lawyers (American Bar Association) and of physicians (American Medical Association), which adopted codes of professional ethics in 1908 and 1912, respectively. The American Institute of Electrical Engineers (AIEE, later to become the Institute of Electrical and Electronic Engineers or IEEE, the largest engineering society in the world) adopted the first professional ethics code in 1912; the ASCE adopted another two years later in 1914. In both cases, the primary duty of the engineer was described as serving as a "faithful agent or trustee" of some employing company – a duty that has been argued to reflect the origins of engineering in the military, where obedience to authority is a primary obligation (see Mitcham 1992).

What was the perceived need that these codes aimed to address? According to Edwin Layton's historical narrative on the sociological development of the American engineering profession, the key element was what he termed *The Revolt of the Engineers* (1971) against subservience to corporate interests. The codes aimed to help engineers resist persistence efforts, both economic and political, to deprive them of their rightful authority over the design and construction of large-scale projects – deprivations that sometimes resulted in dam collapses and bridge failures – and undermined pursuit of the technical ideal of efficiency. Unfortunately the dominance of corporate interests even within the profession forced the codes to stress some version of company loyalty so that the original aim was often subverted. In David Noble's Marxist analysis, American engineering was actually "guided as much by the capitalist need to minimize both the cost and the autonomy of skilled labor as by the desire to harness most efficiently the potentials of matter and energy" (Noble, 1977, p. 34).

What is equally as significant as their functional features is that these early engineering ethics codes were formulated minus any consultation with professors of ethics or philosophy. Instead, despite the presence in American public life of William James, John Dewey, and others, engineering ethics was the product of what might be called folk philosophy.

Initial Engineering-Philosophical Discussions: Germany

The first clear engagements between engineers and philosophers as separate professional traditions took place in Germany. The background for this engagement was emergence in the late 1800s of the first efforts of general philosophical reflection on technology (in German, *Technik*, which can also mean engineering). As Carl Mitcham (1994) has summarized this development of what he terms engineering philosophy of technology, ethics is subordinate to the articulation of an engineering world view. He quotes, for instance, one apology for this world view from the Russian engineer Peter K. Engelmeier, writing in German:

> *Techniker* [technologists or engineers] generally believe they have fulfilled their social tasks when they have delivered good, cheap products. But this is only part of their professional task. The well-educated engineers of today are not found only in factories. Highway and water transport, urban and economic management, etc. are already under the direction of engineers. Our professional colleagues are climbing ever higher up the social ladder; the engineer is even occasionally becoming a statesman.... This extension of the engineering profession not only seems welcome, it is the necessary consequence of the enormous economic growth of modern society and augurs well for its future evolution. (Mitcham 1994, p. 26, modified)

Because of this societal ascent engineers, in order to achieve the social recognition they deserve, should work to articulate their world view as a general philosophy of technology.

According to Mitcham this was most fully realized in the German context in the philosophical efforts of the German inventor and engineer Friedrich Dessauer. Dessauer not only argued his views, like Engelmeier, with conscious reference to such key philosophers and G.W.F. Hegel and Immanuel Kant, but drew on and entered into dialogue with Plato and Aristotle from the classical period of European philosophy and with contemporary Marxist, existentialist, and other thinkers. The result is a general engineering philosophy that sees engineers, through the act of invention, as coming in contact with Platonic forms or closing the gap between Kantian phenomena and noumena. Modern engineering involves "participating in creation" and constitutes "*the greatest earthly experience of mortals*" (quoted from Mitcham, 1994, p. 33, emphasis in the original). Dessauer's philosophical praise of engineering integrated ethics into epistemology, metaphysics, and even aesthetics.

Such a positive philosophical interpretation of engineering was called dramatically into question by World War II, which occasioned pessimism concerning highly developed technology and a tendency to condemn engineers as morally irresponsible contributors to destructive warfare. The actions of German engineers during the National Socialist regime (1933–1944) challenged engineering philosophical idealism. Because many of its members had been compromised by involvement with Nazism, the Verein Deutscher Ingenieure (VDI or Association of German Engineers) undertook to promote a new philosophical reflection among engineers. This led to a more sustained dialogue between engineers and philosophers than had previously taken place in Germany or in any other country.

Prior to World War II, German engineers had attempted to construct an ethics and philosophy of engineering on their own. They were autodidacts who read philosophical texts but did not ask philosophers themselves to think about engineering. After the war they undertook to ask philosophers themselves for help.

In the early 1950s, for instance, the VDI sponsored a series of conferences on "The Responsibility of Engineers," "Humanity and Work in the Technological Era," "Changes in Humanity through Technology," and "Humanity in the Force-field of Technology." In all cases, professional philosophers were invited to discuss the issues with professional engineers. Out of the first conference came "The Engineer's Confession," a Hippocratic-like oath for VDI members, and later the formation of a special Mensch und Technik [Humanity and technology] study group composed of engineers and philosophers. Broken down into working committees on such themes as "Pedagogy and Technology," "Sociology and Technology," "Religion and Technology," and "Philosophy and Technology," the study group produced by the mid-1970s a series of publications focusing on technology and values.

This work in turn led to replacement of the now dated "Engineer's Confession" and to further interdisciplinary engineering-philosophy research, especially on the theory of technology assessment. With regard to professional ethics, one Mensch und Technik working committee report in 1980 proposed simply that "The aim of all engineers is the improvement of the possibilities of life for all humanity by the development and appropriate application of technical means" (VDI 1980, p. x). With regard to the foundations of technology assessment, a second working committee in 1986 identified eight fields of value (environmental quality, health, safety, functionality, economics, living standards, personal development, and social quality), mapped out their interrelations, and developed recommendations for their implementation in the design of technical products and projects. The practice of comprehensive, interdisciplinary technology assessment effectively became a recommended professional ethical obligation for German engineers. The best single introduction to engineering ethics in Germany is a volume edited by philosophers Hans Lenk and Günter Ropohl (1987), which includes as an appendix the "Verein Deutscher Ingenieure, Ausschuß 'Grundlagen der Technikbewertung': Vorentwurf für eine Richtlinie 'Empfehlungen zur Technikbewertung'" [Association of German Engineers, "Foundations of Technology Assessment" Committee: Preliminary draft of a "Recommendations for Technology Assessment" guideline], parts 1–3 of five parts.

This distinctive interaction between engineers and philosophers for the first time created a philosophically rich engineering ethics within a professional engineering society. At the same time, by the early 2000s some philosophers associated with the

VDI began to question the adequacy of their work. Günter Ropohl (2002), for instance, described what he termed "the mixed prospects of engineering ethics." Although a new awareness had emerged among German engineers of endogenous ethical responsibilities, because their work was done mostly in teams under conditions highly influenced by economic and social pressures, they also increasingly acknowledged the extent to which they were subject to exogenous influences. Indeed, in a world undergoing globalization, "the world society – beyond the individual, the corporation and the national state – is appearing as the fourth level of responsibility in technology" (Ropohl 2002, p. 154). Whether and to what extent this is threat or opportunity and how it is to be managed became an issue of continuing discussion. It also points toward the importance, beyond ethics, of politics and political philosophy.

Initial Engineering-Philosophical Discussions: United States

In the United States the stimulus that brought philosophers and engineers together came more from outside than from within the profession. In 1977 the U.S. National Endowment for the Humanities (NEH) created a National Project on Philosophy and Engineering Ethics. Announcing the project in the journal of the National Society of Professional Engineers, one project leader observed, "Philosophers have not yet entered into constructive partnerships with engineers similar to their efforts in the field of medicine." This was not surprising, since "engineers have generally not been aware of the potential contributions philosophers might make [and] philosophers have on the whole failed to appreciate and understand the social and intellectual significance" of ethical problems in engineering. So the NEH, promoting what became known as the "applied turn" in American philosophy, took the initiative. As the announcement went on to explain, "The National Project of Philosophy and Engineering Ethics has been designed to recruit 15–18 professional engineers from both the academic and non-academic engineering communities, who are interested in teaming up with professional philosophers to formulate, develop, and implement projects dealing with the ethical problems in engineering" (Flores 1977, p. 28).

The external initiative also occurred as engineers themselves had been expanding their understandings of the profession. One pivotal event occurred in California in the 1930s when two engineers reported some illegal activities of their supervisors (who were subsequently tried and convicted), but the reporting engineers found themselves expelled from the ASCE for unethically failing to act as a "faithful agent or trustee" of their employer. One of the engineers continued unsuccessfully into the 1950s to seek vindication. Discussion of this and related cases led in the mid-1970s to a fundamental revision of the ASCE ethics code. The first principle of the new code was that "Engineers shall hold paramount the safety, health, and welfare of the public in the performance of their professional duties."

In parallel developments, leading AIEE engineers in the early 1900s had begun to challenge Tredgold's classic definition by conceiving of engineering as focused not just on exploiting the forces of nature for human benefit but of pursuing this end through the management of other human beings as well (see McMahon 1984, chap. 4). Over the course of time, however, it was increasingly recognized that the implementation of any such expanded vision, especially in conjunction with increased recognition of the manifold societal impacts of technology, presented distinctive challenges. As historian Matthew Wisnioski (2012) has richly chronicled, the late 1960s and early 1970s witnessed a blossoming of dissent within the engineering community that both responded to and mirrored public concerns about nuclear weapons, environmental pollution, and the technological transformation of society. In 1974 the IEEE (created in 1963 by merger of the AIEE and the Institute of Radio Engineers or IRE), like the ASCE, affirmed in a revised "Code of Ethics for Engineers" an obligation to "protect the safety, health and welfare of the public." Initially relegated to the fourth of four articles, by 1990 "making decisions consistent with the safety, health and welfare of the public" had become the first of ten principles. Additionally, the 1970s witnessed creation in the IEEE of a new special section to promote reflection on the social implications of technology, and the editor of the flagship journal, *IEEE Spectrum*, began to refer to a "new professionalism" to be "based not only on traditional high standards of technical achievement but that embraces concern for the impact of technological developments on society as well" (Christiansen 1972, p. 17). Again, such concern seemed to invite dialogue with philosophers, some of whom in the applied field had also become critics of technological transformations in society.

One outcome of the NEH project, in collaboration with a new Ethics and Values in Science and Technology (EVIST) Program at the National Science Foundation, was a number of publications oriented toward the teaching of engineering ethics. The textbook that most integrated philosophy into engineering was by the team of Mike Martin (philosopher) and Roland Schinzinger (engineer) titled *Ethics in Engineering* (first edition, 1983). As the authors explained in their introduction, they took engineering ethics to be "the discipline or study of the moral issues arising in and surrounding engineering" and to involve "normative (evaluative) inquiries, conceptual (meaning) inquiries, and descriptive (factual) inquires," with the normative inquires being central. The text itself then developed a challenging notion of engineering as social experimentation that required adaptation of the principles of free and informed consent (from biomedical ethics) and highlighted a primary professional concern for safety. It went on to examine the ways engineering is embedded in organizations and engaged with management and philosophically explicated both the responsibilities and rights of engineers. It concluded with philosophical reflections on career choices. This was the first full book in English that could accurately be described as bringing philosophical ethics to bear in engineering, and through multiple editions (1989, 1996, and 2005) it significantly influenced the field.

Over the course of the 1980s and 1990s engineering ethics courses became increasing features of engineering education. Indeed, the primary philosopher-engineer connection was between philosophy and engineering professors rather than with working engineers. In 2000 ABET (from Accreditation Board for Engineering and Technology), the organization that accredits U.S. engineering programs, began explicitly to list as one of 11 required educational outcomes, "an understanding of professional and ethical responsibility."

The primary way this understanding came to be taught was not so much through the kind of critical philosophical reflection promoted by Martin and Schinzinger as by the teaching of professional ethics codes and case studies. Indeed, professional and public discussion of the case of the space shuttle *Challenger* from 1986, and the way in which engineers had opposed the disastrous launch, helped stimulate ABET accreditation policy. In respect to codes and cases, a second textbook, stimulated this time by National Science Foundation support, became exemplary: Charles Harris (philosopher), Michael Pritchard (philosopher), and Michael Rabins (engineer), *Engineering Ethics: Concepts and Cases* (1995, with subsequent editions in 2000, 2005, 2008, and 2013). Case studies also became important features of subsequent editions of the Martin and Schinzinger text. It is worth noting that code and case study teaching was also typical of the few pre-1970s engineering ethics courses, taught mostly by senior engineering faculty seeking to share their experiences with a younger generation. Post-1970s the case studies just became more carefully developed.

However, in a reflective review of the achievements of engineering ethics, Paul Durbin, a philosophy professor who had also been involved in developing and teaching engineering ethics, questioned whether the philosophy-engineering interaction had realized its promise. It is certainly the case that engineering ethics never became as prominent a discourse as bioethics. Philosopher Stephen Toulmin (1982) once argued that "medicine actually saved the life of ethics" insofar as it had forced philosophers, who had become involved almost exclusively with increasingly abstract questions related to such topics as the language of morals, to begin again to deal with substantive issues of good and bad, right and wrong, in real-life situations. It was unfortunate, Durbin argued, that engineering ethics had not been able to develop into as robust a pursuit. According to, Durbin

> the recent history of engineering ethics in the USA is not a happy one. Philosophical engineering ethics is an almost complete failure, largely because the efforts of engineers and their professional societies are too limited in both scope and impact. With Robert Baum and Albert Flores – in their original hopes for the National Project of Philosophy and Engineering Ethics – I believe that the way to go is through collaborative efforts involving philosophers and engineers. But I would qualify my optimism about the approach by saying that its success depends on significant behavioral changes. The engineers and their professional societies need to broaden their focus, moving beyond a focus on individual misconduct to broader social responsibilities, and to welcome a broader range of people into the dialogue. [Additionally,] philosophers, social critics, reporters and editors, environmental activists, and the like need to be less confrontational and more willing to dialogue. [This would create a better conception] of engineering ethics than a definition that focuses mainly on the potential misconduct of individual engineers and technical professionals. (Durbin, 1997, p. 82)

Only by going beyond a focus on individual responsibility and such issues as whistle blowing, can engineering ethics make an impact on society anyway comparable to the impact made by engineering and technology themselves. Again, Durbin's plea can be framed as calling for a movement from ethics to politics.

Globalization

The American approach to engineering ethics is much less deeply philosophical than the German. But despite the fact that it arrived on the scene two decades after the German version, the American has arguably been the leading influence globally. Engineering ethics as it has been pursued and practiced in developed and developing countries alike has often echoed the American approach of adopting codes of conduct privileging the responsibility of individual engineers to protect public safety, health, and welfare.

To cite relevant examples from the developed world: The ICE in England, after almost 150 years without a code, in 1963 adopted a set of "Rules for Professional Conduct"; the 2010 version of this set of rules equates acting ethically with acting honorably and obligates "All members [to] discharge their professional duties with integrity [, competency, and with] full regard for the public interest, particularly in relation to matters of health and safety." The Conseil National des Ingénieurs et des Scientifiques de France (CNISF), which incorporates the Société Centrale des Ingénieurs Civils (founded 1848), in 2001 adopted a "Charte d'Éthique de l'Ingénieur" proclaiming "engineers are citizens … involved in civic actions aiming for the common good." The Engineering Society of Finland (founded 1880), in 1966 adopted a "Code of Honour" calling on members "to be of service of both [their] country and mankind as a whole." The Institution of Engineers, Australia (founded 1919), in 1981 adopted a "Code of Ethics" which states that "The responsibility of Engineers for the welfare, health and safety of the community shall at all times come before their responsibility to the Profession, to sectional or private interests, or to other Engineers." The Association of Professional Engineers of Ontario, Canada (founded 1922), in 1984 adopted a "Code of Ethics" that states "a practitioner shall regard his duty to public welfare as paramount."

In modest contrast, while professional associations of engineers in developing countries have also created ethics codes, they have more commonly stressed obligations to enhance professional reputation. For example, the Institution of Engineers (India), established in 1920 (royal charter 1935), in 1944 created a code of ethics that stresses how the member "should scrupulously guard his professional reputation and avoid association with any enterprise of questionable character." The Colegio de Ingenieros de Chile (founded 1958), in 1981 adopted a code of professional ethics that aims to promote at once the professional reputation and national subservience of engineering; it is, for instance, contrary to the code "to permit actions or omissions that favor or permit the unnecessary use of foreign engineering for objectives and work for which Chilean engineering is sufficient and adequate."

Finally, transnational or globalizing engineering associations have likewise been influenced by the American model. The Unión Panamericana de Asociaciones de

Ingenieros (UPADI or Pan American Federation of Engineering Societies, founded 1949), in the 1980s adopted a code of professional ethics that stressed professionalism, but then in 2003 created a new code stressing activity that benefited "clients, society and the environment, optimizing the use of resources and with reduced generation of wastes or any types of pollution." The European Federation of National Engineering Associations (FEANI, founded 1951), in 1988 adopted a Code of Conduct that obligates all members "to be conscious of the importance of science and technology for mankind and of their own social responsibilities when engaged in their professional activities." The World Federation of Engineering Organizations (WFEO, founded 1968), in 2001 adopted a Model Code of Ethics in which "Professional engineers shall hold paramount the safety, health and welfare of the public and the protection of both the natural and the built environment in accordance with the Principles of Sustainable Development."

All three types of engineering ethics globalization – engineers in advanced countries other than the United States adopting ethics codes, engineers in developing countries formulating ethics codes, and transnational engineering associations creating ethics codes – reflect influences from the United States, although not exclusively. They highlight in general terms obligations to some version of the common good such as public safety, health, and welfare or professional loyalty to clients or employers. There are few if any codes on the Germany model, in which engineers are obligated to contribute to technology assessment; nor are there any efforts to ground engineering ethics in epistemological or general philosophical reflection.

There are perhaps four reasons for the prominence of American over German influence. First, the fact that English has become a more global language than German makes American discussions more readily communicated. Second, the stigma of German engineering involvement with World War II may continue to exercise some negative influence. Third, ABET is in the process of becoming a de facto global accreditation agency; a number of engineering programs in other countries are now seeking ABET accreditation and thus having to address the ABET criterion for engineering ethics learning. Finally, the fact that the German approach requires engagement with a philosophical tradition of depth and complexity that runs from Gottfried Leibniz through Kant and Hegel to Karl Marx, Friedrich Nietzsche, Martin Heidegger, and Jürgen Habermas makes it inherently more difficult to imitate. Although engineering ethics in the United States has involved philosophers, the kind of philosophy applied (as, e.g., pragmatism) exhibits less historical depth and is simply less demanding than that present in the German tradition. Indeed, in most other countries the engagement of engineers with philosophes has also been minimal.

There are, however, three important exceptions to this generalization; three examples worth noting are Denmark, the Netherlands, and China. In Denmark, initiatives to reform engineering education in the late 1990s led to the introduction of a requirement in 2000 that all technical and engineering curricula at the bachelor level include a course in the philosophy of science for engineers to be implemented no later than 2004. In response, a number of philosophers of technology from the United States were invited for consultations and Danish philosophers themselves undertook to work with engineering educators to develop appropriate courses. One especially remarkable effort in this general area was spearheaded by Steen Hyldgaard Christensen and led to publication of a textbook on *Philosophy in Engineering* (2007). Another created the *Companion to the Philosophy of Technology* edited by Jan Kyrre Berg Olsen, Stig Andur Pedersen, and Vincent F. Hendricks (2009).

The Netherlands is home to what is arguably the most intensive pursuit of the philosophy of technology and engineering in any country. In a nation that is literally an artifact designed and maintained by hydrological engineers, and which appropriately enough has one of the most well developed communities of science, technology, and society (STS) scholars, it was natural that philosophers at technological institutions of higher education would engage with engineers. In 2006 they established the inter-institutional 3TU Centre for Ethics and Technology to bring together the expertise of the philosophy departments from three universities –TU Delft, TU Eindhoven, and Twente University – to pursue the ethics of science, technology, and engineering through interdisciplinary applied research, fundamental research, teaching, and public outreach. The 3TU Centre quickly became the most integrated, interdisciplinary engagement of philosophers and engineers in the world. Among its many publications are Ibo van de Poel and Lambèr Royakkers, *Ethics, Technology, and Engineering: An Introduction* (2011).

In China, as in the Netherlands, initiatives to link engineering and philosophy emerged at universities dedicated to engineering and often in conjunction with STS programs. From the late Qing Dynasty (1644–1911) the modern engineering professional emerged in China as part of an effort to defend against Western colonialism. The first technical school was the Fujian Shipping School, founded in 1866 in response to Chinese defeats in the two Opium Wars (1839–1842 and 1856–1860) in order "to learn the skills of the barbarians in order to fight the barbarians." Because the explicit goal of early Chinese engineering development was to acquire Western science and technology while retaining Chinese culture, technical education necessarily included what might be called a philosophical element. From the founding of the People's Republic of China in 1949 this took the form of technical education that included a significant component of Marxist ideology in order to create "Red engineers." In a study of the premier technological university of China, one scholar describes Tsinghua as

> China's consummate trainer of Red engineers.... [T]he university's party organization is renowned for grooming political cadres [so that] Tsinghua graduates occupy key positions in the upper echelons of the party and state bureaucracies, and one-third of the members of the Political Bureau's Standing Committee ... are alumni. (Andreas 2009, p. 6)

But independent of this ideological version of engineering ethics that promotes engineering loyalty to the Communist Party, a number of technological universities, including Tsinghua, have also created STS and engineering studies programs that promote less ideological engagements between philosophers and engineers. Such engagements are further encouraged by increasing recognition of the historically unprecedented character of technological transformation and the need for general reflection on the contributions engineers are making to the re-making of China.

From Ethics to Politics

Once ethical reflection ceases to focus primarily on guidelines for the behavior of individual engineers or what is best for the engineering profession reflection readily follows the path mapped classically by Aristotle and Confucius, whose discussions of ethics lead into discussions of politics. Such a movement is manifest in multiple ways in North America, in Europe, and in Asia – with the political being given different conceptualizations and expressions in different contexts and traditions.

One general conceptualization might actually be to configure the movement from ethics to politics as another instance of globalization, in a secondary meaning of the term. Most commonly, globalization refers to external processes that lead to ever greater economic, political, and cultural interactions across national borders – something clearly represented by the emergence of transnational engineering ethics codes. However, globalization can also involve expanding some previously narrow perspective into a more holistic one. Taking a global perspective on investing in a new technological innovation, for instance, would involve going beyond the economic interests of shareholders to include multiple benefits and risks related to all shareholders and as well as environmental concerns. There is thus a sense in which engineering ethics can be internally globalized by moving from a narrow focus on what has been called micro ethical issues related to individuals to a broader, more global or holistic focus on macro ethical issues involving engineering organizations or even projects. Globalization in this sense also suggests a need to re-conceive engineering in terms larger than the technical professional occupation.

In North America, for instance, engineering has been mostly understood in the narrow sense as a historically self-defining group constituted by occupation, discipline, and profession. From such a perspective, engineering ethics is equivalent to the professional ethics of engineers. This is the view that has dominated engineering ethics not only in the United States but in other North American countries as well. One problem with this view is that it places a heavy burden of responsibility on individual engineers, often calling on them to exercise moral heroism as whistleblowers in the face of economic, managerial, or political pressures to compromise technical standards in ways that can undermine functionality or safety. In response, a number of scholars have sought to consider some aspect of the broader political context in which individual engineers work.

Engineer-philosopher Joseph Herkert (2001) summarized such efforts using the distinction, original proposed by John Ladd (1980) between micro-ethics (dealing with relationships individual engineers have with each other, their employers, and clients) and macro-ethics (addressing issues of collective social responsibility of the engineering profession as a whole). Herkert compares the efforts of two philosophers (Ladd and Richard De George (De George 1981)), two engineers (G.F. McClean 1993 and Willem Vanderburg 1995), and an STS scholar (Richard Devon 1999) to conceptualize the macro-ethical context and argues that none successfully integrated the micro- and macro-ethical perspectives. Herkert's own proposal was simply for more research on the responsibilities of professional societies as a whole;

specific suggestions are the need for professional engineering societies to establish institutional supports for individual whistle-blowers and to develop statements on public policy issues such as product liability. The idea that professional societies should support individual whistle-blowers had actually been argued for some time by Stephen Unger (1982 and 1994).

While important, Herkert's discussion manifests a thin view of the political. It makes only the most limited reference to the ways in which engineering transforms the political, as articulated by Langdon Winner (1980) or engages with the political through various forms of technocracy. Related efforts to conceptualize engineering in its broad social and political context can be found in work on STS and engineering ethics (Johnson and Wetmore, 2008), humanitarian engineering (Mitcham and Muñoz, 2010), and engineering and social justice (Lucena et al., 2010; Lucena, 2013).

Beyond these isolated efforts, a more systematic approach to the political is found in Philippe Goujon and Bertrand Hériard Dubreuil, *eds., Technology and Ethics: A European Quest for Responsible Engineering* (2001), which expands the engineering ethics perspective on at least two counts: form and content. In regard to form, its collaborative character – with contributions from engineers, philosophers, sociologists, and historians from across Europe – is more extensive than anything previously attempted in North America. In regard to content, this was perhaps the first engineering ethics textbook to attempt a broad contextualization of engineering. In three major parts, the book moves from considerations of (a) problems related to engineers in technical institutions, through (b) technical systems and technical decision making, to (c) technical development as a social issue. Moreover, the tripartite structure of each major section – historical and sociological description, case studies, and philosophical reflection – provides a strong stimulus to think of engineering in more than simple professionalism. Indeed, the very title of the volume suggests a need to link engineering ethics with the ethics of technology, something that, as one summary of the North American field noted, has not been tried.

The distinctive achievement of the European quest is unintentionally highlighted in the introduction to a reprint collection of 57 articles from philosophy and social science journals that would "provide in a single volume the most important essays on engineering ethics in a form that should be useful to a scholar unfamiliar with the field" (Davis 2005, p. xx). The editor notes the absence of an intersection between the philosophy and ethics of technology and engineering ethics. In his words, although "the two fields might seem to have much in common" the fact is that in North America "the philosophy of technology tends to focus on technology itself rather than on those who make it and, even when attending to those who make it, tends to lump engineers with other 'knowledge workers' in the omnibus of 'technologists,' ignoring the special standards of engineering as a distinct profession" (Davis, 2005, p. xvi). The European textbook not only seeks explicitly to connect philosophy of technology and engineering ethics, but also works to bridge multiple disciplines and language communities while placing engineering in broad social and political contexts.

A final expansive effort to move from ethics to politics is that undertaken in China by Li Bocong at the Center for Engineering and Society of the University of the Chinese Academy of Sciences. Li Bocong argues that engineering ethics needs to be complemented by the sociology of engineering and that the professional engineer needs to be understood as but one member of a more inclusive engineering community. The engineering community is, in turn, established by an engineering project.

In the two volumes on *Engineering Education and Practice in Context* to which this chapter contributes, the most common approaches to engineering emphasize engineering as a profession and/or as design. But engineering is broader than any one profession or occupation. This is indicated by the fact that there are "engineers" who are neither professionals nor designers (e.g., persons with engineering degrees who work as managers or investors) and that there are non-engineers who are involved in engineering projects (e.g., persons who have learned from apprenticeship and practice or have degrees in physics, chemistry, or even the social sciences). In recognition of these facts, Li Bocong (2010) argues for anchoring an understanding not in engineering as an isolated activity but an aspect of an engineering project. To do this, of course, suggests the need for some concept of a project, so that an engineering project can be distinguished from, say, a political, economic, or artistic project.

In English the word "project" as a noun commonly references a large undertaking, often involving significant amounts of money, many personnel, and major equipment; it is planned out in advance. As a transitive verb, the term can mean "to propose," "to throw," "to set forth or calculate" (something in the future). As an intransitive verb, it can mean "to extend or protrude," "to use one's voice so as to be heard at a distance," "to produce a clear impression of one's thinking or personality," and in psychology, "to ascribe one's own feelings, thoughts, or attitudes to others."

Simple etymology deepens the appreciation of these straightforward linguistic uses. Its roots are in the Middle English "project(e)," meaning "plan," from the Medieval Latin *projectum*, *projectus*, past participle of *proicere*, to throw forward, extend, from *pro* (preposition, in favor of, for) + *jacere* (verb, to throw). The English word thus connotes a somewhat forceful imposition not just on the future but also into the present.

A political project could be exemplified by the action of establishing a colony on newly explored land or the creation of a political party to seek control of the government. Economic projects are associated with the founding of corporations or investments in money making activities. Artistic projects create not just a single painting or sculpture but a collection of paintings or/and sculptures, buildings with a certain flair, museums. What is remarkable about engineering projects is the degree to which they partake of politics, economics, and material construction, with aesthetic dimensions. To re-think engineering ethics from the perspectives of these various engagements is an effort that is only now emerging – and will need to include politics.

Conclusion

From its pre-philosophical beginnings as an aspect of the politics of the emergence of engineering as a profession, philosophical engineering ethics exhibits a trajectory that runs from discussions in two particular countries to discussions in many countries and suggestions of a need to move beyond ethics to political philosophy. The movement could thus be summarized as one from politics to ethics to political philosophy. Our chapter overview of this trajectory has highlighted the following five theses:

1. Originally engineering ethics did not involve philosophy; instead, it was pursued by engineers alone.
2. The first two collaborations between engineers and philosophers took place in Germany (1950s–present) and in the United States (1970s–present). In Germany the engagement among engineers and philosophers included more than ethics; in the United States engagement tended to focus more narrowly on ethics.
3. The influence of the German approach to engineering ethics has been less influential in other countries than the American approach to engineering ethics.
4. As engineering ethics has become a point of discussion in many other countries, it remains the case that in most contexts engineering ethics has not been significantly involved with philosophers. Three exceptions are Denmark, the Netherlands, and China.
5. Finally, echoing the philosophical connection between ethics and political philosophy as found in Aristotle, Confucius, and others, engineering ethics would benefit from expanding appreciation of the political aspects of engineering.

Coda: Toward a Political Philosophy of Engineering

As we have presented to different audiences and colleagues our argument that the philosophical engagement with engineering needs to expand from ethics to politics, we have regularly been asked for more specifics about how this might work. The following reflections are a provisional response.

Engineering ethics commonly draws on different ethical theories or traditions to help individuals think about non-technical problems they encounter in engineering practice. The result is to produce ethical analyses that involve, for example, consequentialism, deontology, and virtue ethics in traditions associated with Aristotle, Confucius, Jeremy Bentham and John Stuart Mill, Immanuel Kant, and others. As many have noted, however, the results often create new problems. One concerns how individual engineers can live up to various ethical ideals. Another is the challenges that occasionally arise when different ethical perspectives lead in different directions. Such problems are of a political character, the kind often dealt with in political philosophy. Yet most attempts to address ethical problems that call for a shift from thinking about individual behaviors (the primary focus of ethics) to thinking about the behavior of groups and the structures of social institutions (which is the focus of political science) have not yet made much effort to draw on the political philosophical work of the same philosophers and philosophical traditions referenced in ethical analyses.

One example could note how Herkert argues for the importance of a macro engineering ethics that would stress the responsibilities of professional societies not just individuals. Herkert's argument could be advanced by political philosophical reflection on how professional societies are themselves structured. Plato, for instance, argues that the social institution known as the state should be governed by philosophers or those who most embody reason. Aristotle argues that the best structure for a state is a mixture of aristocracy and democracy. Confucius argues that the state should be ruled by virtuous individuals who govern through example more than through law. Bentham and Mill propose a representative democracy that creates laws on the basis of a utilitarian calculus. Might it not be useful to critically examine the structure of professional engineering societies from these various philosophical perspectives?

Another example might examine what Li Bocong calls the engineering community – consisting of engineers but also workers, investors, and others – from the perspective of political philosophy. How might such a community function differently in Plato's Republic, Aristotle's *polis*, the Confucian state, or utilitarian democracy? Moreover, although the sociology of the engineering community will enhance our understanding of engineering and engineering projects, sociology is not a normative science. But the engineering community and its numerous projects need to be subject to normative analysis. The philosophical challenges from engineering concern not only how to conduct engineering projects in the right way but what are the right engineering projects to undertake. The political ideal of justice and the political philosophical traditions of reflection on the nature of justice have implications for engineering. Engineering ethics and the sociology of engineering must be complemented by political philosophical reflections especially when dealing with not just with the construction of dams and cities but especially with confronting the problems of climate change and proposals for geoengineering of the planet.

A final example could address directly the question of individual versus group or collective responsibility. The problem of the diffusion of responsibility in large-scale organizations and complex technological projects is one that repeatedly arises in engineering ethics. Psychologists and sociologists have noted how people are less likely to assume responsibility for a problem when others are present; by themselves individuals who see problems are more likely to take action than when they are members of a group confronted by the same problem. Ethical arguments for collective responsibility of all members of a group for the bad behavior of some of the members, insofar as all ignore or tolerate the bad behavior even if they do not actively collaborate in it, need to be complemented by political arguments for social structures that promote such responsibility. Theories of corporate social responsibility have struggled with how to get firms to do more than simply act in the narrow, profit-making interests of the firm and ways required by law. In all such cases it is reasonable to propose that critical reflection could be advanced by drawing on political philosophy.

What we suggest here, however, is only a beginning. Our basic thesis remains simply that engineering ethics could benefit from recognizing the ways that ethics in philosophy is a prologue to political philosophy – and thus seeking ways that critical reflection on the challenges associated with engineering ethics might be advanced by broadening the scope of discussion to include politics and political philosophy.

Acknowledgments For support in the writing of this chapter, Carl Mitcham acknowledges a sabbatical leave from the Colorado School of Mines; and Wang Nan would like to credit a postdoctoral fellowship from the Hennebach Program in the Humanities at the Colorado School of Mines.

References

Andreas, J. (2009). *Rise of the red engineers*. Stanford: Stanford University Press.

Christensen, S. H., Meganck, M., & Delahousse, B. (2007). *Philosophy in engineering*. Aarhus: Academica.

Christiansen, D. (1972). The new professionalism. *IEEE Spectrum, 8*(6), 17.

Davis, M. (Ed.). (2005). *Engineering ethics*. Burlington: Ashgate.

De George, R. T. (1981). Ethical responsibilities of engineers in large organizations: The pinto case. *Business and Professional Ethics Journal, 1*, 1–14.

Devon, R. (1999). Toward a social ethics of engineering: The norms of engagement. *Journal of Engineering Education, 88*, 87–92.

Durbin, P. T. (1997). Engineering ethics and social responsibility: Reflections on recent development in the USA. *Bulletin of Science Technology and Society, 17*(2–3), 77–83.

Flores, A. (1977). The national project of philosophy and engineering ethics. *Professional Engineer, 47*(8), 26–29.

Harris, C., Pritchard, M., & Rabins, M. (1995). *Engineering ethics: Concepts and cases* (2nd ed. 2000; 3rd ed. 2005; 4th ed. 2008; 5th ed. 2013). Belmont: Wadsworth.

Herkert, J. R. (2001). Future directions in engineering ethics research: Microethics, macroethics, and the role of professional societies. *Science and Engineering Ethics, 7*(3), 403–414.

Johnson, D. G., & Wetmore, J. M. (2008). STS and ethics: Implications for engineering ethics. In E. J. Hackett, O. Amsterdamska, M. E. Lynch, & J. Wajcman (Eds.), *The handbook of science, and technology studies* (3rd ed., pp. 567–581). Cambridge, MA: MIT Press.

Ladd, J. (1980). The quest for a code of professional ethics: An intellectual and moral confusion. In R. A. Chalk, M. S. Frankel, & S. B. Chafer (Eds.), *AAAS professional ethics project: Professional ethics activities in the scientific and engineering societies* (pp. 154–159). Washington, DC: AAAS.

Lenk, H., & Ropohl, G. (Eds.). (1987). *Technik und Ethik*. Stuttgart: Philipp Reclam.

Li Bocong (2010). *Gongcheng shehui xue daolun* [Introduction to sociology of engineering]. Hangzhou: Zhejiang daxue chuban she [Zhejiang University Press].

Lucena, J. (Ed.). (2013). *Engineering education for social justice*. Dordrecht: Springer.

Lucena, J., Schneider, J., & Leydens, J. A. (2010). *Engineering and sustainable community development*. San Rafael: Morgan and Claypool.

Martin, M., & Schinzinger, R. (1983). *Ethics in engineering* (2nd ed. 1989; 3rd ed. 1996; 4th ed. 2005). New York: McGraw-Hill.

McClean, G. F. (1993). Integrating ethics and design. *IEEE Technology and Society, 12*(3), 19–30.
McClelland, E. H. (1917). Engineering ethics. In: Carnegie Library of Pittsburgh (Eds.), *Bibliographies* (Vol. 1, pp. 3–17). Pittsburgh: Carnegie Library: Engineering ethics (1917).
McMahon, M. A. (1984). *The making of a profession: A century of electrical engineering in America*. New York: Institute of Electrical and Electronic Engineers.
Mitcham, C. (1992). *Engineering ethics throughout the world: Introduction, documentation and bibliography. A collection of engineering ethics codes and related materials with introductory analysis and supplementary annotated bibliography. Version 1.0*. University Park: STS Press.
Mitcham, C. (1994). *Thinking through technology: The path between engineering and philosophy*. Chicago: University of Chicago Press.
Mitcham, C., & Muñoz, D. (2010). *Humanitarian engineering*. San Rafael, CA: Morgan & Claypool.
Noble, D. F. (1977). *America by design: Science, technology, and the rise of corporate capitalism*. New York: Oxford University Press.
Olsen, J. K. B., Pedersen, S. A., & Hendricks, V. F. (Eds.). (2009). *A companion to the philosophy of technology*. Malden: Wiley-Blackwell.
Ropohl, G. (2002). Mixed prospect of engineering ethics. *European Journal of Engineering Education, 27*(2), 149–155.
Toulmin, S. (1982). How medicine saved the life of ethics. *Perspectives in Biology and Medicine, 25*(4), 736–750.
Van de Poel, I., & Royakkers, L. (2011). *Ethics, technology, and engineering: an introduction*. Malden: Wiley-Blackwell.
Vanderburg, W. H. (1995). Preventive engineering: Strategy for dealing with negative social and environmental implications of technology. *Journal of Professional Issues in Engineering Education and Practice, 121*, 155–160.
Verein Deutscher Ingenieure. (1980). *VDI: Zukünftige Aufgaben*. Düsseldorf: VDI.
Winner, L. (1980). Do artifacts have politics? *Daedalus, 109*(1), 121–136.
Wisnioski, M. (2012). *Engineers for change: Competing visions of technology in 1960s America*. Cambridge, MA: MIT Press.

Carl Mitcham B.A. and M.A. in Philosophy from University of Colorado, Boulder. Ph.D. in Philosophy from Fordham University. Professor of Philosophy of Science and Technology, Renmin University of China, Beijing; Liberal Arts and International Studies, Colorado School of Mines, Golden, Colorado. Scholarly contributions have been directed toward the philosophy and ethics of science, technology, engineering, and medicine and to science, technology, and society (STS) studies. Teaching areas: ethics, STS, and science and technology policy.

Wang Nan M.A. in the Philosophy of Science and Technology at the Dalian University of Technology. Ph.D. in Philosophy of Science and Technology at the Graduate University of Chinese Academy of Sciences. She is assistant professor at the School of Humanities and Social Sciences, University of Chinese Academy of Sciences. Two English publications are: "Beijing Bauhaus: Art, Engineering, and Socio-Cultural Change", *Design Philosophy Papers*, vol (2), 2011, co-authored with Carl Mitcham; "On Whistleblowing: From the View of Engineering Ethics", in *Whistleblowing: In Defense of Proper Action* (special issue, *Praxiology: International Annual of Practical Philosophy and Methodology*, vol. 18, 2010). Her research areas include philosophy of technology, philosophy of engineering, and engineering sociology.

Chapter 18
Guiding Gulliver: Challenges for Ethical Engineering

Wayne Ambler

Abstract One challenge for ethical engineering is persuading engineers to do the right thing in their day-to-day jobs. More fundamental challenges arise when we consider the vast power of their knowledge in the first place. What ends should this power serve? Is it for health? National greatness? Ameliorating the lives of the poor? Providing entertainment? Generating jobs for the next generation? And hard on the heels of this massive theoretical question is the practical one: in a world ruled by individuals with selfish interests and imperfect wisdom, how can engineering be guided toward its best possible uses? Like Gulliver among the Lilliputians, or Oppenheimer among the Americans, scientific engineering is deeply impressive and larger than life. Nevertheless, understanding its proper goals is a philosophical question of the first order, and guiding its ambiguous power in practice is a constant challenge. Alas, this paper will not solve these twin problems, but it will illuminate them. It will do so by assessing the regulated marketplace as one means of guiding the direction of future engineering innovations. Second, it will note the epistemological challenges brought to the fore especially by David Hume and Max Weber, both of whom stressed the difficulty of ever knowing what *ought* to be done. We can learn from the power of engineering that the knowledge of ends is logically prior to the knowledge of means, even as we marvel at the difficulty of acquiring this knowledge and putting it to use.

Keywords Conquest of nature • Naturalistic fallacy • Engineering ethics • Free economy • Liberal democracy • Moral relativism • Distinction between facts and values • Human well-being

W. Ambler (✉)
Herbst Program of the Humanities, College of Engineering, University of Colorado Boulder, Boulder, CO 80309, USA
e-mail: wayne.ambler@colorado.edu

S.H. Christensen et al. (eds.), *Engineering Identities, Epistemologies and Values*, Philosophy of Engineering and Technology 21,
DOI 10.1007/978-3-319-16172-3_18

The Vast but Ambiguous Power of Modern Engineering

Like Gulliver among the Lilliputians, the power of engineering dwarfs all rivals.[1] Its power cannot be precisely reckoned, for it spans broad areas of human endeavor: power to change the courses of rivers; power to extract liquids and solids from the ground, refine them, and use them to heat and illuminate hundreds of millions of homes, set in motion hundreds of millions of vehicles large and small, and run factories producing goods of all descriptions; power to battle and often defeat diseases and the heart-wrenching consequences of accidents; power to shrink the world by faster communications and transportation; power to make the weapons that modern armies use both for defense and to menace others. One might go so far as to say that the vast and growing power of modern technology rules the world. Although most of us may not often think about it, all who use cars, phones, computers, air conditioners, modern medicine or other such characteristic elements of modern life are deeply dependent upon the successes of modern engineering. We take this success for granted, and, when a little glitch or some major accident deprives us of their services, it seems that life itself is at risk. Indeed it would be if, for example, electric power to cities were lost for an extended period.

So crucial has the role of modern engineering and technology been in ameliorating our lives that we also look to it to solve many of our looming problems, and we do so as if by instinct, with little hesitation. Consider this passage from President Obama's Second Inaugural Address:

> We cannot cede to other nations the technology that will power new jobs and new industries. We must claim its promise. That's how we will maintain our economic vitality and our national treasure, our forests and waterways, our crop lands and snow-capped peaks. That is how we will preserve our planet, commanded to our care by God. That's what will lend meaning to the creed our fathers once declared.

The President is able to count on his listeners to agree without a second thought that technology is of central importance for "powering" new jobs and for protecting "our national treasure, our forests and waterways, our crop lands and snow-capped peaks." It will even "preserve our planet." In short, modern engineering has proven its powers, and great hopes surround it in expectation of further benefits. Partly because his reasons are widely embraced around the globe, perhaps even more enthusiastically in Asia than in the USA, the number of engineering students around the planet continues to rise.[2]

Notwithstanding the great achievements of modern technology and our confidence that more are on the way, it is no big secret that some technologies have been used to do harm and that others might do harm on an even grander scale in the

---

[1] I consider engineering and technology to go hand-in-hand. The former is the application of the natural sciences in such a way as to produce the latter. Without modern engineering there would be no modern technology, and both depend on the Scientific Revolution that was stimulated and defended by such thinkers as Francis Bacon, Isaac Newton, and Rene Descartes.

[2] http://www.nsf.gov/statistics/seind02/c2/c2s4.htm, accessed December 12, 2012.

future, whether by design or by accident. Perhaps dangerous weapons or man-made pathogens come most quickly to mind in this connection, but consider again the passage just quoted from President Obama's Second Inaugural. He hopes technology will protect our "snow-capped peaks" and preserve our planet, but are not the unanticipated effects of modern technology the main reason we worry about the preservation of our glaciers in particular and our planet in general? Modern engineering brought us modern means of production, thus sparking an industrial revolution and world-wide development; along with the many and undeniable benefits from these changes came environmental degradation and, now, global warming, such that it is not entirely hyperbolic to fear for the long-term survival of the planet and "our forests and waterways, our crop lands and snow-capped peaks."

The Bulletin of the Atomic Scientists does what it can to stress that the many benefits of modern engineering have come with risks. On the cover of every issue since 1947 there has been a symbolic clock which seeks to gauge the risks of global disaster. Of the 1,440 min in a day, their clock has never shown more than 17 min remaining before doomsday, represented by midnight.[3] For 60 years the focus in setting this clock was on the risk of nuclear war; in recent years its focus has expanded to include the risk of climate change. In the words of the *Bulletin*,

> The dangers posed by climate change are nearly as dire as those posed by nuclear weapons. The effects may be less dramatic in the short term than the destruction that could be wrought by nuclear explosions, but over the next three to four decades climate change could cause irremediable harm to the habitats upon which human societies depend for survival.[4]

I do not cite this because I claim to know that the authors are correct in making this particular claim, or because I know how close we really are to global catastrophe; I mean to stress only this more limited point, that the great blessings of modern technology have come with great risks.

Further evidence conducive to this line of thought includes the establishment of the new Centre for the Study of Existential Risk (C.S.E.R.) at the University of Cambridge. Its founders and board include distinguished scientists and philosophers, not Luddites of any recognizable description. The first sentences of its website state the following: "Developments in human technology may soon pose new, extinction-level risks to our species as a whole. Such dangers have been suggested from progress in AI, from developments in biotechnology and artificial life, from nanotechnology, and from possible extreme effects of anthropogenic climate change." [5] (Is it a sign of "progress" that C.S.E.R. does not even mention nuclear war?) In short, when we say "technology has vastly improved the way we live," we need to add, "but distinguished scientists think it has also dramatically increased the risk of global disaster."

---

[3] This moment of an unusual sense of security came when the USSR collapsed; but the clock was back in the single digits by the end of the 1990s, as the fear of "loose nukes" grew, India and Pakistan tested nuclear weapons, and the USA and Russia failed to reduce their nuclear stockpiles in a reassuring fashion. http://www.thebulletin.org/content/media-center/announcements/2007/01/17/doomsday-clock-moves-two-minutes-closer-to-midnight, accessed December 15, 2012.

[4] *Ibid.*

[5] http://www.cser.org/

It is in the nature of technology to come with risks, I believe. In developing new technologies, engineers grant us new tools; but tools can be used for many different ends or purposes. A hammer can drive a nail or kill a neighbor. Tools of great power, like those deriving from atomic fission, can have vastly different applications. The same source of power can either illuminate or annihilate a city. And the same electric power that is used to cool homes, factories, and shopping malls is produced in such a way as to contribute to the warming of the planet. Engineering does much good, then, by empowering us to help ourselves in many ways; but we must face squarely its power to harm as well as help. The goodness of modern engineering depends upon its use; goodness is not inherent in the activity itself. (Conversely, that engineering might sometimes do harm does not mean that we can or should do without it. Beauty can also do harm, as Marc Antony learned at a great personal cost, but it does not follow that we should destroy it or wrap it in thick veils.)

As the power of modern technology increases, so too does the gap between the good that it promises and the harm it might do. When discussing justice with his friends two and a half millennia ago, Socrates argued that the person cleverest at guarding against disease would also be the cleverest at producing it (Plato, *Republic* 333e). The doctor who is good at curing malaria, for example, could also be good at spreading it. In general terms, the knowledge required to use a particular power in one way is also the knowledge needed to use it in a vastly different way. But, in Socrates' example of a single doctor, neither the power to cure nor the power to infect compares in magnitude to the powers now generated by modern technology. Socrates noted the principle, which we might call the "moral neutrality" of technical knowledge, but he did not live with technologies whose powers to do good or harm border on the unthinkable. The issue here is far from new in principle, but the stakes are vastly higher. The misuse of a nuclear weapon is more worrisome than the misuse of a hammer.

These are ethical issues, but not in the ordinary sense of on-the-job ethics. I will not preach here that engineers should fulfill their contractual obligations or that they should undergo ethics training to teach them not to plagiarize. I am thinking on the broader scale and want to know whether and how we might become able to judge the benefits and risks of new technologies. My starting point is that everyone wants the products of engineering to be beneficial but realizes that sometimes they are not. Thus we often regulate or even ban particular technologies. How do we do this? What procedural and philosophical problems have to be solved to do this well? How should engineers use the vast powers they call into being and command? To insist on product safety and honesty makes sense, but there are deeper issues.

The Challenge in Practice: Guiding Gulliver in the USA

Gulliver is so powerful we try to govern him even if we do not know how best to do so. Hence, all modern states have policies to regulate engineering and science. Just as in the land of the Lilliputians, Gulliver is both restricted and directed by others "smaller" than he is. In less metaphorical terms, the power to create is not the power to wield: engineers and scientists do not rule the world even if the powers they have brought into being are essential to such rule. If we think of engineering as a powerful "Gulliver," we must add that this Gulliver is not free to use his powers as he sees fit. J. Robert Oppenheimer enjoyed several moments of special influence in discussing the use of and controls on the bomb he helped to build, but his story shows the limits, not the triumph, of the influence of scientists on US policy.[6]

Here in the United States, the main elements of our effort to guide science and engineering are a free economy and a liberal democratic political system. The main case for a free economy is well known, but it is worth a brief quotation by Founding Father Alexander Hamilton to call it to mind:

> The prosperity of commerce is now perceived and acknowledged by all enlightened statesmen to be the most useful as well as the most productive source of national wealth.... By multiplying the means of gratification, by promoting the introduction and circulation of the precious metals, those darling objects of human avarice and enterprise, it serves to vivify and invigorate the channels of industry, and to make them flow with greater activity and copiousness. The assiduous merchant, the laborious husbandman, the active mechanic, and the industrious manufacturer,--all orders of men, look forward with eager expectation and growing alacrity to this pleasing reward of their toils.[7]

With a hint of amusement but none of apology, Hamilton indicates that in a free economy, avarice is politically useful: it may be a vice in moral terms, but avarice leads us all to look forward to the "pleasing reward of [our] toils." What is true of "the assiduous merchant" is true also of the engineer. This is made almost explicit in the provision of Constitution that grants to Congress the authority "to promote the Progress of Science and useful Arts, by securing for limited Times to Authors and Inventors the exclusive Right to their respective Writings and Discoveries."[8] That is, the "exclusive right" to their inventions provides individual engineers and corporations the incentive that in turn leads to the "progress of science and the useful arts." Hamilton neither glorifies this motivation nor denies that other less self-interested motives might be at work, but his main point is clear: in a generally free economy, the desire for individual gain can be used to stimulate the progress in inventions, which in turn is expected to be generally beneficial for society. Hamilton does not quite say it, but in this case the direction of scientific progress will be guided in part by how people spend their money. Whether they are eager to buy farm equipment, video games, or improved medical technology, for example, engineers will work to provide what others seek. If they do not even know what they seek, entrepreneurs will figure it out and give directions. The first or most fundamental practical answer regarding the guiding of Gulliver in the USA is this: the market will do it.

[6] Oppenheimer's hopes for putting the bomb under the control of the United Nations were severely disappointed. I am puzzled by a recent suggestion by Lawrence Krauss that he successfully guided US atomic policy. See http://www.nytimes.com/2013/01/16/opinion/deafness-at-doomsday.html, accessed January 18, 2013.

[7] Alexander Hamilton, *Federalist 12*.

[8] United States Constitution, Article I, Section 8. I believe the US Constitution was the first constitution to call for patent laws as a way of stimulating science and inventions.

But the US Constitution does not call for a completely laissez-faire economy; patent laws themselves require an active Congress to promote science and "vivify and invigorate the channels of industry," as Hamilton put it. Nor are the expressed goals of the Constitution strictly laissez-faire: they are to "establish Justice, insure domestic Tranquility, provide for the common defense, promote the general Welfare, and secure the Blessings of Liberty to ourselves and our Posterity"[9]; the document nowhere hints that maximizing individual economic liberty is the only or best route to these different and lofty goals. However one understands the original Constitution, the current US political system certainly does not just leave science and engineering alone; it stimulates them with contracts, grants, and honors. In other respects, it curtails them either by indifference or outright prohibitions.

Over the last decade the US Federal Government has spent about $60 billion per year to fund engineering and scientific research through agencies like NASA, NIH, DOD, NSF, and the USDA.[10] These allocations have been decided upon by the normal political process, which includes input from the Congress, the Executive Branch, numerous committees in both branches, and agencies outside of the government such as the National Academy of Engineering and a broad array of non-profit organizations with an interest or expertise in how these many billions of dollars are spent. The same is true, of course, for decisions to prohibit certain kinds of research or simply not to fund it.

Note in the first place that all this federal money is not only to achieve humanitarian goals. It is also to achieve national and local goals which may or may not coincide perfectly with the greater good. Everyone wants a cure for cancer, for example, but national and corporate leaders are especially eager that this cure come from their company and their nation. After all, the production and application of the drugs and procedures that bring the cure will result in profit and reputation, and even if anyone should care nothing for reputation in itself, it at least can lead to further profit. Hence in the Inaugural Address quoted above, President Obama implies that technological development is an instrument of national policy: he wants it for the USA in particular. When he says "we cannot cede to other nations the technology that will power new jobs and new industries," he even implies that new technology somehow belongs to the US and is ours to cede or retain. I am tempted to say it is like one of the national treasures to which he refers, but this is not correct: it is a (mere) means to jobs and the protection of our treasures. It has no value in itself, he implies, but only in what it can do for us.

Let us glance now at the main objections to the process for guiding Gulliver in the USA. The two main objections now heard on the public stage are that there is too much influence in the hands of powerful corporations and too little in the hands of scientists. The former argument maintains that huge corporations – Halliburton and Monsanto are two targets frequently invoked – can purchase favorable legislation and regulations by using clever lobbyists and political or personal favors done

[9] United States Constitution, Preamble.

[10] http://www.whitehouse.gov/sites/default/files/microsites/ostp/fy2013rd_summary.pdf, accessed January 2, 2013.

to congressmen and regulators. Thus, it is held, the interests of individuals and the nation as a whole are sold out to help the bottom line of big businesses and the states or districts which benefit most from the jobs they provide and taxes they pay. The argument is of course not unique to science policy but is characteristic of the general critique of capitalism from the political Left.

An example typical of those used to support this conclusion, which I offer not to settle the issue but only to clarify it, is the so-called "Halliburton Loophole." The Energy Policy Act of 2005 amends the Safe Drinking Water Act (42 U.S.C. 300 h(d)) so that it does not cover "the underground injection of fluids or propping agents (other than diesel fuels) pursuant to hydraulic fracturing operations related to oil, gas, or geothermal production activities."[11] Why, one may wonder, should the law exempt so-called "fracking fluids" from consideration when looking at threats to safe drinking water? And, when looking at the Safe Drinking Water Act of 2005 itself, one sees further evidence of care to keep hydrofracking from being held in violation. When it defines "pollutant," the act says quite clearly – even in capital letters! – that "This TERM DOES NOT MEAN ... water, gas, or other material which is injected into a well to facilitate production of oil or gas..."[12] It appears, at least, that the nation's water is put at risk in favor of the interest of producing profits for the giant oil and gas industry: if fracking pollutes the water, it is not "pollution" as understood by the law.[13] (This appearance might be challenged or limited by the observation that many Americans want cheaper sources of domestic energy, so the big oil companies are not really acting without popular support. The counter-argument is that the public is distracted and not fully aware of the long-term dangers that lurk behind the attractive face of immediate economic growth. A further issue is whether the big corporations and other powerful groups share important common interests or whether they might help to "check and balance" one another, as Madison suggested they would in *Federalist 10*.)

Commonly criticized or sneeringly attacked for being too subject to the influence of wealthy corporations, US policy on engineering and science is faulted also for being too little subject to the influence of scientists and engineers themselves. Who, after all, should have more influence over matters of technology and science than scientists and engineers? So why then do we see so few engineers in Congress? This line of criticism is compatible in practice with the view that corporations have too much clout, but it is potentially more radical. The previous point defends democratic principles by sketching the case against an oligarchy of rich corporations; by

[11] http://www1.eere.energy.gov/femp/pdfs/epact_2005.pdf, accessed December 10, 2012. See page 102, section 322.

[12] http://www.epa.gov/npdes/pubs/cwatxt.txt, accessed December 10, 2012. See SEC. 502 [33 U.S.C. 1362] General Definitions (definition #6).

[13] An example regarding Monsanto and other seed companies concerns the surprising law which prohibits independent researchers from publishing their findings on genetically modified seed without the permission of the producer: http://www.nature.com/scientificamerican/journal/v301/n2/full/scientificamerican0809-28.html, accessed September 12, 2010. This gives seed companies the authority to censor published research.

stressing the greater understanding of scientists, this one takes a step away from strictly democratic principles toward a sort of scientific aristocracy or "scientocracy."[14]

As the first sentence of a recent article by distinguished physicist Lawrence Krauss puts it, "To our great peril, the scientific community has had little success in recent years influencing policy on global security."[15] Krauss is confident that scientists know best at least regarding "global security." The last sentence of his essay goes further, for it suggests "science and data" are essential for determining policy in all areas of government: "Until science and data become central to informing our public policies, our civilization will be hamstrung in confronting the gravest threats to its survival." This is a powerful argument, of course, for who could argue against determining policies by "science and data"?

To clarify the preceding implicit critique of democracy, recall one of the most famous passages of Plato's *Republic*:

> Unless the philosophers rule as kings or those now called kings and chiefs genuinely and adequately philosophize, and unless political power and philosophy coincide in the same place … there is no rest from ills for those cities nor, I think, for human kind. (*Republic* 473d)

The hopes expressed here for the benefits of philosophic rule have been transferred in the contemporary argument to the benefits of rule by "science and data," but both share the lament that policy is often formulated out of ignorance.

The first critique of US science and technology policy rests upon an accusation of oligarchy, the second upon an accusation of ignorance. The premise of the first is that the majority should rule (or rule more than they currently do); the premise of the second is that scientists should rule (or rule more than they currently do). Having sketched the outlines of the US constitutional system as well as two critiques of our current practice, let me suggest the following points.

1. The American political system, both with regard to science policy and in general, does not ensure that the best policies are pursued: no political system does or can. Governments are made up of human beings, and human beings do not always act for the sake of the common good, even in the event that they figure out what this common good happens to be. This does not mean that, for example, the institution of checks and balances may not help protect a government of human beings against making the most dangerous mistakes; but it does mean that there is neither a substitute for good judgment nor a political system that can guarantee it.
2. An oligarchy of powerful corporations is indeed a risk in the US system and perhaps in some degree already a reality. But there remain powerful institutions through which this risk can be combatted, at least in principle. Examples include

[14] I do not claim that anyone makes this argument explicit, only that it is implicit in the call for paying more attention in policy making to science and scientists. As for an early and thematic raising of this issue, consider the role of "Salomon's House" in Francis Bacon's *New Atlantis*. A more pessimistic view of the rule of science emerges from the pages of Aldous Huxley's *Brave New World.*

[15] http://www.nytimes.com/2013/01/16/opinion/deafness-at-doomsday.html, accessed January 18, 2013.

the free press, the principle of majority vote, the independent judiciary, and the existence of committees and private organizations that can be insulated from the influence of corporations and vast wealth.

3. It would be foolish to formulate policy without considering such scientific evidence as bears on this policy. Since science shows that greenhouse gasses are causing climate change, it is foolish to pretend that this is not so. If GMO's come with long-term risks of creating super-weeds, lessening biodiversity, or causing cancer, there is no excuse for putting them into circulation just because they might be good for business. But scientists are also human beings and, outside of their areas of expertise, they too have interests and biases. Scientists working on atomic power for Iran seek to understand the same physics as Oppenheimer did, for example, but they need not agree with him on how to use the tool that he first built. Indeed, even in the US, not all the scientists and engineers working on the Manhattan Project agreed about atomic policy, and the later policy clash between Edward Teller and Oppenheimer helped to cost the latter his career. Science is one thing, science policy is another.
4. Global problems need global solutions, but the separate nations of the world generally look first to their own interests. Even if one were persuaded that the USA did a good job of guiding Gulliver – and I do mean to stress the words "even if" – we are only a part of a larger world. As repeated experience confirms, the energy policies of particular countries are not set by the United Nations Framework Convention on Climate Change, or by any other effort to determine the best global policies on this issue; and neither North Korea nor the United States will choose their nuclear policy without considering especially what they think best for their own nation. Whatever degree of satisfaction one might take regarding the science and technology policy of a particular nation, the situation in the larger world is close to anarchic.

Beyond the challenges of trying to determine the best procedures for the formulation of science policy, which entails trying to assess the due influence of various stakeholders, there lies a challenge that is harder to recognize. I call it "the philosophical challenge," and it is the subject of the next section.

The Philosophical Challenge

When thinking about the social context of engineering, we often think of such things as the demands, limitations, and particular traits of the society in which technologies are developed and used. To reverse one of Karl Marx's dictums, the hand mill is the technology suited for feudal society, the steam mill the technology suited for the nineteenth century. Solar power makes even more sense for a dispersed society where there is no electrical grid, and chemical toilets are appropriate if there is no plumbing infrastructure. Engineering, in short, should not look for the best technical solution in the abstract; it must choose the solution best suited for prevailing social conditions. This problem is widely studied and acknowledged, even if it is far from being entirely solved.

The following pages will consider an analogous but distinct issue. Rather than looking primarily at the social context, we will here consider the intellectual context. Just as different societies have different needs and put different pressures on engineering solutions, so they also have different reigning ideas or, to use a popular phrase, different zeitgeists; what looks reasonable according to one zeitgeist might be useless or frightening by the estimate of another. Nuclear power in the hands of a nihilist is likely to be used differently than the same power in the hands of a pacifist.

Beyond the challenge of recognizing that different societies are influenced by different reigning ideas lies the challenge of seeking to guide Gulliver toward what is truly better and away from what is worse. In the end, we want technology to be used for good purposes, not merely for particular national or ideological purposes, so this imposes the not-so-minor requirement that we actually know what is good. This philosophical challenge, the challenge of understanding the good and bad goals of engineering, is easily overlooked, partly because most of us are confident that we already know what it is good most of the time, so no further thought is needed. Curing disease is good; promoting it is bad. Making communications faster and easier is good; living with the horse and buggy or dial phones is bad. Internet searches are fast and easy; having to go to a library and look things up in dated books is time consuming and unpleasant.

Although we often take it for granted that we know what is good, moments of confusion arise nonetheless. What is bad in peace might be good in war; fast communications lead to frequent communications, and frequent communications may distract us from our driving, our friends, and issues which require patient reflection; "information" may become more widespread but yet be more superficial, misleading, or wrong. When we think simultaneously of energy and the environment, we think of the harm done by coal-fired power plants and the risks posed by hydrofracking and horizontal drilling; when we think of energy and the economy, we think of how cheap power can stimulate economic productivity. Thus cheap power is simultaneously good and bad (albeit in different respects). But the greatest cautionary tale rests with the long human history of people who were sure they were doing the right thing only to conclude later that they were not.

The philosophic search for an objective understanding of "good" began long ago, perhaps most dramatically in Plato's *Republic*, where the famous allegory of the cave culminates in "the idea of the good" (517b–c). Aristotle continued this philosophical focus on the good in his *Nicomachean Ethics*, whose first sentence reads, "Every art and every inquiry, and similarly every action as well as choice, is believed to aim at some good." We may add that every application of science and the effort of every engineer aim at something believed to be good, but if we agree with Aristotle, as I do, then we need to wonder whether our beliefs about what is good are well-founded. Even if we possessed a complete knowledge of the future, according to Plato, this would not enable us to judge it well: we also need the knowledge that distinguishes good from bad (Plato, *Charmides* 173d-74d).

If we skip to the modern age, which supplies the intellectual context within which modern engineering must operate, we find that the philosophic climate has cooled to the possibility of knowing "the good," so much so that we often put it in quotation marks to show that we doubt either that it exists or that it can be known. Whereas Plato and Aristotle openly pursued rival views of the human good – might it be pleasure? Justice? Wisdom? Personal advantage? – we moderns more commonly take the asking of this sort of question to be naïve, at least when we are speaking theoretically: why raise questions that cannot be settled in a rigorous fashion? A contributing reason for putting the question aside is the oft-made distinction between facts and values: science and strict methodologies have enabled us to make undisputed progress in disputes over facts, but quarrels over values elude our grasp and still divide the world. As the Romans said of taste, *De gustibus non est disputandum*, we often say of all values. Everyone remains fully entitled to hold their own values; they just need to hold them without believing them to be objectively true or true for others.

The philosopher David Hume is credited with originating and defending this distinction; now it is commonplace.[16] Science cannot successfully defend one value over another. Science does wonderfully well with facts, which it can test and measure in myriad ways, but it can say nothing conclusive about beauty, morality, or, say, the value of honor. To say the same thing in slightly different words, this view holds one cannot ever establish an "ought" from an "is." Science tells us what is; no one really knows what ought to be. This is a consequence of the banishment of final causes, or ultimate purpose, from modern natural science: Aristotle's four "causes" or ways of explaining something have been reduced to two. The greatness of modern science lies in uncovering material and efficient causes, not in discerning ultimate purpose (final causes). If one tries to derive an ought by inferring from what is natural to what is right – for example if one reasons that "human beings are by nature rational animals, so human beings ought to reason" – then one is said to be guilty of the "naturalistic fallacy," the groundless assumption that what is natural is good. Philosophers and humanists still cling to the notion that other, non-scientific ways of knowing might settle questions of value; but science has set high standards for what constitutes a proof, so claims to knowledge in matters of ethics or the human good inevitably look embarrassingly weak by comparison.

If the sharp distinction between facts and values facilitates the acceptance of moral relativism, cultural relativism, its close cousin, makes essentially the same claim at the cultural level: good anthropologists can describe different cultures, but nothing in their scientific training entitles them to rank or judge them. When it comes to ideas of good, bad, just, unjust, noble, and base, neither natural nor social science can determine whether any is truer than another; facts are their domain. I do not wish now to test the truth of this idea, or to ask whether anyone can really live their lives in accord with it, so much as to show its influence and reflect on its consequences.

---

[16] For a powerful statement of his view, see David Hume, *Treatise on Human Nature*, Book III, Part I, Section I (last paragraph).

One sign of the influence of these core ideas is that they have profoundly influenced the shape of the modern social sciences, thanks especially to the work of Max Weber.[17] Now, understandably eager to enjoy the power and prestige of the natural sciences, modern social sciences have sought to imitate them. One visible sign of this is the widespread use of modern mathematics in the study of society; the flip side of this is the effort to exclude values from the subject matter. The modern social sciences are to be "value free" to the extent possible and to avoid the "value laden" questions that would keep them from claiming to be sciences properly understood. But perhaps the most eye- and ear-catching thinker to stress the resistance of values to philosophic justifications was Friedrich Nietzsche, who went so far as to say our preference for truth over falsehood is itself merely a value. As he put it, "It is no more than a moral prejudice that truth is worth more than mere appearance; it is even the worst proved assumption there is in the world."[18]

If we turn from the seminal thinkers who first advanced the powerful distinction between facts and values, it is easy to see evidence of its widespread acceptance today. Even readers of the popular media can be counted on to see values as beyond scrutiny. Hence, for example, when Noble Laureate Eric Cornell wrote a short piece in *Time Magazine* to discourage the teaching of "intelligent design" in science classes, he stressed that the question of ultimate purpose is simply not a scientific question. In his words, science can "teach us nothing about values, ethics, morals, or, for that matter, God." Applying this principle, he concludes, "science can try to predict how human activity may change the climate, but science can't tell us whether those changes would be good or bad." Now the human beings who devote themselves to science will certainly have values, and they may believe in them very strongly. Cornell himself concludes, "My value judgment is that further progress in science will be good for humanity."[19] But strict followers of the distinction between facts and values must admit that their values cannot be scientifically demonstrated to be superior to any others, so the charge made by Nietzsche returns unanswered: the preference for the truth – the preference for science – is no more than a moral prejudice, at least if our intellectual world is rigidly divided between facts, which can be known to be true, and values, which can only be asserted.

Now here, for clarity, is the point: the challenge of guiding Gulliver, which is difficult to begin with, is further complicated by our intellectual climate or philosophical zeitgeist. For those who sense they do not yet know what is good, but think they might if they search, their searching can proceed with greater energy; but if we think such knowledge is simply unattainable, as the distinction between facts and values maintains, then on what solid basis can we give direction to the awesome powers our engineers are summoning into being? Relativism leaves us adrift; and, absent confidence that any higher "value" really means anything, it would not be surprising if its adherents sometimes felt the emptiness of an existence that cannot

[17] Leo Strauss, *Natural Right and History* (Chicago: University of Chicago Press, 1965) 36–78.

[18] Friedrich Nietzsche, *Beyond Good and Evil* (New York: Vintage Books, 1966) 46 [aphorism #34].

[19] All quotations here are from Eric Cornell, *Time Magazine*, November 6, 2005.

profess any special significance. To the extent that no value has a claim over any other, reasons for sacrifice and (painful) dedication grow weaker. Comfort and easy pleasures become their own defense.

I grant, of course, that what I am calling our philosophical zeitgeist is not ubiquitous. I do not think that political leaders have the liberty to announce, for example, that the principles of their countries are, at bottom, mere values. If we glance back at the words of President Obama quoted above, we see these: "…commanded to our care by God. That's what will lend meaning to the creed our fathers once declared." Here, then, are two efforts to claim more solid ground, God and tradition, and surely many feel these to be rocks on which solid principles for action can be based. Yet when it comes to more philosophic discussion among intellectuals, these "rocks" are more commonly subject to debunking than to reaffirmation.

It is also true that we human beings are highly inconsistent, so we can in practice say something like, "that's just my value judgment," even as we act as if it were a law of nature or personal revelation from a superior being. When Eric Cornell says, for example, "my value judgment is that further progress in science will be good for humanity," my guess is that he believes very strongly that he is correct to claim that science is good, even if he labels it a mere value judgment (and hence removes it from the domain of the correct and incorrect). Human inconsistency allows us to believe in one moment what we in other moments hold to be unknowable; this inconsistency is a useful protection against paralysis, so it brings some benefits. (Could this be one reason Nietzsche claimed in the statement above that the truth is overrated, that recognizing its complexity leads to paralysis?) But it is depressingly desperate to place our hopes for vigorous action in the observation that we can sometimes get ourselves to think we really know what we are doing, when in our more reflective moments we deny that this is possible.

Solutions?

I have sketched what I consider to be the problem of moral relativism because I think it complicates the challenge of guiding Gulliver: if we do not think it possible to know how or where best to direct science and technology, it is unlikely we will think well about this absolutely essential issue. We will be experts in increasing our power, children in knowing how best to use it. Let me now mention two very different attempts to solve the problem of moral relativism before I add my own conclusion. I choose these two partly because they show that, powerful though it is, moral relativism is contested; I choose them especially because they suggest how great a gulf divides those who rise to the challenge.

One response, which I mention also for its novelty, first acknowledges the problem but then suggests that our evolved natures manage to solve it. According to Alex Rosenberg, Chair of the Philosophy Department at Duke University, it is true that it is now impossible for us to embrace the old absolutes of traditional religion and moral philosophy: there is no "right" or "wrong" when it comes to questions of values (278). Far from underplaying the moral consequences of this intellectual development, Rosenberg goes even further than I did. He calls it "nihilism" rather than "relativism" (279, 288); for science – or "scientism," as Rosenberg calls is – is not only unable to defend any one value over any other, it also declares all of them to be devoid of meaning and purpose. Science can still describe what is, but its description shows us eternal purposelessness, not a harmonious "universe" or ordered whole. So far, Rosenberg is the radical par excellence; and he is critical of Existentialists, Secular Humanists, and atheists like Richard Dawkins for not having seen just how devoid of meaning and morality our world really is (277–82). Surprisingly, however, even though he sketches what appears to be a troubling picture of a nihilistic world, he also concludes that evolved creatures might still behave decently even in a world without meaning. Hence, in a trice, his ominous nihilism becomes "nice nihilism" (his term), for we are instinctively altruistic or at least "nice," even if the nature of things is utterly pointless (278, 286–87). If we have become hard-wired to behave decently to one another, it does not matter whether there is anything that deserves to be called "justice" or "the good," for it is our wiring that guides our actions, not our ideas. Our evolved natures, now hard-wired into us, will keep us on a path that is sufficiently straight and narrow, even if there is no moral reason for preferring such a path. Whereas I had worried that moral relativism would keep us from thinking well about how to use our vast powers, Rosenberg would persuade us that our powers of choice are far more limited than we realize. Indeed, free choice is a mere illusion, so ideas do not really matter (300).

If I were to pause to examine Rosenberg's line of argument, I would wonder especially how solid the evidence is that we human beings are hard-wired to be consistently "nice." I would wonder not only about our frequent wars against "the other" but also about what look to me like frequent lapses from niceness even in our domestic lives, where one might expect we would be at our nicest. I would happily grant that we have evolved in such a way that we have not yet gone so far as to destroy ourselves, and that a degree of niceness may help to explain this, but I would caution that only a fraction of our evolutionary history has seen us armed with nuclear weapons and other tools of mass destruction. Whatever one may think of the depth and reliability of human "niceness," Rosenberg is useful in dramatizing the possibility that science (or "scientism") is an acid that corrodes every possible moral orientation. Whereas I think and worry that the philosophic or ideological context of engineering is of existential importance, Rosenberg the nihilist advises we respond to such angst by taking Prozac. (Really! He even stresses the point in his title to Chap. 12. See also 275, 281–82, 315).

A more serious response to the problem of relativism, I think, is that by neuroscientist Sam Harris, whose recent book *The Moral Landscape* insists that relativism is both dangerous and wrong: science should not relegate all questions of good and bad, moral and immoral, to the domain of mere non-rational opinion (2010, pp. 1–14). More than any widely-read author I know, Harris calls attention to our need for a reliable moral compass by which to guide our actions, which are frequently empowered by potent technologies, and he correctly identifies moral relativism as, in effect, a tossing of this compass overboard. This is an important achievement. But Harris goes further and even purports to solve the problem both he and I seek to identify and underscore. Far from being excluded from studying the question of the good, science can identify it, according to Harris. The good is the "well-being" of sentient creatures, and the maturing sciences of the brain can even show us when and how far this well-being is being reached (Harris 2011, p. 28). All actions are not equal; good actions promote well-being, and neuroscience can help identify them by studying the states of our brains. In short, Harris promises "a science of morality" to replace the exemption of "values" from scientific scrutiny (27). Such a science could then be employed to guide Gulliver, pointing the giant in the direction increasing "well-being" and keeping him from wandering aimlessly or being misdirected.

If I were to pause to examine Harris's line of argument, I would wonder whether well-being" admits of qualitative distinctions and, if so, whether neuroscience will be able to make them. It is easy to imagine that brain sciences will help quantify the pleasures and pains we and other sentient creatures feel, for example. And this may nudge us in the direction of better treatment of animals, as Harris clearly wishes. But will the measurement of brain states help us decide whether the raptures of a football fan on his couch are similar in value to those of, say, a Mozart at his piano? Can neuroscience teach us how to evaluate the actions of a gourmand in his pleasures in relation to the sacrifices of a hero or martyr? And if comforting illusions register more well-being than hard truths, should we prefer them? If there should ever be conflict between the well-being of one person and the well-being of another, does neuroscience give either person a reason to yield? However one may assess the prospects that well-being might become the foundation of a science of morality, Harris has done a great service in showing clearly the dangers of moral relativism and how bizarre it is that we often do not recognize them.

Conclusion

Modern engineering provides us powerful tools which can do great good and great harm. To guide the use of these tools we need first to know what is good and second to design political processes likely to steer us in the right directions; thus the promise of modern engineering is dependent on both our philosophical orientation and our political institutions.

On the political level, the first challenge is that there is no effective world government, so the policies guiding engineering and technology will be largely subject to national interests (or perceived national interests). Whether we speak of weaponry or air pollution, nations can be trusted to follow policies which generally put their own interests first; and only with great caution will they renounce shorter term interests in favor of those which might hold in the longer term. The opinions or ideals of particular engineers, like those who campaign for open-source software, may show their influence, but it will be seen at the margins. Secondly, and although far from completely free, the market will show where the demand is greatest, and this too will exert great pressure on the engineering profession and the technology it produces.

But we do not interpret or rank our interests independently of our ideas about them; what people think of as their own best interest ultimately depends on such very fundamental ideas as those about pleasure, honor, wealth, justice, and perhaps even divine salvation. Thus the direction engineering takes is dependent not only on the political procedures we follow to shape our science policy but also on the "philosophical zeitgeist" in which these procedures operate. Behind our politics lie our philosophical opinions, and my main point in this regard has been to suggest that relativism is both powerful and disorienting. In leading us to doubt that any course is certifiably better or more moral than another, it invites the sneaking suspicion that we are living blind, while at the same time suggesting it is impossible for us to recover our sight. Consideration of the vast power of engineering can help us see better that the knowledge of ends is logically prior to the knowledge of means, even as we face the difficulties of acquiring this knowledge and putting it to use.

References

Aristotle. (2011). *Nicomachean ethics*. Chicago: University of Chicago.
Bacon, F. (1965). *The new Atlantis*. New York: Macmillan.
Constitution of the United States of America. http://constitutionus.com/. Accessed 10 Dec 2013.
Cornell, E. (2005). What was god thinking? Science can't tell. *Time Magazine*, November 2005. http://www.time.com/time/magazine/article/0,9171,1126751,00.html. Accessed 6 Nov 2012.
Energy Policy Act. (2005). http://www1.eere.energy.gov/femp/pdfs/epact_2005.pdf. Accessed 10 Dec 2013.
Harris, S. (2011). *The moral landscape*. New York: The Free Press.
Hume, D. (1739–40). *The treatise on human nature*. Public domain. Available at http://www.gutenberg.org/files/4705/4705-h/4705-h.htm. Accessed 15 Nov 2013.
Huxley, A. (1946). *Brave new world*. New York: Harper.
Madison, J., Hamilton, A., Jay, J. (2012). *The federalist papers*. Public domain. Available at http://www.foundingfathers.info/federalistpapers/. Accessed 12 Dec 2012.
Marx, K. (1998). *The communist manifesto*. New York: Signet Classic.
Nietzsche, F. (1966). *Beyond good and evil*. New York: Vintage Books.
Obama, B. (2013). *Second inaugural address*. Available at http://www.whitehouse.gov/the-press-office/2013/01/21/inaugural-address-president-barack-obama. Accessed 22 Jan 2013.
Plato. (1968a). Charmides. In *Platonis opera*, Tomus III. Oxonii, Clarendoniano.
Plato. (1968b). Respublica. In *Platonis opera*, Tomus IV. Oxonii, Clarendoniano.
Rosenberg, A. (2012). *The atheist's guide to reality: Enjoying life without illusions*. New York: Norton.
Safe Drinking Water Act. (1977). http://www.epa.gov/npdes/pubs/cwatxt.txt. Accessed 10 Dec 2013.
Strauss, L. (1953). *Natural right and history*. Chicago: University of Chicago.
Swift, J. (1996). *Gulliver's travels*. New York: Dover Thrift.
Weber, M. (1946). Science as a vocation. In Gerth, H. H., & Wright Mills, C. (ed. and transl.). *From max weber: Essays in sociology*. New York: Routledge.

Wayne Ambler B.A. in Government, Cornell University. M.A. in Political Economy, University of Toronto, M.A. in Classics, Boston College, and Ph.D. in Political Philosophy, Boston College. Associate Professor and Co-Director of the Herbst Program of Humanities for Engineers, University of Colorado at Boulder. Scholarly articles and books on Aristotle, Xenophon, and Aristophanes, as well as on the challenges facing modernization through science, technology, and democracy.

Part IV
Competing Contexts in Engineering

Introduction

Matthew Wisnioski and William Grimson

How and why do engineers' actions result in differing socio-material outcomes? The two volumes in this series find answers to this foundational question of engineering studies in the goals and practices of education, the formation of professional identities, the epistemological bases of design, and the norms and values of practitioners. The authors in this final part draw on such domain specific inquiries to address the question at a synthetic level. A mix of engineers, historians, and philosophers from three continents explore the situated practices of engineering by interrogating the contexts that shape engineers' decisions as well as the very idea of "context" itself.

Context long has been deployed as an explanatory frame for the factors that impact (or do not impact) engineers' decisions and products. Over the past half-century, tens of thousands of pages have been written in support of the belief, as one recent American Society for Engineering Education (ASEE) special interest group asserts, that knowing the "context in which engineering is practiced" should be an integral competency of the "global engineer" (ASEE International Engineering Education SIG and IFEES, 2013). Implicit in this statement, and in much of the scholarship and reform that precedes it, is an eagerness to identify better means for developing reflective practitioners.

But context is a slippery concept. As a practical matter, the specific contexts of engineering are infinite and are highly dependent on where and when one stands. Context is an idea that draws heavily on historical analysis, largely because it is

M. Wisnioski
STS Department 0247, Virginia Tech, Blacksburg, VA 24061, USA
e-mail: mwisnios@vt.edu

W. Grimson
Dublin Institute of Technology, 143 Lower Rathmines Road, Dublin 6, Ireland
e-mail: william.grimson@dit.ie

significantly easier to identify causal factors after the fact. Context, however, whether deployed as accusation or justification, often works to cloud causal relationships even as it names them. The authoritative narrator in Bruno Latour's *Aramis, or Love of Technology*, puts it best: "nothing happens by accident; but nothing happens by context either" (1996, p. 138).

Where does this uncertainty over the nature of context leave engineers and intellectuals, both of whom desire robust methodologies for productively accounting for the heterogeneity of technical labor? Authors in this part seek to tame the thicket of interpretations of context by highlighting their competing meanings in two broad ways. First, they show the interactions between contexts and scales of context that come to bear on a project, on what it means to be an engineer, and on where engineering does and should happen. Second, they investigate and in some cases posit competing interpretations of what context means among scholars of engineering studies. Because context is essentially boundless, some important contributing factors are not covered in the chapters. Gender, for example, is not explicitly included. Nevertheless the approaches taken by the authors are not exclusive; gender and a range of additional themes could well be integrated into most of the frameworks that follow.

In Chap. 19, Matthew Wisnioski points out that engineers "make their own context." They mediate the complexity of inputs that contribute to their experience and arrive at a vision of self, society, and technology in which they interpret and proceed with their engineering undertakings. Wisnioski argues that the core issue is not the knowing about context but the very process or activity involved – how engineers contextualize themselves. He revisits Layton's pioneering notion of an "ideology of engineering," but concludes that there is no unitary professional worldview among engineers. In determining what counts as relevant context, engineers arrive at a range of positions for the simple fact that such an exercise is predicated on a set of localized beliefs. Further, since any set of beliefs will have a socio-political dimension it follows that ideologies shift and new convictions emerge as society itself changes. Wisnioski outlines how differing normative visions arise by surveying a selection of engineers' thoughts internationally over the past two centuries and providing an extended case study of American engineers amid the political and cultural movements of the 1960s. The time and place in which an activity is situated, he concludes, has a significant bearing on what ideology is embraced; moreover, the act of embracing or rejecting a socio-political reality is itself an act of contextualization. The challenge to educators here is that of exposing young engineers to navigating the waters of competing ideologies.

Chapter 20 draws attention for two reasons. In its style, this short text appears as a column, commenting and meditating on observations ranging from the construction of a bridge to the life and work of Galileo. And while doing so, the author (Joseph Pitt) questions the mere notion of "context" itself, and hence the whole project of this EEPiC-diptych. The notion of context is so easily used that we think we know what we mean by it ... until one tries to define it properly! In practice, there seem to be many shades of contextuality. This leads Pitt to the observation that "talk of 'engineering in context' is, at best, unhelpful" because it is a "static concept." The lure of context is the desire to identify bounded causes. But how can those boundaries be maintained in ever-evolving local and global changes? Inspired by a Heraclitean approach, Pitt argues that analysts should not focus on context or even contexts, but rather on processes in which people, technologies, ideas, and their interconnections are in constant flux. But Pitt does not want to deploy processes merely as contexts-plus-action. Rather, processes also are interconnected and changing. Instead of trying to identify and implement universal solutions, the focus should be shifted towards local, situated analyses of engineering practice. The value of Pitt's contribution lies in questioning the presumed self-evidence of our talking about engineering in terms of projects and contexts. Using other words (like processes and environments) may at first sight appear a formal detail: yet it is an invitation to reconsider our views both on engineering and on our thinking about it: a philosophical enterprise par excellence.

In Chap. 21, Li Bocong addresses the problem of interconnected scales in engineering projects by arguing for distinctions between micro-, meso-, and macro-contexts. To establish his position, Li Bocong first traces the rise of context as an analytical lens from the mid 1800s to the present. He show how different frames of interpretation emerged to describe the heterogeneous nature of engineering. He then describes engineering action in a micro-context by reexamining the famed Hawthorne experiments to provide a contextualist explanation that accounts for the physical, social, and cultural environment of any local engineering project. In discussing the meso-contexts of engineering, the author connects the local to organizational structures and inter-professional exchanges that emerge from the norm of team-based work in engineering. He notes that engineers are in essence hybrid men and women sharing values and ideologies with scientists and businessmen. Further, engineers shift their ideologies between these two camps depending on the context of the situation. This is not simply an exercise in easy pragmatism. Rather, Li Bocong claims it reflects the contextualist nature of all engineering practice. Moving on to macro-contexts, Li Bocong uses the example of the variable national development of railways to show that industrial policy, economic policy, and political realities together with the underlying cultural environment largely explain why the growth of transportation systems differed between countries. He gives special attention to the situation in China, where for example there was resistance to building railways in the late Qing Dynasty. Li Bocong concludes that, across all these scales, the context of decision making in engineering is not just an interesting theoretical issue but one of great practical importance.

In Chap. 22 Sjoerd Zwart and Peter Kroes aim to tease out the boundaries between what is at the heart of an engineering design problem in terms of its scientific and technical challenge, and what might be considered the external factors surrounding that exercise. In accordance with Li Bocong, the authors note that stakeholders in a technological project inevitably have different perspectives. The corporate director will have the balance sheet as a significant element in his or her context whereas the chemical engineer trying to find an optimum agent for some process will have little need to consider anything other than the technological challenge on hand. Their chapter deploys an extended design example to take a long hard look at what takes place on the ground. Starting with the previous work of Kroes and van de Poel that held it to be impossible to distinguish between technology and its social context in general, here the authors endeavour to dig into the more specific question about the distinction between core and context in *engineering design*. First Zwart and Kroes introduce two pairs of "cores" and "contexts," which they refer to as "substantive" and "procedural." The first core/context pair relates to the factors that help to determine the design brief, and thus the ensuing object of design, whereas the second, the procedural core/context, is aligned with the factors that impinge on the process of design, such as its organisational aspects, the stakeholders and the availability all kinds of resources. To study the distinction between substantive and procedural, Zwart and Kroes apply their framework to the case of a revolutionary water treatment plant design. In their extended technical discussion it emerges that clear distinctions between core and context are not easily made. One reason being that macro- and meso-contexts can conflict, after all it is not uncommon for corporate monetary factors to suddenly and radically change the direction in which a design is headed. Thus the "object" of design itself changes throughout the process. On the evidence of the water treatment example, Zwart and Kroes come to the conclusion that while in theory it is attractive to make a distinction between core and context of a design as object and process, in actual engineering practice such a distinction is problematic; furthermore, gaining a deeper understanding of the interplay of societal, engineering and scientific aspects within a particular engineering design is more crucial than seeking to demarcate sharp boundaries between useful heuristics.

Fittingly, Chap. 23, the last in the two-volume series, is written by an engineer whose career spans industry, the academy, and national and professional service. In what amounts to a microcosm of the larger project, William Grimson maps the status of context in engineering education, individual and organizational professional identities, ethics and values, economic decision-making, law, and grand societal challenges. He laments the paucity of attention paid to contextual thinking in many domains of engineering, while offering successful cases on which to enhance the profession's sophistication in integrating contextual factors into its image and practice. Grimson's contribution is to highlight a central dilemma confronting the reflective professional. On the one hand, the practitioner is tasked to "do it right," and knows that it is impossible to fully account for the gamut of contexts relevant to the definition and solution of engineering problems. After all, hindsight is built into the very defining of contexts. On the other hand, the responsible engineer knows that to "do the right thing" he cannot pretend to unknow his learned awareness that everything he does (and does not do) alters reality.

Taken as a whole, it is clear from this part that while there is a robust set of lenses to view the subject, "context" remains far from tamed in the engineering studies community much less among practicing engineers. In part, this is because there are no academic paths out of Grimson's dilemma, but conceptual tools such as these volumes nonetheless are needed to offer methodological support to aid engineers as they grapple with the many-layered complexities of their labor. "Real world" case studies can provide a firm base on which to build some working principles for broader public engagement and policymaking. One positive example is the so-called Aarhus Convention, which stipulates with administrative and judicial recourse that environmental information must be provided to citizens in a timely and transparent manner so that they can participate in decision-making relevant to their lives (UNECE 1998). One could envisage using this as a model through which other contextual "knots" of engineering are addressed with the aim of underpinning some reasonable level of interaction between the public and relevant technical authorities. In this sense the defining of relevant contexts is a form of governance, and how society deals with context is a reflection on the very nature of democracy.

References

Hundley, S., & Brown, L. G. (2013). The attributes of a global engineer project: Updates, inputs, faculty development considerations. *ASEE International Forum (June 2013)*. Available at: http://www.asee.org/public/conferences/27/papers/8302/view.

Latour, B. (1996). *Aramis or the love of technology*. Cambridge: Harvard University Press.

United Nations Economic Commission for Europe (UNECE). (1998). *Convention on access to information, public participation in decision-making and access to justice in environmental matters* (Aarhus Convention). Available at: http://www.unece.org/env/pp/treatytext.html

Matthew Wisnioski B.S. in Materials Science and Engineering from Johns Hopkins University and Ph.D. in History from Princeton University. Matthew is associate professor of Science and Technology in Society at Virginia Tech where he teaches and researches at the nexus of history of science & technology, American cultural-intellectual history, engineering studies, and the values of design. He is the author of *Engineers for Change: Competing Visions of Technology in 1960s America* (MIT Press, 2012) and an associate editor of *Engineering Studies*. He is working on a book on the role of innovation expertise in the making of technoscientific selves.

William Grimson BAI, B.A., M.A.Sc., Chartered Engineer. An Electronic Engineering graduate of Trinity College, Dublin and the University of Toronto. He has recently retired but is still actively involved in re-thinking engineering education and the relevance of philosophy to engineering and engineering education. He first worked as a R&D engineer for Ferranti Ltd before joining the academic staff of the Dublin Institute of Technology (DIT) where he was at one time Head of the Department of Electrical Engineering. He ended his career as Registrar with overall responsibility for academic quality enhancement. His academic output was and remains eclectic ranging from publications in areas as diverse as plasma physics, clinical information systems, philosophy of engineering, and development issues. He is currently a member of the Executive of the Institution of Engineers of Ireland.

Chapter 19
Engineers Make Their Own Context: Vision-Making in the Profession

Matthew Wisnioski

Abstract Engineering is an inherently normative practice dependent on how engineers understand history. Social vision, however, when mentioned in an engineering context typically brings to mind extremist regimes such as Nazi Germany or Stalinist Russia. But all engineers practice with assumptions about how their interventions in the material world will change society. The historian of American engineering Edwin Layton called this set of socio-political beliefs the *ideology of engineering*. Engineers' beliefs, however, have not been uniform across time and geography. Engineers have worked in specific national, international, corporate, and government contexts that have influenced how they see society's past, present, and future. This essay surveys the historical literature on engineers' social thought and presents a detailed case study of conflicting worldviews in 1960s American engineering to explore how engineers have acted upon differing *normative visions*. I argue that studying how engineers contextualize their world – particularly during moments of historical crisis – provides a source of inspiration and classroom instruction for those concerned with contemporary engineering in a global world.

Keywords Contextualization • New engineer • Ideology • Normative vision • Responsibility • Technological change • Globalization

---

This chapter was first published in *Engineering in Context* (Wisnioski 2009a). It appears here with updates, corrections, and minor modifications with the permission of Academica. The themes it highlights are elaborated in broader detail in *Engineers for Change: Competing Visions of Technology in 1960s America* (Wisnioski 2012).

M. Wisnioski (✉)
STS Department 0247, Virginia Tech, Blacksburg, VA 24061, USA
e-mail: mwisnios@vt.edu

S.H. Christensen et al. (eds.), *Engineering Identities, Epistemologies and Values*, Philosophy of Engineering and Technology 21,
DOI 10.1007/978-3-319-16172-3_19

The Many Shades of New Engineering

Engineers make their own context. They define what it means to be a professional, to be responsible, to be an American, a Swede, or a global citizen. Not anything goes in this act of contextualizing. The laws of physics and nations apply, budget offices reign, contracts must be honored, social norms obeyed. But like the material world, which a skilled engineer bends to his or her will, who engineers are also is malleable.

In the normal state of affairs engineers collapse *is* into *ought*, focusing on solutions to problems without incorporating how a problem came to be, what sets its boundaries, and how reconceptualizing its past has the potential to redefine engineering itself (Alder 1997). This strategy is in part born of necessity. If engineers challenged every aspect of their being, they would not be engineers but rather skeptic philosophers and historians, incapable of resolving their disputations on time and under budget. But it is also a strategy premised on sequestering the social from the technical in constructions of society and self. To question whom or what one serves is to cast doubt on present needs and to challenge the social order.

A minority of engineers, however, dedicate careers trying to make engineering transcend the needs of the day. Not satisfied with engineering in the world as it is, these visionaries seek to articulate and build into reality what it should be. As a consequence of their contextualizing in the United States we currently have the *Engineer of 2020* (National Academy of Engineering 2004) and the *Grand Challenges for Engineering* (National Academy of Engineering 2008). Forty years ago it was the New Professionalism (American Society of Mechanical Engineers 1971), before that a mandate for engineer-economists.

At the edges of the profession, questions of who engineers should be can take on an overtly political cast with the intention of provoking change in the mainstream. Organizations from Technocracy, Inc. to Engineers Without Borders (EWB-USA 2013) have sought to remake engineering to control a failing global economy, to restore human values to autonomous technology, or to bring modernity's fruits to the other 90 % of humanity (Smith 2007).

Those most concerned with recontextualizing engineering often turn to history, literature, or the social sciences to form an ideological basis for implementing change. Assisted by humanist and social scientist mediators, they use these sources not to understand the past on its own terms, but to mine history and culture as sources of meaning in the present and guides to the future, sometimes reaching as far back as the Neolithic Age for a usable past (Wisnioski 2009b).

The engineer as social theorist in action is a phenomenon hardly unique to the United States. Whether in calls for "Bildung in Engineering" or "new Renaissance engineers," a concerned minority from Denmark to Australia, is turning to local heritages to produce flexible engineers for a global world. Indeed, this volume's organizers have argued in a previous transatlantic collaboration that: "it seems justifiable to speak of a general crisis in engineering education calling for 'a new engineer'" (Christensen et al. 2007, p. 14).

By studying how engineers have used visions of self, society, and technology to remake their world, this essay seeks both to interrogate and contribute to efforts to design "new engineers." I first highlight tensions raised by the prospect of engineering in a "global context" by surveying engineers' past engagements with nationalist ideologies. I then turn to a detailed study of another engineering crisis – the crisis of technology in late 1960s America – to show how disputes about the meaning of technology can reshape engineering's dominant images and practices. Finally, I argue that historical accounts of how engineers contextualize their world during periods of conflict offer reformers a more robust usable past than narratives of context.

Vision-Making in Engineering

Do engineers across borders share a common perspective that transcends local tradition and national interest? Edwin Layton (1971) asserted the historical emergence of a common engineering ethos when he posited the existence of an *ideology of engineering*. Engineers, he explained, are hybrid agents, borrowing from the worldview of the scientist and the businessman, yet identifiable as neither.

To create a unique social identity, Layton showed, American engineers in the early twentieth century propagated a set of beliefs about society and self that drew upon a powerful new rhetoric of *technology* as the wellspring of social progress. In a Machine Age rife with class warfare and alienation, engineers portrayed themselves as heroic servants, harnessing man and nature to build technological systems for the greater good.

As engineers crafted their normative vision, they claimed sole authority over the domain of technology. *The Machine* was neither an autonomous historical force nor a collective cultural achievement; it was the product of the engineer's organized intelligence. This creative agency was fused with prevailing notions of masculinity, middle-class respectability, and, for leading reformers, progressive politics. The *professional* engineer was to apply his *scientific expertise* guided by *moral* virtue. His pursuit of *efficiency* would eradicate waste to bring *profit* to all, *bridging* rifts between Labor and the Captains of Industry (Oldenziel 1999).

The ideology of engineering was a resilient set of beliefs despite the fact that its point of origin belied its mythic nature. It cast engineers as autonomous experts, while most were employed in the bureaucratic hierarchy of industrial corporations. But the ideology of engineering maintained its allure even when it failed in practice. Railing against industry's control of professional standards, in the 1910s–1920s a core of reformers drew upon this ideology to challenge the corporate status quo as a first step in remaking society at large. Significantly, however, most of the same forces that led reformers to believe engineers should be at the vanguard of social change via autonomous expertise, proved even more effective in supporting a vision in which the corporation was the necessary agent of social progress. Supporters of this worldview recast rank-and-file service as subordination to the corporation for the greater good, and argued that the rise to top management marked the pinnacle of professionalism (Noble 1977).

A quarter century of historical scholarship has proven Layton's notion of a unitary ideology of engineering to be an Amerocentric perspective, at the same time that it has reinforced the importance of social vision in the lives of engineers. In different national territories engineers have been trained to serve a range of organizational structures and social philosophies that make any single explanation of what it means to be an engineer untenable (Bailes 1978; Hecht 1998; Meiksins and Smith 1996). Currently, there at least 45 distinct paths in Europe alone, from the *Akademiingeniør* (Denmark) to the *Verkfraedingur* (Iceland) (Lucena et al. 2008, p. 438). Moreover, whom engineers should be changes, sometimes dramatically, even within these boundaries. In Germany, for example, engineers who were avid servants of the Nazi project of reactionary modernism became supporters of Communist engineering in the German Democratic Republic (Herf 1984; Augustine 2007).

We might identify these distinct political visions among engineers of different countries to prove that context matters. No doubt it confirms Alder's adage that "engineers are designed to serve" (Alder 1997, p. 86). Alternatively, we might argue that it tells us something *universal* about the character of engineers, that as C.P. Snow once wrote: "In nine cases out of ten engineers are acceptant of any regime in which they find themselves" (Snow 1954, p. 136).

My claim is that if these historical accounts offer a useable past to engineers, it is the recognition of how embracing or resisting a particular political or social "reality" always is an act of contextualization – not only of social goals but also of who engineers are and what they are for. I am primarily interested in the engineers in Snow's remaining tenth, but even the most compliant organization man structures his vision in such a way as to make resistance seem inconceivable. In both extremes engineers rework their social politics hand-in-hand with the work of defining their uniqueness as engineers.

In my book *Engineers for Change*, I introduced the concept of *normative vision* to convey how engineers' interpretations of society and self structure experience and guide future action (Wisnioski 2012). Normative visions are historically grounded explanations that stand at the heart of much engineering decision-making. They explain the present state of affairs and how it came about with either implicit or explicit reference to outcomes. Normative visions are not rigid set of values, rather they are constructed of plastic networks of sometimes contradictory ideas, images, and practices that individuals use to enroll others toward shared goals and a common sense of self. For the minority of engineers who attempt to scale up their societal ambitions, these visions become synonymous with philosophy or ideology. Studying engineering's intellectuals and visionaries proves especially valuable for understanding the dominant images within larger professional norms.

Ideas have consequences and some visions are more flexible or constraining than others. The failure of progressive reformers in America to make good on their claims of technocratic leadership at the same moment that their values were espoused in a different register by the businessmen that defeated them, is indicative of ideology's flexibility and limitations. Ideological concepts, moreover, are often powerful on the local or national level but either fail to extend beyond these borders or are translated and modified when exchanged across them. *Bildung* in engineering, for example, may be an effective strategy in Germany, may spark contestation in debates about making *European* engineers, and likely will find little traction at all across the Atlantic.

Making the Local Universal

Accepting the local character of vision-making, how then can we account for the obvious similarities among engineering cultures internationally; particularly since what it means to be an engineer appears to be converging as a consequence of transformations in the global economy?

For one, engineers have never lived in hermetic national containers. Whether in frontier expeditions, the development of colonial infrastructures, humanitarian missions, or corporate multinational outsourcing, engineers, their practices, and their devices have been among modernity's most mobile actors. While most engineers have been fiercely protective of local norms and standards, they nonetheless have altered and recontextualized their practices and ideas in this international circulation. The nationalist policymakers who worry about falling behind accelerate this exchange in their desire to know what competitors are doing to define what it means to compete.

At the same time, the expansion of multinational corporations and the globalization of design and manufacturing have brought the role of circulation in engineering to the fore of international consciousness. But employers' visions of what engineers are and whom they serve are sometimes at odds with those of professional societies, national policymakers, and engineering educators. Consequently, one finds both employers and professional organizations expressing fear about autonomous social changes that engineers are failing to master.

Two powerful keywords – *technology* and *globalization* – are at the center of local efforts to contextualize these macroscopic economic, material and social changes. Recognizing how these concepts have come to structure engineering's normative vision is of more than mere academic interest, because "while 'technology' expands its rhetorical reach, that of 'engineering' shrinks" (Williams 2002, p. 17). But the hazardous concepts (Marx 1997) of *technology* and *globalization*, as well as the visions of historical change they support, are not spread by their own volition. They have been and continue to be appropriated, modified, and promoted by individuals and organizations – including engineers – in support of political and material projects.

The historical period I know best, the American 1960s, offers vital insight into the processes by which engineers make coherent their changing world by contextualizing the meaning of technology. In the cultural revolutions of the late 1960s factions within American engineering struggled to enlist the profession's majority in normative visions that challenged dominant narratives of the engineer's societal role. It provides not only a case from which to generalize about vision-making in engineering cultures, but also an understanding of the undercurrents behind the current clamor for "new engineers."

Competing Visions of Technology and Engineering in 1960s America

In the late 1960s, American engineers' compression of *is* into *ought* exploded in stunning fashion. Confronted with critiques of technology from without and changes in labor and knowledge practices from within, engineers' semantic control over technology threatened to unravel. Decrying an onslaught of "New Luddites," professional society officers beseeched engineers to "look back in history and recognize the frailty of the philosophical foundations upon which our acceptance of technology rests" (Marlowe 1970, p. 12). At the same time, radicalized engineers preached "revolutionary engineering" for "countertechnology" (Aquarius Project 1971).

The heyday of the Cold War scientific state – from roughly 1950 to 1969 – brought about the largest transformation in the engineering profession since the rise of the corporation. Engineering expanded rapidly to maintain domestic, military, and cosmic supremacy. Depending upon who was counting, engineering manpower in the United States increased four-fold from 250,000 after World War II to over 1,000,000 by 1965. The consolidation of America's largest corporations was equally impressive. Nearly 75 % of the engineering workforce was employed in just 1 % of all firms. For engineers, however, the most visible change was the expansion of government patronage. When corporate contracts for federal R&D were taken into account, the United States government had become the profession's largest employer, supporting the work of 45 % of the nation's engineers (National Science Foundation 1968) (Perrucci and Gerstl 1969).

In the political economy of Cold War, engineers' self-image evolved from system-builders to servants of the system. The cultural ascendancy of the scientist, contract-based work on giant teams, and the compartmentalization mandated by secret research strained the tenability of narratives of autonomous expertise or entrepreneurship for the rank-and-file, prompting disgruntled engineers to describe themselves as "high-class migrant labor" (Wakeman 1970, p. 70). The success of a handful of systems-entrepreneurs such as Simon Ramo, director of the American ICBM program, and engineering-scientists like Vannevar Bush, head of the Office of Scientic Research and Development during World War II and author of *Science the Endless Frontier*, provided templates for new identities, but did little to combat the image of engineers as organization men.

Instead of an America of unbridled technological optimism, engineers found themselves in a hostile intellectual world. The rise of civil rights, environmental, and antiwar movements challenged the authority of the institutions that engineers served at the same time that they reified technology's agency in social change. Technology, which had been the de facto marker of societal achievement, became a specter of out-of-control power. Texts like Lewis Mumford's *Myth of the Machine* (1967, 1970), Jacques Ellul's *Technological Society* (1964), and Herbert Marcuse's *One Dimensional Man* (1964) portrayed society as totalitarian, based on rationalizing, destructive values. According to these intellectuals, unless the existing system was dismantled, there was no hope for a future in which technology enhanced the "human center."

By the late 1960s, engineers of all stripes concluded that the world was in the midst of a crisis of modernity, characterized by men on the moon and children on fire; campus research centers occupied by student dissidents; environmental hazards resistant to linear solutions; an aerospace industry subsidized by the Department of Defense; a creed of individualism and company loyalty in a system that normalized mass layoffs; and a popular culture in which journalists wrote proudly that "There is no stopping the engineering mentality, we can only try and stop the Engineers" (Marine 1969, p. 46).

It would be a mistake to recognize these changes inside engineering and trace how the *context* of the late 1960s impacted the profession in one way or another. Rather, given the multitude of challenges, this historical moment offers a rich case of how engineers *contextualize* ideas and events in order to negotiate who they are and what they do. While the proximate causes of the crisis were new demands for pollution control, ecological harmony, and the conversion of techniques of warfare to welfare, above all else engineers experienced it as an existential and intellectual threat. As *technology* expanded its rhetorical reach, engineers feared that they were losing their identity as its masters.

A tiny minority, no more than a few thousand engineers, responded by turning to the growing genre of *Technology & Society* literature, written by American and European intellectuals, in an attempt to retool engineering's normative vision. This literature encompassed a range of beliefs, but two competing positions dominated how partisan intellectuals talked about technology.

The political theorist Langdon Winner (1977) described the first position – advocated in the writings of critics like Ellul and Mumford – as an ideology of *technological politics*. This view had two defining tenets: (i) technical decisions were inherently political, and thus technological systems embodied political philosophies; and (ii) once the dominant system was sufficiently advanced, it would become autonomous. Analysis between society's *is* and its *ought* would drive reasoned action to liberate man from totalitarian rationality. In the absence of blueprints for a human-centered society, one's first task was to become aware of his place in the system.

A small core of engineers – generally professors and disaffected defense industry employees – was catalyzed by an ideology of technological politics. They strove to integrate the lessons of technology's critics into their vision of engineering service. John Boyd, a mechanical engineer at WPI, for example, summarized Ellul in his *Journal of Engineering Education* essay "Science is Dead – Long Live Technology!" He claimed that *technique* rendered even engineers into cogs, concluding that engineers needed "to teach how man can use technology rather than be used by it" (Boyd 1972, p. 895). Underground corporate newsletters and alternative journals like General Electric's *GE Resistor* and the Committee for Social Responsibility in Engineering's (1971) *Spark* were venues for bringing together those interested in recontextualizing engineers as collaborative partners with the public rather than as servants of the "military-industrial complex."

This worldview did not *require* reading critical texts, but doing so provided engineers with intellectual authority. It appealed because it named the complex in which they worked – its control of the economy, employment patterns, Vietnam – and most of all because it confirmed their alienation and offered a way forward. Nor were these engineers guided solely by the beauty of their ideas. In addition to constructing human technology, they saw their criticism as a means for gaining control over their labor.

An ideology of technological politics thus was an attractive proposition because it created new and successful modes of political engagement for engineers. Reform movements within the professional societies and engineering schools pushed for all engineers to better understand technology's critics. Its converts, however, could become so engaged in recontextualizing their profession's past, present, and future that they could lose sight of the goal of engineering it. Indeed, there were few projects where the principles of technological politics were implemented into actual projects. Nonetheless, from the perspective of these engineers, success seemed just over the horizon.

But talking about a revolution extended beyond dissenters and reformers. Alarmed by the outsized impact of the critical minority, the nation's engineering deans, society officers, and top management also strove to restore progressive meaning to engineering through a new vision of technology. They conceded that technology created real problems, but a culture of protest was no solution. With the help of establishment intellectuals they crafted a robust counter-ideology to technological politics in a worldview that following Winner and Williams I call an *ideology of technological change*.

An ideology of technological change posited that technology was neither good, nor evil; neither was it neutral. Technological change was a semi-autonomous force that was accelerating rapidly, outracing the ability of social institutions could adjust. It produced tremendous opportunities, but also social dislocations, alienation, and the threat of nuclear holocaust. Through rational management, however, technology's negative *unintended consequences* could be *minimized* and its positive capacities *maximized*.

Engineers circulated three variants of this worldview. The first called for a new "socio-technologist" – an engineer that would be equally versed in the natural sciences and socio-humanistic learning. Ramo, then one of the nation's best-known engineers, was a vocal advocate of this strand as a means of restoring the profession's heroic luster. In two monographs and over a hundred articles and speeches, he argued that systems engineering would be the *Cure for Chaos* (1969) in a *Century of Mismatch* (1970) between technology and society. The second placed the onus for socio-technical decisions on the social science and policy think tanks that gave an ideology of technological change its academic credibility (Harvard University Program on Technology and Society 1972). The third variant targeted the rank-and-file, arguing that the engineer's primary responsibility was to technological change itself. While elite experts resolved the complicated human problems of technology, the majority of engineers were to keep pace with accelerating change.

Advocates of both an ideology of technological politics and technological change in the late 1960s and early 1970s attempted to implement their ideals in multiple venues from professional member societies to develop projects across the globe, but they targeted the nation's engineering schools as the key battleground for reform. That engineering schools were such an important focus of attention comes as no surprise. It was in pedagogy where "new engineers" could be formed before students were disciplined into existing norms. It was also where engineers had greatest access not only to texts in the *technology & society* genre, but in many cases to the authors themselves.

At least since the aftermath of Sputnik, educators had struggled to establish a distinct identity for engineers amid Cold War transformations, particularly the new hegemony of science. Exacerbated by campus unrest, critical theories of technology redirected institutional reform. The same organizations that had pushed to overhaul the nation's curricula to produce engineering-scientists now explored how to use "liberal education" to distinguish the "genuine engineer" from the irresponsible scientist and the menial "technician" as a man who could envision the "system of the future as a whole" (ASEE Humanistic-Social Research Project 1968, pp. 11, 20, 21). One survey identified over 200 schools revamping their curricula (Knepler 1973).

Some engineering educators found themselves empowered in the rush for change. The active mediation of humanists and social scientists made pedagogical reform possible. The historian Lynn White, for example, called engineers "the chief revolutionaries of our time" and suggested that humanists and engineers join forces in the creation of a "new humanism" and a "global democratic culture" (White 1967, pp. 375–376).

Still, educators conceived of what engineers were for differently, and thus read history differently. A handful of faculty and programs sought to use the past to train humility and alternative power relations in technological decisions. Harvey Mudd College, for example, pursued a model in which classic texts were used to recognize that problems of human nature could not simply be engineered out of existence (Waldman 1971). The great majority of educators, however, cultivated an ideology of technological change. They used historical texts to instruct that the consequences of rapid technological change had been unforeseen, but now that its logic was identified, it could be managed by incorporating history and culture as variables in formal design methodologies (Rosenstein 1968).

At the center of these differing interpretations were assertions of whether the engineer was a "conscious agent of social change," or a tool to be directed by change's "manipulators" (ASEE Humanistic-Social Research Project 1968, p. 4) (De Simone 1968, p. 6). However, once establishment intellectuals, top management, society officers, and many educators contextualized a vision that both explained technology's ill effects and absolved engineers for the "unintended consequences" of their work, the force of alternative visions dissipated. Re-fusing the divide between *is* and *ought*, American engineers in the early 1970s internalized an ideology of technological change as a self-evident reality rather than a socio-political worldview.

This moment of rupture in American engineering offers two significant insights about engineering formation – one hopeful, the other cautionary:

1. During moments of expressed crisis engineers – especially engineering educators – are more likely to turn to texts, practices, and human collaborators outside the accepted realm of engineering to recontextualize what engineering is and who (or what) it should serve.
2. Challenged by alternative models of engineering, those vested in the dominant practices of engineering do not simply dismiss technology's critics, they craft alternative visions, which sometimes amount to *counter-ideologies*, to recontextualize the reformers' claims to weaken their impact.

In other words, the story of engineers in the late 1960s ultimately is not about alternative conceptions of whom engineers are for. Rather, it is an account of the contextualization of the hazardous concepts that structure the lives not only of engineers, but of most citizens of the globalized world.

Conclusion: Change Without Change?

The historical inquiries into vision formation presented here offer those concerned with educating "new engineers" in the age of Microsoft and Royal Dutch Shell a pedagogical resource by suggesting that we present history to engineers as more than mere context. Since even the most unrepentant humanist will recognize that anything we teach engineers is likely to perform a normative function, engineering educators should emphasize processes of contextualization to help students make meaning of their world, now and in their future careers. To do so values method over genealogy. This is far less likely to lead to mythology or, worse, strong normative control in the workplace (Kunda 1992), because it gives students the resources to recognize the ways in which dominant images are constructed.

In this regard, studies of conflicting visions in engineering are especially valuable. Cases of contradictory worldviews are pedagogically useful because they make visible the processes that take place implicitly in everyday engineering. Attention to opposing conceptions of what engineering is for emphasizes the mechanisms of authority and the reality of alternatives.

Educators might conclude that introducing nascent engineers to the history of conflict in their own ranks will lead to cynicism. While the perhaps quixotic escapades of radicalized engineers in the 1960s fade into obscurity, the meaning of technology developed in reaction has become the background ideology of modernity (Habermas 1972). Indeed, if anything, the rhetorical structure of *globalization* has further likened engineering service to keeping pace. To read Thomas Friedman (2005) is to read the ideology of technological change in a new register. This vision is the impetus behind many campaigns for "new engineers," which emphasize making engineers maximally appropriate for accelerating global change. What is to be gained, these skeptics might charge, by showing students that the beat goes on?

Thankfully, Bruno Latour reminds us that to simply conclude that "plus ça change plus c'est pareil" is to practice crude scholarship (Latour 1996, p. 131). Engineers are not preordained to reproduce the status quo, and a particular technological future is not inevitable.

Educators – who are critical mediators for helping new engineers contextualize what it means to be a "new engineer" – might apply the lessons of the 1960s to think not about engineers in the *context* of a global world, but rather how they *contextualize* that world in ways that constrain and afford engineers' service. Before asking what competencies make someone maximally appropriate for the global economy it is worth questioning what one assumes *globalization* to be and what engineers should be doing to make it something else. As engineers become involved in global teams, for example, they might focus on where there is room to learn from other groups and where differences are important. By doing so students can learn how to recognize the limitations of the conceptual framework of *technological change* and *globalization* and how those concepts structure their labor.

At the very least, when student exercises aimed at demonstrating the difference between *context* and *contextualization* become integrated into engineering pedagogy, the majority of engineers who do not dedicate their lives to questioning first principles might come to see that there are assumptions involved in their everyday practices – and that in attempting to solve problems they always recontextualize what is given, and thus what is possible.

Acknowledgments The author thanks the editors for suggesting that the chapter be updated for a new audience, Jongmin Lee for his assistance in revisions, and Gary Downey for the spirit of communitas.

References

Alder, K. (1997). *Engineering the revolution: Arms and enlightenment in France, 1763–1815*. Princeton: Princeton University Press.

American Society of Mechanical Engineers. (1971). Goals: Basis for action programs, 'Making Technology a True Servant of Man'. *Mechanical Engineering, 93*(4), 16–19.

Aquarius Project. (1971). Revolutionary engineering. *Spark, 1*(2), 12.

ASEE Humanistic-Social Research Project. (1968). *Liberal learning for the engineer*. Washington, DC: American Society for Engineering Education.

Augustine, D. L. (2007). *Red prometheus: Engineering and dictatorship in East Germany, 1945–1990*. Cambridge, MA: MIT Press.

Bailes, K. E. (1978). *Technology and society under Lenin and Stalin: Origins of the Soviet technical intelligentsia, 1917–1941*. Princeton: Princeton University Press.

Boyd, J. M. (1972). Science is dead – long live technology! *Journal of Engineering Education, 62*, 892–895.

Christensen, S. H., Meganck, M., & Delahousse, B. (Eds.). (2007). *Philosophy in engineering*. Århus: Academica.
Committee for Social Responsibility in Engineering. (1971). *Spark* 1, 1.
De Simone, D. V. (1968). *Education for innovation*. New York: Pergamon Press.
Ellul, J. (1964). *The technological society*. New York: Knopf.
Engineers Without Borders USA. (2013). http://www.ewb-usa.org
Friedman, T. L. (2005). *The world is flat: A brief history of the twenty-first century*. New York: Farrar, Straus and Giroux.
Habermas, J. (1972). Science and technology as ideology. In B. Barnes (Ed.), *Sociology of science* (pp. 353–375). Baltimore: Penguin.
Harvard University Program on Technology and Society. (1972). *A final review, 1964–1972*. Cambridge, MA: Harvard University Press.
Hecht, G. (1998). *The radiance of France: Nuclear power and national identity after world war II*. Cambridge, MA: MIT Press.
Herf, J. (1984). *Reactionary modernism: Technology, culture, and politics in Weimar and the Third Reich*. New York: Cambridge University Press.
Knepler, H. (1973). Engineering education and the humanities in America. *Leonardo, 6*(4), 305–309.
Kunda, G. (1992). *Engineering culture: Control and commitment in a high-tech corporation*. Philadelphia: Temple University Press.
Latour, B. (1996). *Aramis or the love of technology*. Cambridge, MA: Harvard University Press.
Layton, E. T. (1971). *Revolt of the engineers: Social responsibility and the American Engineering Profession*. Cleveland: Press of Case Western Reserve University.
Lucena, J., Downey, G., Jesiek, B., & Elber, S. (2008). Competencies beyond countries: The re-organization of engineering education in the United States, Europe, and Latin America. *Journal of Engineering Education, 97*(4), 433–447.
Marcuse, H. (1964). *One dimensional man: Studies in the ideology of advanced industrial society*. Boston: Beacon.
Marine, G. (1969). *America the raped: The engineering mentality and the devastation of a continent*. New York: Simon and Schuster.
Marlowe, D. (1970). The new luddites. *Mechanical Engineering, 92*(3), 12–13.
Marx, L. (1997). Technology: The emergence of a hazardous concept. *Social Research, 64*(3), 965–988.
Meiksins, P., & Smith, C. (Eds.). (1996). *Engineering labour: Technical workers in comparative perspective*. London: Verso.
Mumford, L. (1967). *The myth of the machine: Technics and human development*. London: Secker & Warburg.
Mumford, L. (1970). *The myth of the machine: The pentagon of power*. New York: Harcourt Brace Jovanovich.
National Academy of Engineering. (2004). *The engineer of 2020*. Washington, DC: The National Academies Press.
National Academy of Engineering. (2008). *Grand challenges of engineering*. Available at http://www.engineeringchallenges.org
National Science Foundation. (1968). *Employment of scientists and engineers in the United States, 1950–1966 (No. 68–30)*. National Science Foundation in Cooperation with US Department of Labor, Bureau of Labor Statistics, Washington, DC.
Noble, D. F. (1977). *America by design: Science, technology, and the rise of corporate capitalism*. New York: Oxford University Press.
Oldenziel, R. (1999). *Making technology masculine: Men, women and modern machines in America, 1870–1945*. Amsterdam: Amsterdam University Press.
Perrucci, R., & Gerstl, J. E. (1969). *Profession without community: Engineers in American Society*. New York: Random House.

Ramo, S. (1969). *Cure for chaos: Fresh solutions to social problems through the systems approach.* New York: D. McKay Co.

Ramo, S. (1970). *Century of mismatch.* New York: D. McKay Co.

Rosenstein, A. B. (1968). *A study of a profession and professional education: The final publication and recommendations of the UCLA educational development program (EDP 7–68).* Los Angeles: School of Engineering and Applied Science University of California.

Smith, C. E. (Ed.). (2007). *Design for the other 90%.* New York: Cooper-Hewitt, National Design Museum.

Snow, C. P. (1954). *The new men.* London: Macmillan.

Wakeman, R. E. (1970). Letter to the editor. *Mechanical Engineering, 92*(11), 70.

Waldman, T. (1971). Quest for commonwealth, an applied humanities course. In *Applying humanities to engineering* (pp. 1–11). Claremont: Harvey Mudd College.

White, L. T. (1967). Engineers and the making of a new humanism. *Journal of Engineering Education, 57*, 375–376.

Williams, R. H. (2002). *Retooling: A historian confronts technological change.* Cambridge, MA: MIT Press.

Winner, L. (1977). *Autonomous technology: Technics-out-of-control as a theme in political thought.* Cambridge, MA: MIT Press.

Wisnioski, M. H. (2009a). Engineers make their own context: Vision-making in the profession. In S. H. Christensen, B. Delahousse, & M. Meganck (Eds.), *Engineering in context* (pp. 403–415). Århus: Academica.

Wisnioski, M. H. (2009b). "Liberal education has failed": Reading like an engineer in 1960s America. *Technology and Culture, 50*(4), 753–782.

Wisnioski, M. H. (2012). *Engineers for change: Competing visions of technology in 1960s America.* Cambridge, MA: MIT Press.

Matthew Wisnioski B.S. in Materials Science and Engineering from Johns Hopkins University and Ph.D. in History from Princeton University. Matthew is associate professor of Science and Technology in Society at Virginia Tech where he teaches and researches at the nexus of history of science & technology, American cultural-intellectual history, engineering studies, and the values of design. He is the author of *Engineers for Change: Competing Visions of Technology in 1960s America* (MIT Press 2012) and an associate editor of *Engineering Studies*. He is working on a book on the role of innovation expertise in the making of technoscientific selves.

Chapter 20
Context Versus Processes

Joseph C. Pitt

Abstract An examination of the difficulties in identifying the conditions for asserting that something is a context reveals the problem of not having a principle of selection for contexts. In this chapter it is proposed that rather than talk about engineering in context, we look to a broader understanding of what is involved in an engineering project, replacing "context" with "process" or perhaps "environment".

Keywords Engineering environment • Fundamental values • Political considerations • Multiple factors • Principle of selection

Introduction

It's all very complicated (Margorie Grene)

It is a commonplace that all engineering is done in some context or other. This is not to say that all engineering education is attuned to context, since the more engineering strives to be science-like, the more it seeks for general non-contextually bound principles. But despite the fact that we think engineering projects take place in a context, there really are no contexts. What there is instead is a process occurring over time in which the components and their relationships are constantly changing. An engineering project is basically the transformation of some set of ideas, designs and materials into something else. It takes place over time and in the process that unfolds the only constant is change. The situation is similar to that of science, in which the individuating of contexts is terribly difficult. The problems scientists work on have a history and change over time as players come and go, so that when talking about science it would be better to concentrate on problematics, which are historically extended investigations (Pitt 2007).

J.C. Pitt (✉)
Department of Philosophy (0126), Virginia Tech,
220 Stanger St., Blacksburg, VA 24061, USA
e-mail: jcpitt@vt.edu

S.H. Christensen et al. (eds.), *Engineering Identities, Epistemologies and Values*, Philosophy of Engineering and Technology 21,
DOI 10.1007/978-3-319-16172-3_20

In engineering, there are multiple stages in the process. We begin with the charge to build X – call it a bridge. The kind of bridge we are to build depends on multiple factors: its site, resources (i.e., budget), availability of materials, deadlines, local politics, cultural factors and personalities and leadership. Each of these factors will play a different role at different times. (Although not an engineer, I did build my own house and so I am drawing on some of my own experience and extrapolating, in addition to drawing on knowledge of what happens at an engineering site.)

The Site

The site is the initial context; how much work does it need? As we grapple with that problem the site itself recedes into the background; questions of what kinds of machinery we are going to need to transfer to the site comes into focus. Not only does the question arise as to whether or not the relevant machinery is available, but is the relevant expertise available? As an example, consider what happens when the geology of a site is not fully explored. When Virginia Tech was constructing its new $90,000,000 Center for the Arts, excavations for the stage area under the theater revealed a geological fault that took over a million dollars to accommodate. What kind of a timeline are we looking at and when can we get going on other facets of the project, such as ordering and storing materials? Are there problems in getting to the site? Do we need to build access roads, will they be permanent or are we going to have to plan on regrading when done? Now these problems are not tackled by a single team. The project is broken down into component parts – with smaller teams dealing with smaller problems – but they need coordination. And whoever does the coordinating brings one problem into prominence and then another.

The Social Environment

Sometimes everything has to stop while delicate negotiations with the locals take place – questions about who will be hired to work on the project and where will they be housed and are there appropriate support facilities in the local community don't seem like engineering problems, but they are part of the context of the project and they too change over time. If a support community needs to be built, what happens to it once the bridge is completed and it is no longer needed? These are concerns that undergird the project and perhaps help situate part of the context, and they can make or break a project. So someone needs to be an expert on local relations – a local or a so-called "expert" from outside, which presents its own set of problems. The local may smooth the way in the community, but not be familiar enough with the ins and outs of the project to be a good liaison between project and local community, with the reverse problems coming with bringing in an outsider. For example, if we return to the Virginia Tech Arts Center, we find a serious clash between the needs of the construction crew to have spaces to park large trucks carrying oversized materials and the faculty and staff of the university where parking is already in serious short supply and there is insufficient public transportation. From this perspective the site is not merely the physical space where the components of the bridge will eventually be placed. It is a broader, more amorphous geo-cultural-political domain, the components of which are in constant play.

Complicating the Context

If we turn to the physical components, they too share this kind of constant motion in relation to one another: earthmovers and drills were really important at first, but once the footers are set, their roles recede as the cranes move in. So the once simple site has now become a beehive of activity, and it is not clear what the context of the project is because there is no single, simple context. Consider who and what is involved in this particular engineering project. We began with the site, but the site can't be the context, because the site is in a location, a place, which may or may not be a simple geographical location. But no geographical location is simple. There are questions of subsurface geology, access to the area, ecological impacts, visual impacts and more to be considered.

Depending on whether the project is for a private firm or a governmental agency, a variety of other factors will make a difference to the context. And depending on what kind of governmental agency is involved, there will be different issues to be considered. If, for example, a state or federal agency wants to put a building in a city or town over which it does not have jurisdiction (e.g., the U.S. Government controls directly certain aspects of Washington DC) not only do the relevant state and federal regulations need to be considered, but the local zoning laws and city ordinances also factor in.

There are further significant ambiguities when considering "context". One we often do not think of in the engineering frame, is the political climate surrounding a project. Thus an oil pipeline from Canada to the United States comes under fire because of environmental concerns such as global warming and the dangers of spillage. But such concerns will impact design factors and may initiate further regulations, transforming the context yet again. Or consider the construction of a memorial to someone such as Martin Luther King Jr. Dr. King is still a polarizing figure in some parts of the United States. The context here is political in its most fundamental sense and involves some fundamental values in American society. And in the case of the King Memorial in Washington, DC, there is a problem determining who is responsible for proposing it, for getting it funded, and for securing approval from the appropriate commissions and committees. Each of these actions may be the responsibility of different persons. The impetus for the memorial came from a variety of different sources and, likewise, was resisted by a variety of forces. The decision to build the memorial was not made the same way a decision to build a bridge is made. And while political considerations can clearly impact a bridge project, consider the issues surround the location of the Tappan Zee Bridge across the

Hudson River in New York state; these were political but not in the same sense as those surrounding the King Memorial.

Looking for a Better Term

Consider a slightly more complicated example. Let's assume a U.S. corporation decides to build a plant to manufacture its smart phone in China. Some of the factors involved are cost of manufacture of the phone, U.S. market needs, international market needs, international trade protocols, copyright issues, transportation in the United States such as port access, rail access or truck networks, Chinese regulations, negotiations with the various relevant Chinese domains, and transportation issues in China. Which is the engineering context? Aren't they all engineering issues to some degree or other? But if they are all to be considered, do they all make up the context? Or do we need a better term like, for example, "environment"?

We need some term that captures the complicated range of factors that we have been discussing. So we should ask, would shifting from "engineering in context" to "engineering environment" do a better job? Well that depends on what the job is. And that is, in fact, one thing we have yet to consider. What job is supposed to be done by talking about engineering in context? At the beginning of this essay, I mentioned attempts in the teaching of engineering to make present engineering more like science, seeking general rules and procedures like scientific laws and the scientific method. It is not clear how this meshes with talk of engineering in context. Perhaps what we have here is a clash between the more theoretical and the more practical minded teachers and practitioners. Or is the reference to engineering in context merely to put an emphasis on the fact that practicing engineers deal with specific projects? But if this is the case, then the specific project is, as we have seen, not a simple thing. And here I am not referring to the complexity of constructing a bridge. From the first suggestion of building a bridge *here*, there enter most of the issues we have been discussing. What is the need (this can be a political or economic or combined concerned)? What are the problems to be faced (economic, political, geographical, geological, manpower, materials)? What social factors need to be considered? And more.

And it is not clear that all of this is managed by one person or at least not the same person throughout the process. At some point the site manager is in control, at another the finance people, at another, the architects – and it goes on like this. There is no one context. Rather, there is a complicated environment that is constantly in flux.

Impact on Engineering Education

That being said, there are some consequences that follow from this realization that what we once glibly referred to as a "context" is now a complicated environment that bear primarily on engineering education. For what we have seen is that the various people in charge of one phase or another of a project need to have multiple skills. They need to be able to work in political, social, engineering, and economic environments. They need to know some history, especially if they are involved in an international project, and some psychology. They need management skills and people skills. And I think it not unkind to observe that the standard engineering curriculum is not designed to educate such persons. (One useful approach is systems engineering, where a system is recognized as a system of systems. But elaborating this idea would take us in another direction.)

Talk of "engineering in context" is, at best, unhelpful. For even the smallest engineering project has multiple dimensions. One of the first to note this is Larry Bucciarelli. We would all do well to go back and reread his *Designing Engineers* (1994). Having said that, it is important to note that Bucciarelli's approach sounds in places similar to the context approach. He speaks of different "worlds" – the design world, the manufacturing world, etc. So while he breaks the process down – it still looks like he has contexts – plural, to be sure – in mind.

Some Historical Considerations

If enough of a case has been made to conclude that whatever happens, undertaking an engineering project is complicated in multiple ways, we still need to address the question of the value of talking about process or environment rather than context. We can start to make the case by looking at one of my favorite engineers, Galileo, and asking about the context in which he worked.

There is a case to be made for approaching the history of philosophy contextually. In so doing one runs the risk of offending defenders of the idea of a perennial philosophy – where philosophical questions are seen as having remained the same over the ages and the search for answers to them is a search for a universal response, one that will be good for all times and places. Not wanting to offend, but also committed to historical realism, I respectfully suggest that there are no perennial philosophical questions. Some questions may sound the same when asked today as when they were posed in Socrates' Athens. One such question is "What is the good life?" My reason for claiming a different question is being asked today than 2,500 years ago is simple. What would have satisfied the Athenians as an answer would not satisfy us today. Consider Socrates' Athens, a city-state that practiced slavery and where women did not have the vote. Today we abhor slavery and seek constantly to ensure equality for all. Our good life is a different one from Socrates' – although one would hope that if pushed hard enough he might come to the same point of view we share today. If the answer is different, the question cannot be the same.

Now for Galileo. The point of historical contextualization is to try to understand the nature of the problems Galileo worked on as he understood them, not as we understand them today. This is to reject the Whig approach to history.[1] However, if we push the question regarding the context in which Galileo worked we are going

[1] See Butterfield (1931).

to find ourselves in a mess. For there are several ways to characterize the context in which Galileo worked or, to put it somewhat differently, Galileo worked in many contexts. He was by title Court Mathematician and Philosopher to Cosimo II. He was a catholic who was also considered a heretic. He was a lover, a father, head of his family but never married, constantly obsessed by money. He was a Platonist and an Aristotelian and a Copernican and still again something new as a methodologist. He was a musician and he designed things like the military compass and his telescope and microscope. How are we to understand the context in which he labored? Was it the court? Was it that part of Italian society dominated by the Church? Was it his family? Was it his love of puzzles? Was it his commitment to mathematics and measurement? A major problem with contexts it seems is that there is no principle of selection that allows us to identify the proper context.

I want to argue that while there was no singular context in which to place Galileo, he was a product of all of those contexts and probably others not mentioned. To understand Galileo, the man, his work, his passions, requires knowing a lot about a lot of different things and how they all fit together. To try and find one or even more contexts in which to place him would be to create unnecessary problems. For example, was he a Platonist or an Aristotelian? Does it really matter? Why can't we allow him to have started his career in an Aristotelian mode (having been educated in that fashion) and through his increasing commitment to the role of mathematics in problem solving drawn into a Platonic mode only to adopt a Euclidean methodology and end up creating two new sciences. If we really want to understand Galileo, we need to see him developing and responding to the various forces by which he was buffeted. Over his lifetime he was constantly changing, as we all do. I object to historians of philosophy who insist that a philosopher's views must be consistent over time. Why must this be the case? Does anyone seriously believe that we should hold a 75 year old thinker to ideas she first propounded in her 20s? Of course not. Engineering contexts are much the same – there is no one context in which to frame an engineering project, there are many and they struggle for prominence depending on personalities, places and times.

Conclusion

The value of abandoning contexts in favor of processes is that it allows us to obtain a greater understanding of the dynamics of the environment in which engineering projects take place. In their introduction to *Engineering in Context* (2009), the editors, explaining the project, noted,

> Thus an important aim of this book is a better understanding of the contexts in which engineering activities are situated within the larger realm of human activities and the culture which surrounds them at the micro, meso and macro levels. (Christensen et al., p. 5)

This is a lofty goal and one worth pursuing. But to achieve it we need to rethink one premise, that it can be achieved by focusing on contexts. For "context" is a static concept. It suggests some kind of identifiable set of boundaries. But in the real world, boundaries are fluid and constantly shift in response to varying forces and changing priorities. As noted earlier, in our bridge example, different individuals take on leadership responsibilities depending on where the project is at any point in time. Appreciating the flow of the process where sometimes *this* is being done and *this* person is the one to ask about something and then something else needs to be done and others take charge is to come closer to understanding what really happens. Talk about context may make things simpler, but simplicity does not always yield true comprehension. In this context seeking simplicity is like looking for universal answers to local questions. It just may be time to take the next step and look for local answers in all their complexity and to abandon interfering artificial boundaries.

References

Bucciarelli, L. L. (1994). *Designing engineers*. Cambridge, MA: MIT Press.

Butterfield, H. (1931). *The whig interpretation of history*. New York: Norton, 1965.

Christensen, S. H., Delahousse, B., & Meganck, M. (Eds.). (2009). *Engineering in context*. Aarhus: Academica.

Pitt, J. (2007). Seeing nature; origins of scientific observation. In J. Gayon, R. Burian, & M.-C. Lorne (Eds.), *Conceptions de la Science: Hier, Aujourd'Hui et Demain. Hommage à Marjorie Grene* (pp. 272–289). Brussels: Ousia.

Joseph C. Pitt AB College of William and Mary. M.A. and Ph.D. in Philosophy from University of Western Ontario. Professor of Philosophy, Virginia Tech. Founding Editor of *Perspectives on Science: Historical Philosophical, Social*. Editor-in-Chief of *Techné: Research in Philosophy and Technology*. Latest book: *Doing Philosophy of Technology*. Founding Director of the Center for the Study of Science in Science, which has transformed over the years into the Graduate Program in Science and Technology Studies. Major interests: the impact of technological innovation on scientific change. Current Projects: Editor, *The Routledge Companion to the Philosophy of Technology* (in progress); *Seeing Near and Far, a Heraclitean Philosophy of Science.*

Chapter 21
Engineering Action in Micro-, Meso-, and Macro-contexts

Li Bocong

Abstract Context initially referred to linguistic context of language texts and discourse in the fields of linguistics and communication. But philosophers of engineering should research the context in which engineering practitioners both speak and act. Engineering action means not only an individual's action, but also a collective action participated in by many kinds of engineering practitioners. Modern engineering action is usually undertaken by an enterprise as a special kind of community. The context of engineering action can be divided into three levels: micro-, meso-, and macro-levels. For a particular engineering decision-maker and a particular engineering action, the boundary between action and context is to some extent may be changeable, but it does not mean that there is no boundary between action and context. The problem of context is not only a theoretical one, but also a practical one.

Keywords Engineering action • Engineering community • Context • Micro • Meso • Macro

The publication of *Engineering in Context* (Christensen et al. 2009) marked an important advance in the study of engineering. In its preface, the editors referenced the book *Science in Context: Readings in the Sociology of Science* (Barnes and Edge 1982) as a classic in the sociology of science. Their own volume sought to bring engineering under the same contextualist perspective. As a further contribution to this approach, the present chapter in a new collection on engineering in context reviews the emergence of context as a general principle of understanding and then explores its application to engineering at three levels.

L. Bocong (✉)
College of Humanities and Social Sciences, University of Chinese Academy of Sciences, 19A Yuquan Road, Beijing 100049, People's Republic of China
e-mail: libocong@ucas.ac.cn

S.H. Christensen et al. (eds.), *Engineering Identities, Epistemologies and Values*, Philosophy of Engineering and Technology 21,
DOI 10.1007/978-3-319-16172-3_21

The Emergence of the Theory of Context

The theory of context was brought forth in the later years of the nineteenth century. Studies on various issues of context in such academic fields as linguistics, philosophy, anthropology, and communication, have now become common.

In the history of science, Gottlob Frege (1848–1925) is the earliest scholar who placed context at the core of theory. He used the concept of "context" for the first time in *Die Grundlagen der Arithmetik* (The Foundations of Arithmetic) (Frege 1980[1884]), and brought forth the famous "context principle", which urges people never to ask for the meaning of a word in isolation but only in the context of a sentence. For various reasons, Frege was virtually unknown when alive and did not become famous in the fields of linguistics, philosophy, logic, and analytic philosophy until the second half of the twentieth century, when he was recognized as the founder of modern logic and a major figure in analytic philosophy. Because his primary status was as mathematician and logician, his views on context exerted an influence mostly on logicians, philosophers of mathematics, and analytic philosophers, and were unfamiliar to scholars in other fields. For a long time and even now, scholars especially in the fields of social science failed to appreciate Frege and his work.

Many scholars regard Bronislaw Malinowski (1884–1942) as the founder of the theory of context. Malinowski proposed the concept "context of situation" in 1923 and used it in translating and understanding utterances. In 1935 he put forward the new concept "context of culture" and applied this idea in anthropological studies. While Frege focused only on context of words, Malinowski focused not only on words but also on larger units of language, sentences, passages, and so on. He paid close attention to both written and verbal language, and to linguistic and non-linguistic contexts. As a result, Malinowski has been influential well beyond anthropology.

In the second half of the twentieth century, the theory of context evolved rapidly. Many linguists, such as J.R. Firth, M.A.K. Halliday, D. Hymes, J. Lyons, Wang Zhanfu, and Wang Jianhua, developed their own linguistic theories of context (Zhou Shuping 2011, pp. 13–49). As a result, linguistics became an important field in which the study of context bore rich fruits. At the same time, scholars expanded its scope. As the study of context entered new fields, contextualism emerged as a general approach. Differing from those who interpreted text as language text, and context as linguistic context, the French philosopher Paul Ricoeur (1913–2005) expanded the understanding of text and context. Ricoeur proposed that not only book text and verbal language can be regarded as a text, but action can be seen as another kind of text. Ricoeur pointed out that not only the context of a language text, but also the context of an action, can and must be studied as such (Ricoeur 1981, pp. 197–221).

While in the fields of linguistics, communication, and linguistic philosophy scholars focused attention mainly on the context of written language, verbal language, discourse, and book text, philosophers of science focused attention on both the contexts of written scientific papers, on the one hand, and of scientific experiments as a special kind of practice, on the other.

Studies of context have been further advanced in the philosophy of technology. For example, Anthonie W.M. Meijers has argued that the properties of technical artifacts can be divided into three types, one of which is context-dependent properties (Meijers 2001, p. 83). In addition, there is more attention to and conscientious historical study of context in the history of technology. The publication of the book *In Context: History and History of Technology* (Cutcliffe and Post 1989) demonstrated how historians of technology appreciate the importance of the study of context.

Context with Regard to Engineering

Insofar as context can now be applied to the study of engineering, there are three points to be noted and underscored. The first is related to the object of study in context. Different from scholars in the fields of rhetoric, linguistics, and linguistic philosophy who take the context of language text as the object of the research, philosophers of engineering should lay emphasis not only on the context of written and verbal language of engineering papers but also on the context of engineering action as a particular kind of text. In other words, the crux of the matter lies in the context in which an individual or an enterprise acts rather than in the context in which an individual speaks.

The second point is that following economists, ethicists, and sociologists who put forward the hierarchical distinction between micro-, meso-, and macro-levels of both action and analysis, philosophers of engineering should analyze and investigate the context of engineering action based on this same three-level framework (Li Bocong 2012, pp. 31–35).

A third point concerns the Chinese translation of these terms. When studying context in the field of philosophy of engineering, Chinese scholars must translate the English word "context" into Chinese. However, there is a problem here. In linguistics, rhetoric, and philosophy of science, "context" is generally translated into the Chinese word 语境 (*yǔ jìng*) that refers to linguistic context or the environment of discourse or utterance clearly and strictly in Chinese. Yet when studying the context of engineering action, the context obviously refers not only to the environment of discourse or utterance, but more importantly, to the social, economic, and natural environments of engineering action. But such a meaning cannot be included in the Chinese 语境. Engineering action is not just a linguistic action, but a kind of action creating artifacts. In order to express these meanings, "context" must be translated by another Chinese word, 场境 (*chǎng jìng*) which has a wider meaning than 语境. Therefore, while the same English word can be used to express both the environment of a text such as novel, poem, or scientific paper, and the environment of an action, two different Chinese words 语境 and 场境, which should not be confused, must be used respectively to express the meaning of the context of language and the context of action, that is, the environment of language and of action.

Practitioners' Engineering Action in the Micro-context

The divisions of micro and macro in physics, economics, and sociology have been supplemented by economists's addition of the concept of meso, thus establishing a micro-meso-macro framework. The three levels of the framework are obviously inter-related and connected, although the micro level is in some sense fundamental.

While the scientific community consists of scientists as homogeneous members, the engineering community consists of heterogeneous members, including workers, engineers, managers, investors, and other stakeholders (Li Bocong et al. 2010, pp. 27–29). Consequently, engineering action inherently involves multiple actors including workers, engineers, managers, investors, and so on.

When analyzing the activities of engineering practitioners who are the micro subjects of engineering action, one discovers an interesting phenomenon: different members of engineering communities have reciprocal relationships with each other between action and context. For instance, engineers, managers, investors, and some other stakeholders become the context of activities of workers. At the same time, workers become an important factor of the context for managers, engineers, investors, and other stakeholders. What follows touches on some issues related to the context of activities of workers and engineers.

As we know, many scholars focused their studies on workers's activities. Social scientists and management experts proposed various theories to interpret workers's activities. A good example is the Hawthorne experiments of Elton Mayo (1933 [1960]).

The Hawthorne experiments, which were carried out at the Western Electric Company's Hawthorne plant in Cicero, Illinois, in the 1920s and 1930s, have had a far-reaching influence. The initial purpose of the experiments was to determine the relationship between the situation of the physical workplace and productivity. At the beginning of the experiments, researchers did not obtain results they expected. However, as a result of subsequent involvement Elton Mayo and Fritz Roethlisberger, the experiments underwent dramatic changes. In particular, a significant breakthrough was made in the theoretical explanation of the experimental results.

Admittedly, different researchers have given different interpretations to the same results by approaching them from different perspectives. Some scholars interpreted the Hawthorne experiments from a humanistic perspective. From this perspective, workers should not be regarded merely as "economic men" but also as "social men". It was argued that after the Hawthorne experiments, management theory entered the "era of social men". To a certain extent, this can be thought of as a human nature oriented theory.

But there is another, context-oriented theory that differs from the human nature oriented theory. Although the interpretation that focuses on the social nature of human beings is to a great extent justifiable, it would be a mistake to believe that the Hawthorne experiments demonstrate only the importance of human nature in engineering action and that the context of the workplace is of little importance.

Roethlisberger, who participated in the Hawthorne experiments as an assistant of Mayo, generalized their findings in *The Elusive Phenomenon* as follows:

> 1. Work conditions have more effect on production than the number of workdays in the work. …
> 3. The supervisor's method is the single most important outside influence. Home condition may affect the worker and his work. However, a supervisor who can listen and not talk can in many instances almost completely compensate for such depressing influences.
> 5. The most surprising result came toward the end of the experiments, … when the researchers returned to the original forty-eight-hour week without rest pauses. Once again, productivity rose! Yet again, it seemed that the workers were responding to the positive concern of the experiments rather than to the physical work conditions (quoted from Gabor 2000, pp. 113–114).

Obviously, the Hawthorne experiments never denied the importance of context. But they definitely reject any view that regards context as only the physical work conditions or the material environment. And the experiments fully revealed that workers work in contexts with various factors.

When mentioning the content and research method of context, Andrew Jamision (2009) indicated that the context of engineering includes economic, social, and cultural contexts, and Sylvain Lavelle (2009) pointed out that context may be studied from analytic, phenomenological, and pragmatic perspectives. The Hawthorne experiments do not mean that only humanity or morale is important while context is unimportant for practitioner's activities. From the contextualist point of view, the Hawthorne experiments have two important theoretical implications:

1. Context includes various aspects, such as the physical, social, and cultural environment. So the colorful and plentiful contents of context should not be simplified.
2. The Hawthorne experiments demonstrated that worker morale is affected by the cultural context. It is the special cultural context created in the experiments by Mayo that inspired the worker morale. Without the special cultural context created in the experiments, the results would not have been obtained.

Mayo's theory has been widely regarded as the foundation of human relations management. Strangely and even paradoxically, the two theories – the theory of human relations and the theory of human nature – amount to two different theoretical orientations. The latter focuses more on the properties of the subjects – especially the properties that are not easily affected by an external environment. The former focuses more on the correlation and interaction of different subjects, and on the properties and features of different subjects that are likely to be affected by an external environment and external relationships. Judging from this, human relations theory is imbued with a strong sense of contextualism.

While Mayo focused on workers' actions, some other scholars focused on engineers' actions. The work of engineers is important and complicated in engineering action; the tasks of engineers complex and mixed. "The position of engineers, partially as labor and partially as managers, prompted Herbert Shepart to call engineers marginal men; part scientist and part businessman, sharing value and ideologies with both camps" (Beder 1998, p. 25).

Shepart's opinion is initially surprising. However, on reflection, many people, even engineers themselves, may accept Shepart's thesis. It is certain that the role of an engineer is different from that of a scientist. Engineers more often than scientists undertake difficult and contradictory tasks. Sometimes, if the demands of labor or scientific factors put great pressures on engineers, engineers will tend to be partial to one over the other. If the pressures of capital and business become stronger, engineers will lean toward satisfying the needs of capital – although this does not always happen. These shifts in behavior do not imply that engineers have a kind of ethical disorder. They simply reveal the fact that engineers, whose positions and functions are quite different from those of workers and investors, are on the horns of a dilemma, which should be analyzed from a contextualist perspective.

Enterprise Engineering Action in the Meso-context

Generally speaking, engineering action involves collective action. In other words, it is carried out by a team, a group, or an organization rather than an isolated individual. In the contemporary world, an isolated individual, for example, an isolated manager, an isolated engineer, or an isolated investor, could by no means engage in actual engineering action. Especially, those who have the same profession or occupation, such as workers or engineers, cannot by themselves initiate engineering action. Engineering teams or groups must be composed of different kinds of members: engineers, workers, investors, managers, and other stakeholders.

Many scholars regard engineering activity as what engineers do. There is no doubt that this is to some extent true. But it is only part of the truth. The complete truth is that engineering activity consists of what engineering teams do, including what engineers do, what workers do, what investors do, what managers do, and especially, what the team as a whole does. If there are only engineers, without the participation of investors, workers and managers, such engineers could not take engineering action at all. In fact, engineers only engage in some part of engineering action. A complete and actual engineering action must be completed by an engineering team composed of engineers, workers, investors, and managers. Adopting the concept of social reality put forward by J. R. Searle (1995), we have every reason to believe that an enterprise is a particular type of social reality (Li Bocong 2009).

Engineering action is impossible without the activity of individuals, including workers, engineers, investors, and more. At the same time, engineering action is a collective action. The two points mentioned above are not contradictory because any human activity must, in the final analysis, be carried out by individuals. If it is not the case, there will be no engineering action to speak of. However, without a team or a collective organization, for example in an enterprise, there would be no engineering action either. What is the bridge that connects one point with the other?

The answer lies in division of labor and cooperation, work in cooperation with an appropriate division of labor.

We must pay attention to the fact that when individuals do something in an engineering team or an enterprise, usually, they cannot do it just on their own initiative or for their own sake but only for the sake of others. For instance, designers design for investors and users instead of for themselves and engineers and workers also do their duties for the sake of others. In an enterprise, workers and engineers cooperate with their partners. What they do must be suitable to what their partners do. For instance, the general manager of an enterprise signs an agreement on behalf of the enterprise, which means that people must differentiate an individual acting in some role within a team from that the same individual acting on his or her own. More importantly, people should not confuse an enterprise with some isolated individual or even with all its members. There is a distinction between an individual's action and an enterprise's action, which leads to a conclusion that when studying context of engineering action, we should study not only the contexts of individuals but also the contexts of enterprises.

When engineering action is taking place, the context in which an enterprise acts is different from the one in which an individual as a member of an enterprise acts. So there are two kinds of context which correspond with two kinds of action, an individual's action and an enterprise's action. How should we distinguish these two kinds of context? Are there still other kinds of context?

As some economists argue, an economy can be divided into three levels: the micro-economy, the meso-economy, and the macro-economy. Similarly, the contexts of engineering action can be divided into three kinds of context: the micro-context, the meso-context, and the macro-context. Generally speaking, the meso-level is at the place which is situated between micro- and the macro-levels. Usually, an individual acts in micro-context and an enterprise acts in meso-context.

Obviously, the meso-context is more complicated than a micro-context. When studying the meso-context in which an enterprise acts, scholars must pay attention to why and how technological, economic, institutional, cultural, and social environments influence the structure and function of an enterprise, and why and how meso-contexts play important roles in an enterprise's decision-making and its engineering actions. Because an enterprise can keenly understand and experience meso-contextual influences which may profoundly affect its development, the enterprise must carefully and prudently take into account meso-contexts such as the regional economic situation, cultural traditions, institutional environments, and business prospects.

Silicon Valley in the United States is an attractive region for many enterprises, especially for those in the field of information and electrical engineering. Many entrepreneurs hope they can build their enterprises there. What makes Silicon Valley more attractive than some other regions is that it is blessed with special technological, economic, cultural, and political contexts which are superior to other regions in their receptivity to and promotion of innovative activity. The gist of AnnaLee Saxenian's *Regional Advantage: Culture and Competition in Silicon Valley and Route 128* (1994), for instance, is to analyze and emphasize the importance of context for business and engineering development, especially the importance of cultural context.

Nobody can deny the huge impact of some advantageous contexts on enterprises and engineering action. A good context can significantly facilitate business and engineering development. By contrast, a context with all kinds of disadvantages will surely impede engineering development. Considering such situations, many countries and governments set up "industrial parks" with a view to create an enabling context in which enterprises can act and develop smoothly.

Engineering Action in the Macro-context

Macroeconomics focuses on the structure, function, and trends in an economy as a whole. In the global era, macro refers to not only a national but necessarily as well to international contexts. A similar interpretation applied to macro in the fields of philosophy and sociology of engineering.

In different countries, different enterprises and different kinds of engineering act in different macro contexts. Now, consider railway engineering as an example. Many countries have constructed railway systems during the modernization process. However, even three leading modernized countries – the UK, the US, and France – are quite different from each other in the railway development process, which cannot be attributed only to technical factors. The railway network in the UK developed initially at a fast pace but ran up against problems such as redundancy of transportation capacity and low construction quality on some lines with resultant replacement construction. In the US, the construction of a railway network was faster and on a larger scale. But during the great development of railway engineering in the US, some companies overestimated the rate of return of engineering investment and went bankrupt. In addition, some railway lines in America were defective in design and construction. The characteristics of the construction of French railway network were that the government took the leading role in the railway network. The French railway system was constructed much slower, which helped prevent unnecessary needs for replacement construction, and established a railway system that in the end had reasonable design and high quality. Frank Dobbin's *Forging Industrial Policy: The United States, Britain, and France in the Railway Age* (1994) carefully and thoroughly analyzed the marked and serious differences in the railway construction in these three countries. He pointed out that the root cause lay in the vast differences of the three countries in industrial policy, economic policy, political condition, and cultural environment. To put it simply, it is the different macro context of the three countries that led to the different processes of railway construction and resulted in different railway networks in the three countries.

Although many developing countries shared the same view that they must develop their own railway systems, railway construction processes in developing countries too were based in different economic, political, and cultural environments. Among them, China seems to be a very special case. The initial railway development in China might be the most difficult case in the world history of railway. After the first railway line was constructed in the UK in 1825, the US, France, Germany, and Russia rapidly initiated their own national railway construction projects in 1830, 1832, 1835, and 1837, respectively. While there was almost no resistance to construction of railways in many countries, there was violent and stubborn resistance to building railways in the late Qing Dynasty. The resistance was so strong that many events in the history of railway in the late Qing Dynasty are unimaginable to later generations. In 1865, a British businessman built 1-li-long railway line outside Xuanwu gate, Beijing, without authorization. But the roaring sound of the train triggered a terrible shock to the common people. The railway was rapidly demolished by the government. In 1876, the Woosong Road Company, which was established by British and American businessmen, built a railway line which stretched for ten miles in Shanghai. However, the Chinese government bought it for 285,000 taels of silver and then demolished it.

In 1878, a 3-li-long railway line was exclusively built for the Empress Dowager Cixi along the bank of Beihai Lake in the Forbidden City after a negotiation between Li Hung-Chang and several British businessmen. The railway was designed to provide convenience for the Empress Dowager Cixi to have meals, take rest, and enjoy the scenery of the imperial garden in order to show the advantage of railway to the Empress Dowager Cixi. But because the Empress Dowager Cixi disliked the roaring sound of the locomotive, the exclusive line in the imperial garden was pulled by eunuchs with ropes instead of by a locomotive (Ji and Kang 2011, pp. 8–15).

Objection to railway construction in China came from China's traditional cultural, political, and social conditions in the late Qing Dynasty. For instance, railway construction may lead to misfortune according to *fengshui* (geomancy). It should be underscored that the obstacles were not thrown mainly by some particular individuals but by the Chinese cultural and political tradition as a whole. Chinese officials engaged in fierce debates over railway policy for some 20 years. Only after eliminating many obstructions did the government in the late Qing Dynasty finally and officially began railway construction in China. Such was the macro context in which railway construction began to take place in China.

Time passes on like an arrow. At the end of the twentieth century, the macro context in China was strikingly different from that in China a hundred years earlier. The twenty-first century witnesses the large-scale construction of high speed railways in China. Why does China stand at the forefront of the construction of high-speed railway in the world at the beginning of the twenty-first century? The main cause resides again in the macro context. Although high speed railway technology was mainly invented in developed countries, the construction of a high speed railway system in developed countries has fallen behind that of China. Why did this case take place? The answer lies in the different macro contexts.

Conclusion

Context is an external factor in engineering activity from the perspective of philosophy. When analyzing and studying context, we must pay attention to the different micro-, meso-, and macro-levels. In addition, we should attend to interactions among different aspects and between different levels. Different aspects and levels in context interact in complex ways. Contextualism in the field of philosophy of engineering can be considered an overall perspective that studies various issues of context. Scholars should analyze engineering action, engineering practitioners, and the engineering communities including various sub-communities such as enterprises, engineering teams, and engineering institutions from a contextualist perspective.

We must admit that context is to a certain extent relative. However, such relativity does not necessarily mean that we can take context as an imaginary issue or an issue that can be neglected. There is no doubt that the issue of context cannot be eliminated. It is crucial for any particular individual and any particular research task. More importantly, as for a particular decision-maker or a particular enterprise, the boundary between text and context cannot be drawn arbitrarily. A decision maker or an enterprise must draw contextual boundaries correctly.

To sum up, context of engineering action, including the context of decision-making, of designing, of manufacturing, of maintenance, and of using products is not just an important theoretical issue, but also an important practical issue. To decision makers, to engineering practitioners, and to managers of an enterprise, there are many particular contextual problems they must analyze and treat in a practical manner all the time. To philosophers, to ethicists, to sociologists, to psychologists, and to management experts, there are many theoretically contextual problems in their fields to be analyzed and treated in theoretical ways.

References

Barnes, B., & Edge, D. (Eds.). (1982). *Science in context: Readings in the sociology of science*. Milton Keynes: Open University Press.

Beder, S. (1998). *The new engineer: Management and professional responsibility in a changing world*. South Yarra: MacMillan Education Australia PTY Ltd.

Christensen, S. H., Delahousse, B., & Meganck, M. (Eds.). (2009). *Engineering in context*. Aarhus: Academica.

Cutcliffe, S. H., & Post, R. C. (Eds.). (1989). *In context: History and history of technology: Essays in honor of Melvin Kranzberg*. Bethlehem: Lehigh University Press.

Dobbin, F. (1994). *Forging industrial policy: The United States, Britain, and France in the railway age*. Cambridge: Cambridge University Press.

Frege, G. (1980[1884]). *The foundations of arithmetic: A logico-mathematical enquiry into the concept of number*. Evanston: Northwestern University Press.

Gabor, A. (2000). *The capitalist philosophers*. Chichester: Wiley.

Jamison, A. (2009). The historiography of engineering contexts. In S. H. Christensen, B. Delahousse, & M. Meganck (Eds.), *Engineering in context* (pp. 49–60). Aarhus: Academica.

Lavelle, S. (2009). Technology and engineering in context: Analytical, phenomenological and pragmatic perspectives. In S. H. Christensen, B. Delahousse, & M. Meganck (Eds.), *Engineering in context* (pp. 75–95). Aarhus: Academica.

Li, Bocong. (2009). Lue lun shehui shizai: Yi qiye wei fanli de yanjiu [On social reality: A study on the enterprise as a model]. *Zhexue yanjiu* [Philosophical Research], *5*, 104–110.

Li, Bocong. (2012). From a micro-macro framework to a micro-meso-macro framework. In S. H. Christensen, C. Mitcham, Li Bocong, & An Yanming (Eds.), *Engineering, development and philosophy: American, European and Chinese perspectives* (pp. 23–36). Dordrecht: Springer.

Li, Bocong, et al. (2010). *Gongcheng shehuixue daolun: Gongcheng gongtongti yanjiu* [An introduction to sociology of engineering: A study of the engineering community]. Hangzhou: Zhejiang daxue chubanshe [Zhejiang University Press].

Li, L., & Kang, B. (2011). *Wang qing tie lu* [The railway in the late Qing dynasty]. Beijing: Zhongguo tiedao chubanshe [China raiway press].

Meijers, A. W. M. (2001). The relational ontology of technical artifacts. In P. Kroes & A. W. M. Meijers (Eds.), *The empirical turn in the philosophy of technology* (pp. 81–96). Amsterdam: Elsevier Science.

Ricoeur, P. (1981). In: J. B. Thompson (Ed. & Trans.). *Hermeneutics and the human sciences: Essays on language, action and interpretation.* Cambridge, MA: Cambridge University Press.

Saxenian, A. L. (1994). *Regional advantage: Culture and competition in Silicon Valley and route 128*. Cambridge, MA: Harvard University Press.

Searle, J. R. (1995). *The construction of social reality*. London: Penguin Press.

Zhou, Shuping. (2011). *Yu jing yanjiu* [A study of context]. Xiamen: Xiamen daxue chubanshe [Xiamen University Press].

Li Bocong M.A. in Philosophy. Professor in Philosophy at University of Chinese Academy of Sciences, Vice-Director of Research Center for Engineering and Society of UCAS and Editor-in-Chief of *Journal of Engineering Studies.* He has published numerous books in Chinese; among the most relevant are (with translated English language titles), *An Introduction to Philosophy of Engineering* (2002), *Selection and Construction* (2008), *An Introduction to Sociology of Engineering* (co-author, 2010), and *Theory of Engineering Evolution* (co-author, 2011). He is also the co-editor of the English language volume *Engineering, Development and Philosophy: American, Chinese and European Perspectives* (2012). Research areas: philosophy of engineering, philosophy of science, sociology of engineering, history of engineering.

Chapter 22
Substantive and Procedural Contexts of Engineering Design

Sjoerd D. Zwart and Peter Kroes

Abstract Kroes and Van de Poel (Problematizing the notion of social context of technology. In S. H. Christensen, B. Delahousse, & M. Meganck (Eds.), *Engineering in context* (pp. 61–74). Aarhus: Academica, 2009) maintain that distinguishing between technology and its social (intentional) context is impossible, because social phenomena are definitive (constitutive) for technology. This raises the problem of differentiating between the social processes that are internal (definitive) and those that are external (contextual) to technology. To explore this problem we distinguish instead between the core and the context of design as object and as process, and we apply them to a case study of the design and development of a new technology for sewage water treatment to find out whether these distinctions make sense in real life engineering practice. Despite the *in abstracto* plausibility of this distinction between core and context, our analysis reveals that its application may turn out to be very problematic in actual engineering practices. The same holds for characterizing particular design features as being the result of either internal (technological) or external (social) factors.

Keywords Product of design • Process of design • Social context • Substantive context • Procedural context

Introduction

In their analysis of the notion of social context of technology Peter Kroes and Ibo Van de Poel (2009, p. 71) come to the conclusion that "independently of whether technology is interpreted as a process or a product … it is not possible to draw a

S.D. Zwart
Departments of Philosophy, Delft and Eindhoven Universities of Technology,
Jaffalaan 5, Delft 2628BX, The Netherlands
e-mail: S.D.Zwart@tudelft.nl

P. Kroes
Department of Philosophy, Delft University of Technology,
Jaffalaan 5, Delft 2628BX, The Netherlands
e-mail: p.a.kroes@tudelft.nl

S.H. Christensen et al. (eds.), *Engineering Identities, Epistemologies and Values*, Philosophy of Engineering and Technology 21,
DOI 10.1007/978-3-319-16172-3_22

demarcation line with technology on the one side and its social context on the other. The reason is that the definition of technology as a process or a product involves reference to social phenomena. Social phenomena are conceptually definitive of technology (or, in ontological terms, constitutive of technology)." This means that "it is not possible to treat all social phenomena as belonging to the context of technology, since some social phenomena are definitive or constitutive of technology." Regarding technology as process the authors interpret engineering (design) practices as social practices to show that social phenomena are more than just part of the context of technology, and regarding technology as product they refer to the dual nature of technical artifacts according to which intentional (social) features are constitutive of technical artifacts.

In this chapter we intend to follow up on this analysis of how to distinguish (social) context from technology and to further problematize the distinction between technology and its social context. We focus on the notion of context with regard to engineering design practice. We identify and analyze *in abstracto* two different kinds of contexts, referred to as 'substantive' and 'procedural' contexts. Both contexts appear to be operative in engineering design practice in the sense that they may influence the outcome of engineering design projects (section "Substantive and Procedural Contexts of Design"). To confront these abstract distinctions to real engineering design practice, we describe in detail an actual design and development project (section "Example: GSBR Technology in Nereda Wastewater Treatment"). Finally we examine whether the distinction between substantive and procedural contexts may be applied to this project and so may be of help in classifying factors that influence the outcome of a design and development process as technological (internal) vs. contextual (external) (section "Discussion").

Substantive and Procedural Contexts of Design

Engineering design may be considered as a *process* and as a *product*, viz., that which is developed during this process and finally is presented as its outcome (Kroes and Van de Poel 2009). To delineate these concepts more precisely we follow (Dorst and Overveld 2009). Engineering design as a process is a human activity in which plans are developed to create an artifact that helps the user attain certain of his/her goals and therefore has value for the future users. An engineering design project has many intended, non-intended, known, and unknown outcomes some of which are directly related to (the plans of) the prospective artifact. The (intermediary) artifact related outcomes of an engineering design project we will call the *object of design*. This object of design in its final form consists of at least the following descriptions, viz., a description of the "artifact itself", a description of the interface between this artifact and the outside world, and an outline of how and in which contexts the designer has imagined the artifact should be used. The latter has also been called the artifact's "use plan" (Houkes et al. 2002). The description of the artifact itself should at least cover the *form* of the artifact, its *function* and its *working principle*. Our delineation of the object of design is still very general and allows many kinds of descriptions, which may include blueprints, texts written in natural language, all kinds of mathematical, iconic and structural models, etc. Moreover it does not exclude the object of design to be a process (for instance, a service), in which case the emphasis of the artifact description is on a series of actions, which may or may not include reference to use plans depending on whether or not some actions imply the use of artifacts.[1] Here we do not focus on the distinction between object of design as object or process; whenever in the following we refer to the object of design as an object it is intended as shorthand for object or process. We will concentrate on the distinction between *design as a process* and *design as a product* (which, thus, may be either an object or a process). Note that our general delineation of the object of design does exclude the object of design to be the physically realized artifact itself. Thus the object of design is not the specific building at its unique place as the solution of a design problem. It is the description of the building and of plans for how to build it. So the object of design is an abstract object.

Following the distinction between the object of design and the design process we may distinguish between the *substantive* and the *procedural* contexts of design. Roughly, the first consists of all factors that influence the object of design and the second all factors that influence the process of design. Regarding the former, it is the context that plays a role in determining the design problem that is to be solved. Here we are dealing with contextual factors that have a direct influence on the object of design in the sense that these factors determine what kind of object is to be designed and the list of requirements or specifications it has to satisfy. The procedural context, however, plays a role on the level of the design process and is the set of factors that determine the time frame and resources available for solving a design problem, which, of course, may indirectly influence the object of design. Clearly both contexts 'shape' the object of the design (and thus, a technical artifact that is an embodiment of that design) that is proposed at the end of a design project. Before we turn to a more detailed discussion of the substantive and procedural contexts of engineering design, it is necessary to delve somewhat deeper into the meaning of the notion 'context' that we are using.

Intuitively, the notion of context of something implies that it is possible to distinguish between what belongs to the 'inside' of that thing and what to its 'outside', its environment or context. However, how the distinction between what belongs to the inside and what to the outside is made, may depend heavily upon how the thing under consideration is conceptualized: "The notion of the context of something has a well-defined meaning only from a certain perspective, one which determines what kind of conceptualizations are adequate or useful and which ones not" (Kroes and Van de Poel 2009). With regard to design this means that the distinction between design and its context depends to a large extent on the conceptualization or framing of what design is.[2] For instance, from the point of view of a project manager the

[1] Note that if the object of design is a process (such as a service) then a use plan may be associated with this process.

[2] Note that from this perspective the borderline between objects (or processes) and their contexts is merely a conceptual and not an ontological affair.

context of an engineering design project differs from the context of the 'same project' in the eyes of other stakeholders such as a corporate director, a design manager or a design engineer. For a corporate director the context of the design project may consist mainly of balance sheets, for a project manager a potential new product line may be an important element from its context, for a design manager the patent position of the firm and the search for new patents and for an engineer designer, finally, anything that is not technically relevant for solving the design problem at hand.

Besides differences in framing or viewpoint there are also differences in resolution of a design project and its context (Hales 1993, Chap. 1; Hales and Gooch 2004, pp. 21–23). From the point of view of a manager of an engineering design project that project may be situated primarily within the context of his/her company or within the context of the company's local or national market which in turn may be situated in the context of a global or international market. Which level of resolution is chosen in analyzing a design project and its context (macro- and micro-economical, corporate, project or even personal level) depends, of course, on the specific problem about the design project and its context that one is dealing with.

As have been argued by Bucciarelli (1994), even at a very detailed resolution level, individual design engineers frame their design often very differently, even if they collaborate in one design team. He refers to these framings or viewpoints as "object worlds". After years of cooperative observation in corporate multi-disciplinary design teams, Bucciarelli draws the conclusion that the participating engineers see the object of design differently; they live in distinct object worlds. Engineers with different backgrounds such as mechanics, electrotechnics, fluid dynamics, thermodynamics and biotechnology, live in their own object worlds, which are characterized by branch-specific instrumentation, standards, codes, and quantitative instrumental rationality. They tend to concentrate on the "hard stuff" of the design, free of any context in which people and societal values play a role, and are brought up to solve single-answer problems using quantitative methods. Bucciarelli's ethnographical approach shows that today's real-world engineering design practices are characterized by multi-disciplinary teamwork in which many negotiations between individuals from different object worlds are taking place. His ethnographic observations show that there is only partial mutual understanding of the individuals having different conceptions of the object of design. Because there is not one single object of design and because of the negotiations between different design engineers, Bucciarelli concludes that today design is first and foremost a social process. For our purposes the most important lesson to be drawn from Bucciarelli's work is that even at a very fine-grained level of resolution differences in perspective or framing of a design project or object and its context may play a crucial role.

In order to further clarify the role of various contexts in shaping technical artifacts (technology) we will now have a closer look at how contexts may be conceptualized in the case of design as a process and as an object. With regard to the object of design, the distinction between what belongs to its context and what belongs to *engineering design proper* we will take to correspond roughly to the distinction between the kinds of factors and considerations that play a role in fixing the constraints and specifications and those that play a role in fixing the physical structure of the future technical artifact (the design as a blue print for production). More in particular, we assume that design decisions that fix this physical structure on the basis of factors and considerations of a technological or scientific nature are internal to the core of engineering design. Other factors and considerations, that influence design decisions, we take to be part of the context of the object of design. We have been referring to this context as the *substantive context* because it is the context that determines the substance of the design effort, namely what kind of technical artifact is going to be designed.

As far as design as a process is concerned, what belongs to the *design process proper* we take to be primarily those actions that have to be performed in order to bring about an adequate solution to the given design problem in a systematic or methodic way where 'adequate' means that a particular design solution meets the lists of specifications, irrespective of the means and resources available or employed to arrive at this design solution.

These kinds of actions and their order have been studied extensively by design methodologists. They have proposed numerous general flow diagrams that prescribe the various steps that have to be taken in order to solve a design problem in a methodically justified way. We take the *procedural context* of the design process to be the factors that determine the conditions concerning available means, resources and time under which actions can be taken to solve the given design problem. All this is summarized in Table 22.1.

A first comment to be made with regard to Table 22.1 is that the distinctions between object and process and between core and context for engineering design are not crisp or clear cut. This is important to keep in mind. *Prima facie* Table 22.1 suggests that the influence of (social) factors from the procedural and substantive contexts on the object of design are not on a par, since only factors from the substantive contexts define the 'essential' features of the object of design, that is, the features as defined by the list of specifications. Factors from the procedural context appear to have only a contingent influence on the object of design since its influence does not affect the set of features that is definitive for the object of design. Suppose that the same design brief is given to two design teams, that have different resources

Table 22.1 Substantive and procedural contexts of engineering design as product and as process

	Design as a product	Design as a process
Core	Scientific and technological considerations that fix the physical structure of the technical artifact to be (the design as a blue print for production) = 'factors internal to engineering design proper'	The kind of actions (and their order) as prescribed by the flow diagrams of design methodology
Context	The kinds of factors and considerations that play a role in fixing the constraints/ specifications (substantive context)	The factors that determine the means, resources and time available for a design project (procedural context)

available or that have developed different design cultures. Due to these differences in resources or design cultures each design team may come up with a different design solution that satisfies the list of specifications. In that case the influence of factors from the substantive context on the object of design is the same for both cases, but the influence of the procedural factors varies.[3] However, the situation becomes more complicated as soon as procedural factors make it necessary to adapt or revise the list of specifications (for instance, because it turns out not to be possible to meet certain specifications with the available resources). In such situations factors from the procedural context may have a direct influence on the object of design that no longer can be characterized as a contingent influence. In other words, whenever the list of specifications is adapted during the process ("on the fly") for reasons related to constraints put on the design process the above *prima facie* difference in influence of factors from the substantive and procedural context appears to break down, because the distinction and relation between object of design and process of design becomes more complicated than suggested by Table 22.1. As is often remarked, in real life design practices, feed-back loops that end up in revisions of the design specifications during the design process are more the rule than the exception. In so far these feed-back loops find their origin in reasons related to process constraints they undermine the simple picture of the influence of substantive and procedural contextual factors of Table 22.1.

A second comment on Table 22.1 concerns the 'visibility' or 'traceability' of social influences, whether stemming from the substantial or procedural context, on an object of design or technical artifact. Consider the following series of design tasks, ranging from designing a raw material, to designing components, up to designing an end-user product:

1. The design of some steel with properties X, Y, and Z.
2. The design of a valve with that steel for an engine of a certain type
3. The design of an engine for some type of car
4. The design of a car.

It seems that the closer we get to end-user products the easier it is to trace the social influences operative in shaping the technical artifacts and to determine their functional features. When confronted with a specimen of steel with properties X, Y and Z it may be immediately clear that we are dealing with an artifact, something made purposely by humans, since we have never come across steel in nature. But it may be more difficult to trace the specific social influences that shaped this material into what it is and to determine its use-plan than in the case of an end-user product like a car. The steel has only physical micro- and macro-properties and does not, so to speak, carry a use plan with it. One might be tempted to say that a raw material, a component and an end-user product differ in the extent to which the artifact "carries with it its use plan". Nevertheless we have to realize that all these designed objects have a particular intentional (social) history and it is this intentional history that

---

[3] Of course, the outcome of the two design projects may be such that after all one design is to be preferred above the other. But that is not the point at issue here; here the question is which contextual factors have a definitive or contingent influence on the object of design.

makes them different from objects with the same physical properties but lacking this intentional history (for more details on the constitutive role of intentional features for being a technical artifact, see Kroes 2012).

This difference in visibility or traceability of social features raises the question to what extent design engineers may bracket social influences and concentrate on the 'purely' physical aspects of the technical artifacts they design.

Generally speaking it may be the case that the closer the artifact is to ready-made consumer goods the more difficult it is to bracket the impact of social factors from design practice. But much depends on how well all societal constraints on the object of design have been translated into functional requirements and in specifications that can be stated clearly in physical terms. Suppose that the properties X, Y and Z of the steel to be designed can be stated in purely physical terms, then the design engineers may forget about the functional requirements from which these properties were derived. Then the social (intentional) context of the new type of steel, its intended use, can easily be bracketed or cloaked (for a detailed discussion of cloaking either social-intentional or physical aspects, see (Vermaas and Houkes 2006)). This will be more difficult when the object of a design is, for instance, a car, since it will be much more difficult to express (or operationalize) all functional requirements of a car in purely physical terms. But even if in the case of steel the properties X, Y and Z may be expressed in clear physical terms, social factors may enter the scene, so to speak by the backdoor, because some of the chemicals used to meet the specifications for properties X, Y and Z may be poisonous, expensive, politically problematic or bad for the environment.

Here ends our analysis *in abstracto* of the role of substantive and procedural contexts in engineering design. Its main result is summarized in Table 22.1. We have already noted that the distinction between the substantive and procedural context may become blurred in case the definition of the object of design is changing during the design process. In the following section we will present a description of a real life research/design process in which a change in object of design actually took place. In the final section we will then discuss whether our analysis of substantive and procedural contexts may be of help in understanding the role of social factors in this particular example.

Example: GSBR Technology in Nereda Wastewater Treatment

Having introduced the notions of context and core for engineering design as product and process, we now turn to a real life example of an engineering design project from biotechnology. It concerns the introduction of a new and successful wastewater treatment technology.[4]

[4] Much of the information contained in this section has been collected by one of the authors and colleagues during ethical parallel research on the design and development of this new waste water treatment technology during the period 2004 until 2006 (De Kreuk et al. 2010; Van de Poel and Zwart 2010; Zwart et al. 2006).

On May 8, 2012 the world's first full-scale municipal wastewater treatment plant using Granular Sequencing Batch Reactor (GSBR) technology was put into operation in Epe, the Netherlands. Contrary to more traditional biological activated sludge wastewater treatments, this brand new technology treats domestic and industrial wastewater using *aerobic granular sludge*. In the Netherlands other Water Boards have ordered similar plants, which are under construction. The consultancy firm DHV, which has been the commercial driving force in scaling up the technology from laboratory to full-scale, is building comparable plants in the Stellenbosch region, South Africa, and in Ryki in the southeastern part of Poland. It baptized the new technology *Nereda*. At the opening ceremony, Joop Atsma, the Dutch State Secretary for Infrastructure and the Environment, claimed: "The development of this technology stands as a perfect example of what can be achieved when the public sector, universities and the private sector come together to develop smart solutions." (DHV 2012).

A drawback of traditional biological wastewater treatment plants is their large footprint in terms of space. In these plants water is purified using bacteria flocks, the so-called activated sludge. The low average biomass concentration and the low settling velocities force traditional plants to use large settling tanks. Besides these settling tanks, the plants need other tanks to accommodate the various steps for nitrogen, COD and phosphate removal, with large recycle flows and a high total hydraulic retention time. Moreover to process the surplus sludge from municipal wastewater plants it needs to be thickened and filter-pressed. In the newly developed aerobic GSBR, biomass grows in dense aerobic granules. This means increased biomass concentration in the reactor tanks and improved separation efficiency. The time needed for the sludge to sink to the bottom at the end of each cycle is substantially diminished, which increases the throughput of the installation. The new technology is based on a batch process in which the bacteria that treat the waste water pass through a cycle consisting of a phase of nutrition under anaerobic conditions and a phase of growth under aerobic conditions. This cycle has been chosen in order to promote the formation of stable granules by the slow-growing bacteria.

Apart from their improved settling characteristics, the aerobic granules can cope with nitrogen, COD and phosphate removal in one tank due to their unique layered structure. Because of diffusion gradients inside the granules, the various process conditions usually found in different tanks are now satisfied inside the granular sludge – the plant-in-the-granule concept. The technology uses effectively only one tank without the need for large recycle flows. Theoretically these granules can reach high removal rates, namely 100 % organic carbon removal, 90–95 % of phosphate removal and 90–95 % of total nitrogen removal with 100 % ammonium removal (De Bruin et al. 2005; De Kreuk et al. 2005). Feasibility and design studies showed that the required land area of traditional waste water treatment plants can be reduced by 80 % and the energy needed can be decreased more than 30 % because of a decrease in construction material and energy needed during building and operation (De Bruin et al. 2004).

Before we delve deeper into the history of this technology, let us present a brief overview of the main parties that have been involved in the design and development of Nereda. The GSBR technology has been developed by Mark van Loosdrecht and his colleagues at the Department of Biotechnology (Kluyver Lab.), Delft University of Technology, the Netherlands. After successful laboratory experiments, Van Loosdrecht, approached various governmental and private organizations to gather funds for the further development of the technology. STOWA, the Foundation for Applied Water Research in the Netherlands, proved willing to invest in the scaling-up of the three litres laboratory reactor to an outdoor pilot plant of 1.5 m^3. STOWA is an organization of the water boards, the local authorities responsible for sewage treatment in the Netherlands. STOWA finances research on new treatment technologies. Van Loosdrecht also acquired funds from STW, a governmental agency stimulating and promoting innovative academic research, for a Ph.D. research project that was carried out in parallel to the pilot plant research. Finally, DHV, an international engineering and consulting firm, with water management technology as one its main domains, showed interest in the commercial exploration of the GSBR technology. DHV was in charge of the research at the pilot plant.

The history of research on GSBR technology goes back at least to research on anaerobic sludge for waste water treatment during the 1970s. Anaerobic Granular Sludge was known to be formed in Upflow Anaerobic Sludge Blanket (UASB) reactors used to produce methane while treating wastewater using an anaerobic process. A blanket of granular sludge was formed which starts to reach maturity after 3 months and suspends in the tank. This blanket contains dense compact granules with a particle size larger than 0.75 mm. Lettinga from Wageningen University in The Netherlands is well known for his late 70s work on UASB reactors. Expanded granular sludge bed (EGSB) digesters are waste water treatment systems similar to the UASB reactors. At the start of the 1990s *aerobic* sludge research was given a boost by the hypothesis of Mishima and Nakamura claiming that also aerobic filamentous bacteria could mutually entangle into aerobic granules. It turned out that aerobic granules could be formed but the explanation of the process remained controversial. For anaerobic granules it was suggested that bacteria stick together into granulates because of mutual exchange of indispensable nutrients. Aerobic bacteria are autotrophic and perfectly capable to live on their own. So why would they agglutinate into granules? At the end of the 1990s researchers started to theorize about the answer to the latter question (we will return to the issue of aerobic granulation below). At the same time researchers started setting up the first lab-scale aerobic granules experiments (Morgenroth et al. 1997; Beun et al. 1999; Dangcong et al. 1999; Etterer and Wilderer 2001), and from the start of the twentieth century interest in aerobic granule research increased and pilot scale studies started to be carried out. The history of the GSBR technology and of Nereda is to be placed against the background of this scientific and engineering interest in aerobic granular sludge in waste water treatment.

Van Loosdrecht and his colleague Sef Heijnen became interested in what later became the GSBR technology after a 1996 visit to colleagues in Munich who were working on Sequential Batch Reactors (SBR). They thought about how they might combine their own experiences with airlift reactors with SBR's. SBR's are processing tanks for a five stage treatment of batches of wastewater. They were not primarily directed towards the production of granular sludge. However, SBR's were suitable to make smooth rounded particles so perhaps they could be used to create aerobic granular sludge in a SBR.

After their Munich visit two issues became prominent on the research agenda of Van Loosdrecht and his colleagues, one of a scientific, the other of an applied nature. The scientific issue concerned the biological explanation of the development of granular sludge. Anaerobic granulation was biologically explained: anaerobic bacteria stick together because of their mutual exchange of indispensable nutrients. But aerobic and autotrophic bacteria were also shown to granulate. How can this be explained biologically as they are perfectly capable of living on their own? According to Van Loosdrecht the formation of aerobic granules undermined the biological collaboration theory of granulation for anaerobic granulation; at least it could not provide a general explanation of granulation. Another explanation attempt, which focused on the extracellular matrix of proteins in the granules, could be used to explain granulation for the aerobic and anaerobic case. But according to van Loosdrecht this matrix is indeed important for the structure of the granules but not for the granulation process. Van Loosdrecht wanted to show that even the fastest growing (aerobic) organisms could granulate and that mechanical shear forces in a reactor are decisive for granulation. For this reason, he set up a research project (performed by Beun, and financed by NWO) with the aim of comparing granular formation in an airlift and a bubble column in which the shear forces are different. The outcome showed that fast growing organisms granulated in an airlift but resisted granulation in a bubble column. So, this research showed a clear influence of shear on growth rate of the organisms: the lower the growth rate gets, the less shear is needed to produce granules. Moreover it showed on a laboratory scale the possibility of an aerobic granular sludge airlift reactor.

The second, application-oriented issue concerned the structure of the granules of the sludge. People were producing aerobic sludge with autotrophic nitrifying granules but they did not succeed in producing aerobic granules with an anaerobic core of heterotrophic bacteria. Van Loosdrecht and his colleagues wanted to develop granules with a layered structure, with an aerobic outside layer where nitrification could take place and an anaerobic zone in the center taking care for the denitrification by heterotrophic organisms. With the help of such granules it would be possible to combine two stages in traditional waste water treatment.

After the possibility of an aerobic granular sludge airlift reactor had been demonstrated in the laboratory, Kreuk carried out another research project that aimed at scaling-up the aerobic granular sludge technology from the laboratory acetate set-up to pilot plant scale using real waste water (in the period 2000–2004). It was funded by STOWA and STW. In fact, work on scaling-up issues had already begun earlier, when the laboratory work of Beun was still on its way. Van Loosdrecht approached several engineering firms as a result of which a major engineering firm in the Netherlands, DHV, became involved in the scaling-up research. An application for a STOWA grant was written in the second half of 1998. STOWA combined the Delft aerobic GSBR proposal and another proposal stemming from the University of Wageningen into a compact reactor innovational research incentives scheme. Doing so, they allowed DHV to carry out a more precise and specific tentative feasibility study. By changing the technology somewhat and calculating the costs, DHV managed to put the technology in a financially more attractive perspective. Accordingly, the engineers were allowed to pursue the project and thus the GSBR connection between DHV and the Kluijver Laboratory came about. Because STOWA and DHV were both most interested in household sewage water treatment, the latter became the main focus of the aerobic GSBR project. This focus was not fixed at the outset.

In 2002 the STOWA compact reactor scheme came to an end and was finalized with a full-blown feasibility study for a real scale aerobic GSBR household sewage plant. The study was positive about the real scale possibilities of the new technology. For this reason STOWA wanted to proceed with the development on pilot scale and provided DHV with financial means to do so. The pilot started in 2003 in Ede. STW was not involved in financing the pilot, which was too practical for their standards, but they did finance the research of Kreuk, the Kluyver Lab participant in the project who was dealing with practical and theoretical up-scaling questions on a daily basis.

As was more or less its standard practice, STOWA installed a Supervisory Committee (SC) to monitor the progress of the project. Its members were representatives from STOWA, STW, the Kluyver Lab, HDV, the water boards as being potential users of the GSBR technology, and members of other engineering firms. The SC was to act as a forum where actors from the network, and some stakeholders from outside the network, could meet to have discussions and make relevant research decisions. The committee did not have a formal decision-making procedure in place, but it influenced research and development decisions by providing a forum for negotiations between the actors. All in all, the SC had three functions. First, it was to control the quality of the research and the progress of the project. Second, it provided oversight so that, besides the scientific knowledge acquired, the practical and applied knowledge was also published in a clear and explicit way. Third, the SC was to function as a critical sounding board.

One of the issues discussed in the SC concerns the problem of the production of granules in the GSB pilot reactor. The engineers/scientists had produced granules in the Kluyver Lab on small scale (3-liter reactors), at room temperature using acetate and not real sewage water, and using an airlift reactor. At the pilot circumstances were very different. Besides the difference in volume, the working temperatures were significantly lower and the substrate was real sewage wastewater instead of acetate. As a result of these different circumstances the granules in the pilot were showing up very slowly. Moreover, the people involved in the scaling-up project tended to talk in terms of granules but these were of such small dimensions that the external specialists only saw flocculated sludge. This question became an important issue at the end of 2003 when the Sludge Volume Index (SVI; the sludge volume index is an important measure of sludge settleability and thus for granule formation) were difficult to measure because of sludge flotation (STOWA report of januari 8, 2004). This flotation was not a sign of firm granules. The lagging behind of the granulation process provoked serious discussions about changing the criteria and specifications of the GSBR technology. DHV and the scientists were asked to provide a new version of these criteria. Among other specifications the first proposed version regarding granulation stated: "stable and significant lower SVI" than measured before (SC minutes, Jan 15, 2004). Of course, the researchers did not want to put themselves in a straightjacket during the process. The SC was not satisfied, however, and they started a discussion about the definition of the granules. This discussion resulted in go/no criteria which only concerned granulation. The criteria decided upon at the SC of May 18, 2004 read:

1. The fraction of dry matter should be at least 15 kg/m^3
2. The SVI after 5 min should be around 50 ml/g
3. Half of the substance sludge/granules should consist of granules with minimal diameter of 200 μm (SC minutes May 18, 2004)

Although the lower limit of the granules seems to be modest relative to the granules produced in the laboratory, the go/no go episode and its definition of granulation clearly illustrate that the SC took its role seriously.

Actual work on the pilot installation started in 2002 and ended in 2005. In 2002 the pilot set-up in Ede started with one airlift and one bubble column reactor to compare their performances. In June 2003, however, the set-up of the airlift reactor was transformed into a second bubble reactor and today's full-scale technology is based on bubble column and not on air-lift reactors. Thus, the decision to change the core of the technology into bubble column reactors has had a decisive influence on the final design of the waste water treatment installation. Interestingly, the choice between the two types of reactors is closely related to at least four different issues or factors, viz., the scientific experiments about the role of shear forces in granular growth; the working principle of the granulation; considerations of financial and energetic costs, and finally the network-of-actors involved in the collaboration.

Let us first have a closer look at the difference between a bubble column and an airlift reactor. A bubble column reactor is just a container in which air is pumped in from the bottom. An air-lift reactor is a bubble column reactor with an additional (internal or external) feedback loop for the liquid. The air bubbles force the liquid to rise from the air inlet at the bottom of the reactor and to go down in the feed-back loop where no air bubbles forces it to rise. The feedback loop ends again at the air inlet. Obviously, an airlift column is more difficult to build but has better circulation and oxygen transfer characteristics. Moreover it provides higher and more equally distributed shear forces between the granules and the substrate in the reactor. At first sight, the design decision between the two working principles depends primarily on the balance of the extra costs of an airlift reactor against the advantages of better oxygen and shear-force characteristics. A somewhat closer look, however, reveals interesting interdependencies regarding the four issues mentioned above.

We have already come across the issue of the role of shear forces on granular growth. Van Loosdrecht initiated research into this issue because he had serious doubts about the prevailing explanation of anaerobic granular growth. Experiments performed at the Kluyver Laboratory showed that shear forces in airlift reactors were high enough to impede fast growing aerobic bacteria to grow in flocks and to stimulate the growth of granules. Since they stimulate granules production, and shear forces in an airlift column are higher than in a bubble column reactor, scientifically airlift reactors are preferred to bubble columns for granulation. However, the issue of the influence of shear forces on granular growth rate did not in the end decide the choice between and airlift or bubble column reactor.

Another issue of paramount importance was the oxygen concentration in the reactor. Most successful experiments with aerobic granules at the start of the twenty-first century operated under relatively high oxygen concentration and constant aeration. For nitrogen removal, more specifically for denitrification, and for lower energy costs during operation low oxygen concentration was necessary but these low concentrations rendered the granules unstable (Mosquera-Corral et al. 2005). Earlier experiences with biofilms had shown that slowly growing organisms had a stabilizing effect on biofilms. Consequently, Kreuk and Van Loosdrecht argued that for the full-scale reactor to work under low oxygen concentrations the total organism growth rate should be decreased during one cycle. They lowered the growth rate by ingeniously letting the "feast phase", where growth is on an external substrate (nutrients), be preceded by a "famine phase", where the organisms feed on nutrients that are internally stored. The introduction of the famine phase in the bubble reactor was successful. Kreuk showed granulation to occur in bubble columns if the phase of aerobic growth started with a famine phase, that is, a period of anaerobic feeding. Still the airlift reactor outperformed the bubble column since the first produces granules already after 5 days whereas with the latter it took a month before granules occurred. As the famine-feast regime, which became one of the operating principles of the final design, selected organisms with a lower growing rate, the influence of the shear-growth-rate principle established by the research of Beun implied that the shear was an important ingredient for granulation in aerobic systems.

In spite of these laboratory results, an airlift and a bubble column reactor were put in parallel to compare their performances at the pilot plant in Ede. Apparently the researchers and scaling-up engineers did not know how these laboratory results would translate to pilot-scale conditions. In the pilot set-up with two reactors of 6 m high and 0.6 m diameter granule growth turned out to be very disappointing. Moreover the SVI measurements of the airlift and bubble column were comparable although those of the first were somewhat better than those of the latter. These outcomes made the SC decide to transform the airlift reactor into a second bubble column and to concentrate on granule formation. This decision was decisive for the final Nereda aerobic GSBR design.

Finally, according to Van der Roest, the project manager at the engineering firm DHV, the decision to opt for the bubble column reactor was made by DHV and DHV had to convince Van Loosdrecht to abandon the airlift reactor, who did so only reluctantly. From the point of view of Van der Roest, if it had not been for DHV to abandon the airlift reactor the aerobic GSBRs would never have come to the commercial market. According to Van der Roest, he had to challenge the scientists to adapt the process for practical purposes. He had to make the scientists aware that on real scale the oxygen concentration would be lower than in the laboratory because of restrictions on pump capacity. From the perspective of the Kluyver Lab it was evident that bubble column reactors were less complicated and thus less expensive than airlift reactors. DHV calculations clearly showed the extra costs of airlift technology. However, in the eyes of the scientists DHV had not gone far enough in technologically optimizing the standard airlift construction and to adapt it to the new GSBR technology. The extra start-up costs could have been reduced. The crucial question, however, would remain whether these extra costs balanced the savings in operational costs regarding energy and after treatment. This was almost impossible to predict especially if one realizes that the after treatment is very expensive.

Discussion

In this final section we will use the GSBR example as a test bed for our interpretations of the notion of context of design as process and as object. Before we apply our conceptual framework to this case, we have to take into consideration the specific nature of the Nereda project. The Nereda technology is an example of what Vincenti (1990) calls *radical design*: it is a design based on a new working principle and its development was strongly research driven. In that respect it is different from normal design, which is involved in run of the mill (industrial) design projects in which (minor) variations on existing designs are developed and in which research plays only a minor role. We expect that if our conceptual framework is of help in understanding the bearing of contextual factors on the object of radical design, as in the Nereda case, then it may also be fruitfully applied to cases of normal design. This expectation is based on the fact that the pivotal distinction of our conceptual framework, namely between process of design and object of design, forms the basis of almost all schematic diagrams of the design processes developed and employed by engineers, and the fact that these diagrams are primarily intended to cover cases of normal design. But, of course, this expectation would have to be borne out by further research.

In line with the characterization of Nereda as an example of radical design, the case study strongly suggests a differentiation between at least *two* kinds of objects that scientists and engineers were working on. In the first place, there is the sewage water treatment plant as object of design in the sense of section "Substantive and Procedural Contexts of Design". There we characterized the object of design as descriptions in terms of blueprints, texts or all kinds of models of at least (1) the artifact itself, i.e., its structure, its function and its working principle, (2) its interface with the outside world, and (3) its 'use plan'. We will refer to (1–3) as the *final design* of an artifact. Besides this object of the treatment plant design, there is also an *object of research*, namely the working principle on which to base the sewage water treatment plant. In the course of the development of Nereda both objects played an important role.

Let us first have a look at the object of design in the Nereda case. In hindsight, the object of design is the final design that was implemented in the first full-scale operating plant using GSBR technology. This 'backward' looking determination of the relevant object in the Nereda case is, however, rather one-sided. It is not a characterization of the object of design that may fruitfully be applied to the early stages of the development of this technology. In view of the fact that these early stages were strongly research driven, it might be more appropriate to introduce the second object of the Nereda development mentioned above, viz. the object of research during the early stages, which gradually was transformed, co-elaborated or accompanied by the object of design.

How are we to distinguish more precisely between the object of design and the object of research in the Nereda case? We are primarily interested in characterizations of these objects that are valid from a 'forward' looking perspective, which means that they may function as goals driving the development of Nereda (where different stakeholders may have had different interpretations of these goals). One of these goals is the production of aerobic granules with aerobic organisms at the outside and anaerobic organisms in the core, using a batch process which might be used to purify wastewater in one reactor; we will refer this goal as the *GSBR-working principle*. Purifying here means oxidation of organic matter and ammonium, nitrate reduction, and biological or chemical phosphate removal etc. Another one is the engineering-scientists' goal of the proof of concept of the GSBR-working principle, the feasibility of which was proven in the laboratory. Finally, there was the practical goal of the engineering firms and water boards who wanted reliable, effective and economic – less energy and land use – wastewater treatment plants based on the GSBR-working principle. This difference in goals made STW subsidize the Ph.D. proof of concept research in the laboratory and STOWA finance parts of the pilot-plant research. In the following we take the object of research in the Nereda case to be the proof of concept of the GSBR-working principle, and the object of design to be the final design for reliable and effective sewage water treatments plants based on the GSBR-working principle.

Interestingly, the notion of the object of research still leaves open whether to use an airlift or a bubble column reactor in the object of design. The GSBR-working principle as the core of the object of research fixes a number of important design characteristics and parameters (for instance, use of aerobic granules and one reactor tank). Defined in this way, the core of the object of research may be taken to constrain the 'technological space' within which the object of design has to be developed. This research object determines the GSBR-technology but not the final design (blueprints) of a Nereda plant. Within the technological space defined by the working principle a design based on airlift and bubble column reactors are still possible. This, however, does not preclude that further scientific and technological considerations may decide the choice for one of these types of reactor in the final design (building plans).

Given our analysis of design as object and design as process in section "Substantive and Procedural Contexts of Design" and our interpretation of the object of research and the object of design, how can we fill in Table 22.1 for the Nereda case? As always, real life turns out to be much more complicated than our abstractions of it. Our case description clearly illustrates the difficulties of projecting our abstract concepts onto this real life engineering design case. Nevertheless, we will make an attempt.

Let us focus first on the kind of factors that played a role in the substantive context of the final design of Nereda. One of the main ingredients of the substantive context is the decision to develop and use the GBRS technology. This decision, which was taken by the various parties involved in the Nereda project, was made in a network of collaboration, without centralized power relations, between scientists, engineering firms, users and subsidizing partners. The reasons to opt for the GBRS technology, and thus to constrain the object of design to this technology, are directly related to the proof of concept and the possibility created by GSBR technology to reach the practical goal of a sewage water treatment facility that was smaller, less energy consuming and at least as effective and reliable as traditional treatment plants. Within the network of collaboration the SC played a key role in the communication between the various parties. As the design project was on the way, negotiations between these different stakeholders in the SC led to various modifications in the object of design. The SC added new criteria and modified existing specifications. The setting of the go/no go criteria serves as a paradigmatic example of fixing the constraints or specifications of the object of design. All decisions and developments regarding design criteria belong to what we have called the substantive context of the design object, including the decision to try to implement the GSBR-working principle for waste water treatment. The fact that laboratory experiments had shown the feasibility of the working principle (the object of research) did not by itself imply that a design project should be set up.

Now let us turn to the object of design, it is the final design of a Nereda plant. What features of this design may be considered to be determined by engineering design proper and therefore belong to the core of design as product, or to engineering design proper? In our opinion, these are all design features that may be fixed on the basis of scientific and technological considerations given the constraint of using the GSBR technology and of coming with an effective and efficient final design for a sewage water treatment plant.

It should be noted that the distinction between the object of design and its substantive context is more intricate than suggested above. Take the decision to use a bubble column reactor and not an airlift reactor. This has been an important decision for the final design. Is this decision to be interpreted as a contextual factor, a factor that influenced the design of the GSBR technology from the 'outside', or as a decision that was taken from 'within', that is, within the technological space and that was based on technological considerations. Van Loosdrecht's opposition to this decision may be interpreted as finding its origin in his idea that there were convincing internal scientific or technological reasons to opt for the airlift reactor, and consequently from his perspective the decision to use the bubble column reactor was forced by reasons originating in the (substantive) context of the object of design. According to Van der Roest (DHV), however, the design decision was based on practical purposes; in his opinion a design based on an airlift reactor would never have reached the commercial market. At first sight, these market considerations may be considered to be of a contextual nature but that remains to be seen. The whole GSBR project was intended to be a practical alternative for traditional technologies used to treat waste water and from that perspective all constraints that derive from this goal of being a practical alternative are definitive of the object of design, also the constraints derivable from market considerations. Van der Roest might, therefore, argue that given all constraints on the object of design the bubble column was the only scientific or technologically feasible option. From that perspective the decision to go for this type of reactor becomes a decision from within the technological space. Consequently, to interpret the bubble column reactor decision as a contextual or a scientific/technological design decision depends heavily on how the object of design is conceived, that is, which factors are taken to be relevant for, or go into the definition of the design object/problem.

Let us turn to the process of the Nereda design and first ask ourselves: What is the core of this process, that is, what are the main actions and considerations that were believed to bring about an adequate solution to this design problem? The core of the Nereda design process, as far as its substantive content is concerned, has been the scaling-up strategy from laboratory scale, via pilot plant scale to full-scale systems development. It was believed that if the proof of concept of the GSBR-working principle in the laboratory succeeded, the concept could successfully be scaled up to full-scale and ensuing actions were undertaken. Depending on the specificity of the working-principle formulation, we may claim that in the course of the up-scaling the working principle changed somewhat from airlift to bubble column reactor. However, this substantive content is not to be confused with the core of the design process as defined in section "Substantive and Procedural Contexts of Design". There this core was defined as the kind of actions (and their order) as prescribed by the flow diagrams of design methodology. This notion of core of the design process may be applicable to cases of normal design but seems hardly applicable to this case of radical design in which research and design activities are so closely intertwined. Nevertheless, some remarks about the procedural context of the Nereda design process may be made. All decisions about the means and resources to solve the research and design problems belong to this context. Clearly, everything that had to do with fund raising and finding interested commercial partners to develop the laboratory technology to full scale is part of the procedural context. In addition, decisions by the main scientists to devote research capacity in the laboratory and at the pilot plant to carry out feasibility and scaling-up research belongs to the procedural context. Also the installment of the SC belongs to this category.

Although the distinction between design process core and context is difficult to make, the following shows that it does play a role in design practices. Design engineers and methodologists have written numerous books and articles that discuss various flow diagrams about how to structure design projects such that design problems may be solved in a systematic way. The basic idea behind these flow diagrams is that there are good and bad ways to try to solve a design problem. These flow diagrams may be considered to describe the core of design as a process. Whether or not there actually is such a core (or only one core/design method, or several) is a matter of controversy. Nevertheless, most design engineers would probably subscribe to the following remarks by Hales (1993, p. 17):

> One of the most frustrating things about being a design engineer or design manager is the way projects are manipulated by those who have very little to do with the design process

itself. One minute everything is extremely urgent and the next minute the project is no longer required or the money has run out. More and more influences affect the course of design projects.

Hales' remark clearly suggests that many influences on the design process are experienced as coming from outside the world of design and "have little to do with the design process itself". So, somehow a distinction may be made between what legitimately belongs to the inside or core of a design process and what to its outside or what we have called its procedural context, even if it is in fact very difficult to spell out the specific details of this core. Some of the frustrations referred to in the Hales quote can be found in the words of van der Roest when he claims: "The first Nereda purification plants could have been up and running years ago if a guarantee fund had been available" (Wassink 2011). It may at least be safely concluded that often considerations of the funding of design processes belong primarily to its procedural context.

It may be rather problematic to become more specific about core and context regarding the daily developments in the Nereda case because the whole process did not start with a design brief or an assignment of some client. Undoubtedly in the final stages of the development of the full-scale plant, there will have been some process that started with a design brief and for which some kind of method for solving that design problem was used. But whether our distinction between the core of design as a process and its procedural context can be fruitfully applied to this design project remains an open issue.

Let us briefly summarize our main results. Kroes and Van de Poel (2009) have argued that it is not possible to make a neat distinction in general between technology on the one hand and its social (intentional) context on the other since some social phenomena are definitive (constitutive) for technology. This leaves open the question whether it is possible to delineate those social processes that are internal (definitive) or external (contextual) to technology. In order to explore this problem we have introduced a distinction between core and context of design as object and as process. To see whether these abstract distinctions make sense in real life engineering practice we have tried to apply our distinctions to the case of the design and development of a new kind of sewage water treatment technology. Our analysis makes clear that while *in abstracto* a distinction between core and context of design as object and process may seem plausible, it may be very problematic to apply this distinction to actual engineering practice and to characterize a particular design feature as the result of internal (technological) or external (social) factors.

References

Beun, J. J., Hendriks, A., van Loosdrecht, M. C. M., Morgenroth, E., Wilderer, P. A., & Heijnen, S. J. (1999). Aerobic granulation in a sequencing batch reactor. *Water Research, 33*(10), 2283–2290.

Bucciarelli, L. L. (1994). *Designing engineers*. Cambridge, MA: MIT Press.

Dangcong, P., Bernet, N., Delgenes, J.-P., & Moletta, R. (1999). Aerobic granular sludge – A case report. *Water Research, 33*(3), 890–893.

De Bruin, B. M. M., de Kreuk, M. K., van der Roest, H. F. R., Van Loosdrecht, M. C. M., & Uijterlinde, C. (2004). Aerobic granular sludge technology, alternative for activated sludge technology? *Water Science and Technology, 49*(11–12), 1–9.

De Bruin, B. M. M., van der Roest, H. F., de Kreuk, M. K., & Van Loosdrecht, M. C. M. (2005). Promising results pilot research aerobic sludge technology at WWTP ede. In S. Bathe, M. K. de Kreuk, B. S. Mc Swain, & N. Schwarzenbeck (Eds.), *Aerobic granular sludge*. London: IWA Publishing.

De Kreuk, M. K., Heijnen, S. J., & Van Loosdrecht, M. C. M. (2005). Simultaneous COD, nitrogen and phosphate removal by aerobic granular sludge. *Biotechnology and Bioengineering, 90*(6), 761–769.

De Kreuk, M. K., van de Poel, I. R., Zwart, S. D., & Van Loosdrecht, M. C. M. (2010). Ethics in innovation: Cooperation and tension. In I. R. van de Poel & D. E. Goldberg (Eds.), *Philosophy and engineering: An emerging Agenda* (pp. 215–226). Dordrecht: Springer.

DHV. (2012). *Dutch prince Willem-Alexander opens first municipal Nereda sewage treatment plant.* Press Release 22 May 2012. Available at http://www.dhv.com/Newsroom/Press-Releases/2012/2012-05-22-Dutch-Prince-Willem-Alexander-opens-fir?ui=/Markets/Water/Water-Treatment/Water-treatment---Wastewater/Nereda/Nereda

Dorst, K., & van Overveld, K. (2009). Typologies of design practice. In A. Meijers (Ed.), *Philosophy of technology and engineering sciences* (pp. 455–487). New York: Elsevier.

Etterer, T. J., & Wilderer, P. A. (2001). Generation and properties of aerobic granular sludge. *Water Science and Technology, 43*(3), 19–26.

Hales, C. (1993). *Managing engineering design*. Harlow: Longman Scientific & Technical.

Hales, C., & Gooch, S. (2004). *Managing engineering design*. London/Berlin: Springer.

Houkes, W., Vermaas, P. E., Dorst, K., & de Vries, M. J. (2002). Design and use as plans: An action-theoretical account. *Design Studies, 23*(2), 303–320.

Kroes, P. (2012). *Technical artefacts: Creations of mind and matter*. Dordrecht: Springer.

Kroes, P., & van de Poel, I. (2009). Problematizing the notion of social context of technology. In S. H. Christensen, B. Delahousse, & M. Meganck (Eds.), *Engineering in context* (pp. 61–74). Aarhus: Academica.

Morgenroth, E., Sherden, T., Van Loosdrecht, M. C. M., Heijnen, S. J., & Wilderer, P. A. (1997). Aerobic granular sludge in sequencing batch reactor. *Water Research, 31*(12), 3191–3494.

Mosquera-Corral, A., De Kreuk, M. K., Heijnen, J. J., & Van Loosdrecht, M. C. M. (2005). Effects of oxygen concentration on N-removal in an aerobic granular sludge reactor. *Water Research, 39*(12), 2676–2686.

Van de Poel, I., & Zwart, S. D. (2010). Reflective equilibrium in R&D networks. *Science, Technology & Human Values, 35*(2), 174–199.

Vermaas, P. E., & Houkes, W. N. (2006). Technical functions: A drawbridge between the intentional and structural natures of technical artefacts. *Studies in History and Philosophy of Science, 37*(1), 5–18.

Vincenti, W. G. (1990). *What engineers know and how they know it.* Baltimore: John Hopkins University Press.

Wassink J. (2011). Purely based on character. *Delft Outlook 3,* 6–11. Available at http://tudelft.nl//en/current/university-magazines/delft-outlook/former-editions/2011/2011-3/achtergrond/purely-based-on-character/

Zwart, S. D., van de Poel, I., van Mil, H., & Brumsen, M. (2006). A network approach for distinguishing ethical issues in research and development. *Science and Engineering Ethics, 12*(4), 663–684.

Sjoerd D. Zwart Assistant Professor in the Philosophy of Technology and Engineering Sciences at the Delft and Eindhoven Universities of Technology, The Netherlands. He has studied Mathematics and has an M.A. in the Formal Philosophy of Science (1989) and wrote a Ph.D.-thesis on *Verisimilitude Distance Measures on Lindenbaum Algebras that bear a Considerable Similarity to Classical Belief Revision* (University of Groningen 1998). He has been teaching courses in logic, argumentation, the philosophy of science and technology, and engineering ethics, mainly for engineering students. His recent research focuses on methods and techniques in the engineering sciences (modeling, scaling, measurement and causality), engineering ethics (responsibility, just design and norms in modeling) and practices in engineering sciences such as tacit knowledge. Relevant publications are 'Scale Modeling in Engineering: Froude's Case' in A.W.M. Meijers, *Philosophy of Technology and Engineering Sciences*, pp. 759–798, 2009; and 'Reflective Equilibrium in R&D Networks' *ST&HV* 35(2), pp. 174–199, 2012 (together with Ibo van de Poel).

Peter Kroes Professor in Philosophy of Technology at Delft University of Technology, The Netherlands. He has an engineering degree in physics (1974) and wrote a Ph.D. thesis on *The Notion of Time in Physical Theories*, University of Nijmegen, 1982. He has been teaching courses in Philosophy of Science and Technology and Ethics of Technology, mainly for engineering students. His research in Philosophy of Technology focuses on the nature of technical artifacts and engineering design, the modeling of socio-technical systems and the nature of technological knowledge. His most recent book publications are: *Technical artefacts: creations of mind and matter*, Springer, 2012, *A philosophy of technology; from technical artefacts to socio-technical systems* (together with Pieter Vermaas, Ibo van de Poel, Maarten Franssen and Wybo Houkes, Morgan and Claypool, 2011) and *Functions in biological and artificial worlds; comparative philosophical perspectives* (editor with Ulrich Krohs, MIT Press, 2009).

Chapter 23
The De-contextualising of Engineering: A Myth or a Misunderstanding

William Grimson

Abstract Engineers impact on how people live, where they live, and the physical environment in which they live. The interaction between society and engineering has a long history but it remains both complex and problematic. Complex because of the many factors involved including the political and economic dimensions. Problematic because it is not always clear how individuals, groups of individuals or society at large negotiate with engineers to ensure the right 'product' is created. One of the ways engineering deals with the problem is through context – the set of circumstances in both the foreground and background of any project. Understanding what counts as valid context and then formulating appropriate responses is something that is encountered to varying degrees first in educational programs and then through multiple processes as engineering is practiced. Some of these processes have legislative force and others are established as best practice. Ethics as the basis of making sound decisions is directly related to how contexts once understood result in appropriate action. And in that sense engineers reflect the norms of the society in which they work.

Keywords Context • Context awareness • Context sensitivity

Introduction

> Do the right thing. It will gratify some people and astonish the rest. (Mark Twain)

To do the 'right thing' one must understand and be responsive to the context surrounding whatever enterprise is being undertaken. But it is neither easy to understand fully what constitutes a context that is relevant to a particular situation nor is it always obvious how to take that context adequately into account. Like any other

W. Grimson (✉)
Dublin Institute of Technology, 143 Lower Rathmines Road, Dublin 6, Ireland
e-mail: william.grimson@dit.ie

S.H. Christensen et al. (eds.), *Engineering Identities, Epistemologies and Values*, Philosophy of Engineering and Technology 21,
DOI 10.1007/978-3-319-16172-3_23

practical profession, such as medicine, engineering is continually faced with the challenge of addressing context. It is not unreasonable to say that the majority of engineers aim to meet the twin objectives of 'doing the right thing' and 'doing it right'. In the latter case, 'doing it right', the challenge is essentially a technical one of working within a set of constraints be they legal, financial or of a scientific and technological nature. In the former case, 'doing the right thing', is in many respects more complex and often contentious. One only need mention a few examples to demonstrate this last point. Nuclear power stations have had an uneven history as regards their public acceptability; wind farms with numerous tall turbines, despite following a green or sustainability agenda, are not always welcomed in a rural community; fracking by which shale gas is retrieved is a current hotly contested matter in many countries; devices and equipment for military purposes such as cluster bombs and anti-personnel mines face ethical questions. All of these examples and more raise the issue of whether the 'right thing' is being created or produced. As a counter another list of 'things' that have high public acceptance with little or nothing in the way of controversy could include: medical devices; rehabilitation engineering; restoring vision or hearing; low-energy or green buildings; improved access to clean water and better sanitation; safer transport systems. These and many more would support but not necessarily prove the proposition that the 'right thing' has been developed. A number of questions immediately spring to mind. Who decides what is right as in giving approval to the development or creation of 'the right thing'? Stating that it depends on the context is only part of the answer. And in any case, what might be considered 'right' in one context may well be thought wrong in another context. Context, then, is identified as the critical factor in determining whether or not a work of engineering is deemed appropriate and right, assuming in the first place the adequacy of all the associated technical aspects of the undertaking. Two points need to be stressed here. First, in the practical world of engineering there is no simple and satisfactory way of drawing a boundary around what constitutes context. Second, there is obviously a difference between the context considered at the time of design and implementation, and a more complete context that only emerges in time. It is not just a question of unintended consequences: value systems change with time and what appeared reasonable at one time can be found to be unacceptable later. Before proceeding some definition needs to be established as to what is meant here by context.

Context, from the Latin *contextus,* means a joining together or connection. So in a given text the passage leading up to a particular word establishes a background by which a fuller understanding of the use of that word can be achieved by the reader. In this chapter context is taken to be the circumstances constituting a background in which something, largely engineering in nature, is to be placed. The circumstances, or set of circumstances, are potentially anything but typically would include the following: economic, social, cultural, political, and environmental factors together with ethical considerations. Time is an additional factor: for example circumstances involving famine, war, drought or any major adverse event inevitably change what might otherwise have been a settled context.

Because engineering impacts on our physical world it must always be situated within a context: engineering cannot be context free. It follows then that the profession is obliged to address the many challenges associated with the set of circumstances surrounding a given engineering event – be it positioning a new bridge, opening a new runway at an airport, locating a new hospital, or installing alternative energy systems such as wind farms. The central questions asked in this chapter focus on the degree to which engineering is sufficiently context sensitive and context responsible. An intriguing question not asked nor answered is whether engineering is better or worse in this respect compared to other professions such as architecture, medicine and law. The book *Engineering in Context* (Christensen et al. 2009) demonstrates the complexity of the subject of its title and not surprisingly offers diverse views. The main themes in the book are: Contextualism in Engineering, Engineering Education in Context, Engineering Design, Engineers in workplaces and institutions, and Engineers in Civil Society. Joseph Herkert claims that engineering codes of ethics focus on microethical (individual) responsibilities but are weak on macroethical (collective) responsibilities (Herkert 2009). And the more general charge has been made that engineering has become de-contextualized. Indeed it is claimed that 'engineers are often unaware of, and sometimes even trained to explicitly ignore, the broader contexts of their work' (Fisher and Miller 2009; Bucciarelli 1994). This might well be the case in some instances but such training runs counter to what is expected of engineering educational programs as discussed later in this chapter. In terms of ethical behavior or duty, closely related to addressing issues of context, there is the perceived duty of engineers, *plus respicere*, which can be broadly interpreted to mean 'go the extra mile' and thus take more into account (Mitcham 1994).

Another initial point that can be made is that individuals work within social environments, including their workplace, and their response to context will inevitably vary according to the circumstances. On the matter of the individual as distinct to the corporate engineer, Li Bocong examines a micro-meso-macro framework and in which ethical stances can be understood and hence positioned with respect to general issues of context (Li Bocong 2012). From a very different perspective Christelle Didier has explored the intersection between religious and political values and their transformation into an engineering ethos (Didier 2012). What is abundantly clear is the complexity of how an engineer, just as any other citizen, develops a worldview from which ethical stances evolve together with an understanding of the relevance of context in whatever situation they are faced.

To unravel the strands of the central question posed the remainder of this chapter looks briefly in turn at types of engineering educational programs, value systems in engineering, identification of grand engineering challenges, the role of professional institutions and academies of engineering, and text books in attempt to gain a understanding of how engineers and engineering address context. In addition some views are expressed as to how dialogue between society and the engineering profession can be enhanced and formalized when it comes to major undertakings.

Engineering Education

Undergraduate engineering programs exist within a wide orbit of models each of which in principle is eligible for approval by accreditation bodies based on published criteria that include amongst other items program learning outcomes. A number of authors have commented on the highly constrained nature of engineering curriculum where there is pressure from all quarters to accommodate additional material and not just technical subjects (Williams 2002). So it could be expected that context like any other aspect of engineering has to be specially pressed if it is to have adequate exposure. It is instructive then to examine the extent to which context is explicitly or implicitly included in the learning outcomes defined for engineering programs. Institutions of engineering have promoted the global harmonization of program learning outcomes and for the purposes of this chapter the work of the European Network for Accreditation of Engineering Education is used (ENAEE 2009). The Network authorizes accreditation and quality assurance agencies to award the EUR-ACE® label to accredited engineering degree programs (EUR-ACE 2008). Within the EUR-ACE standards framework document there are six sets of learning outcomes as follows:

- Knowledge and Understanding
- Engineering Analysis
- Engineering Design
- Investigations
- Engineering Practice
- Transferable Skills

Under the Knowledge and Understanding heading the document states that 'Graduates should demonstrate their knowledge and understanding of their engineering specialisation, and also of the wider context of engineering … and awareness of the wider multidisciplinary context of engineering'. It is left to the colleges offering engineering programs to interpret what is meant by the 'wider context' when an explanatory footnote in accreditation documents on what the accrediting body intends by 'wider context' would be helpful to engineering schools and review panels alike. One of the learning outcomes under Transferable Skills states that graduates should 'demonstrate awareness of the health, safety and legal issues and responsibilities of engineering practice, the impact of engineering solutions in a societal and environmental context, and commit to professional ethics, responsibilities and norms of engineering practice'. The intention then is clear. However the real strength or otherwise depends on how, first, engineering colleges respond to the stated criteria and second, the diligence of accreditation panels in ensuring the criteria are met. Bearing in mind that accreditation panels usually consist of only engineering academics and engineering practitioners some concern arises as to whether the societal aspects of engineering are adequately scrutinised (Grimson and Murphy 2013). Nevertheless in the first instance the onus resides within engineering schools to ensure that addressing societal aspects are incorporated into programs.

A question can be asked as to whether different types of program make it more or less likely that context, societal, ethical and directly related topics are given sufficient attention. It is not the purpose here to describe in detail the structure of engineering programs, rather the intention is to give sufficient information to allow some comments to be made on their suitability to deal with context. Whilst there is a wide range of curriculum implementations the following ones are the dominant ones. First, what might be called the conventional engineering education model with its mixture of mathematics, science and technology covers engineering principles applied to a limited field. Such programs were originally general in character in that the first and second years prepared for the introduction of a range of sub-disciplines in the latter stages of the course. In various colleges a liberal studies element was also included. The breadth of coverage however came at the price of limiting the depth of material. Analysis dominated though design did feature in such programs but was not overly emphasized. Most engineers over the age of forty would have graduated from such a program. In time specialism became a feature of many engineering programs with say electronic engineering being the target subject throughout the entire course of study. On one hand the graduates of such programs were technically more competent in their chosen field than heretofore, but on the other hand they were less flexible and often lacked a wider engineering vision.

The second type of approach resulted in what became known as engineering science. In such programs the focus is primarily on the science concerned with the physical and mathematical basis of engineering. Correspondingly less attention is given both to engineering practice, which often has its roots in craft, and the use of approaches such as heuristics as described by Billy Koen with great force (Koen 2003). Further, it is claimed that design is marginalized in engineering science. This is not a necessary consequence of adopting such an approach but the greater emphasis on science does come at a cost. To counter what is effectively the lack of a holistic approach where many factors have to be taken into account, engineering schools developed a third way generally called systems engineering: MIT in particular were early adopters. One of the key characteristics of systems engineering is that it is intrinsically interdisciplinary. The NASA Systems Engineering Handbook states that 'systems engineering is a robust approach to the design, creation, and operation of systems. In simple terms, the approach consists of identification and quantification of system goals, creation of alternative system design concepts, performance of design trades, selection and implementation of the best design, verification that the design is properly built and integrated, and post-implementation assessment of how well the system meets (or met) the goals' (NASA 1995). The final important variety of program is the one where it is project based. Here the project comes first, with a small team of students who 'discover' what knowledge they require to complete their project. It is the team working aspect plus the opportunity to learn about what to learn that gives project based programs their main advantage. Whilst each type of program has its defining characteristic it must be noted that they all have much in common. And all are capable of meeting the program accreditation criteria in jurisdictions where such schemes are in operation.

Returning to the question of whether context is more likely to receive adequate attention in one type of program compared to another. Whilst context can always be introduced by an instructor in any of the program types outlined above there can be little doubt that both the systems engineering and project based approaches lend themselves to dealing with context in a natural or organic manner. Ruth Graham has explored this question in a white paper and notes that project based learning facilitates the 'greater emphasis on embedding sustainability and ethics within the project context' and 'the creation of new cross-campus multi-disciplinary projects, centred on engineering challenges' (Graham 2010). Sustainability and ethics are not necessarily the totality of context in any given situation but it illustrates the point that project based learning is a good vehicle for its inclusion in the engineering process. Likewise the multi-disciplinary dimension can only enhance context having greater visibility.

One mechanism by which a student can be made aware of context is through work placement. Work placements that focus on narrow technical areas where specialised knowledge and skills can be developed are not necessarily good candidates for developing context awareness and context sensitivity. At the other end of the spectrum, placements abroad as part of, say, humanitarian efforts are likely to instil an awareness that simply cannot be gained in the classroom. Work placement has as a basic objective the aim of showing an undergraduate engineer what it 'feels' like to be a 'real' professional, that is to say one working in the real world. To that end placements that present broadly based opportunities are likely to be of the most benefit. In addition organisations such as *Ingénieurs sans Frontières* have as their ideal the act of addressing societal concerns that contribute to the general context in which engineering takes place.

Textbooks deserve a special mention. Most textbooks for engineering undergraduates do not set out contexts in which the technical material that follows might apply. It is not hard to see why this might be the case, for the treatment of subjects aims to be as general as possible and is therefore not so much context free as context neutral. This is especially the case for introductory textbooks where it is clear that the basics must first be established. As an analogy the pianist would be restricted to learning piano scales and then embracing the Czerny exercises before being let loose on real music however simple it might be. In Electronic Engineering signal processing is heavily dependent on Fourier series and Fourier transforms yet few introductory textbooks seem to be prepared to set out in advance the overwhelming rationale for their use. A few exercises at the end of each chapter are a poor substitute for a rounded discourse on the rich set of applications of this subject material. Is it then just left to the lecturer or instructor to provide some context? The situation in general improves though when graduate course textbooks are encountered. The use of case studies helps greatly in anchoring the technical discussion in realistic settings.

One last point related to the engineering curriculum, bearing in mind the importance of context, there is merit in having mandatory courses in the History of Engineering, Science and Technology. Ideally such courses would include students from other disciplines which would enrich class discussions and expose the engineering students to other viewpoints. Coupled with Philosophy, which promotes right and critical reasoning, the case for including these subjects as part of a Liberal Studies component in the curriculum is strong (Grimson et al. 2008). The notion of banishing engineering to some form of a boot-camp cut off from the ideal university would hardly serve the needs of engineering to be context aware (Robert Wolff 1992). Indeed the idea of humanists and engineers working together to form a 'global democratic culture' has great appeal (White 1967).

Attributes of an Engineer

A number of bodies have conducted exercises to establish the essential skills of an engineer and these are discussed by Ela Krawczyk and Mike Murphy as part of reviewing the challenges in educating engineers. Professional bodies, business, and new graduates all have their favorite priorities. From the lists examined 'context' or the ability to appreciate context does not appear (Krawczyk and Murphy 2012). Further it would take a degree of imagination to choose some of the attributes listed to at least point in the direction of 'context'. Perhaps this gets to the core of the matter: engineering is inherently contextual but context is in many respects only implicit in what transpires. As is demonstrated elsewhere in this chapter there are manifold 'hooks' or opportunities by which context is or can be made explicit. Perhaps in the future a list of attributes will include an item such as 'contextual awareness'! Interestingly the same authors rank the desirability of skills and competences for three very different scenarios: one, where there is a growing libertarianism; one where a balance is strived for between civic society, governments and business; and one where after a long economic stagnation and a fragile socio-political environment people turn back to their local communities and cultural roots for comfort. A brief analysis of the desirability/scenario matrix shows that Social and Ethical Awareness, and Cultural Awareness (close relatives of context awareness) are both ranked low. This reinforces, one would think, the point that issues surrounding context need to be made explicit.

Value Systems

Engineers and engineering have inherited value systems which can be thought of as a set of ethical values that are consistent and which are derived in a personal and corporate manner. It is probably true though that such value systems are more varied than in any other profession, such as medicine, partially due to the wide divergence across the many and varied sub-disciplines of engineering and the range of levels at which it is practiced. In addition at least some of the value system must reflect norms within any given country and culture. Nevertheless engineering institutions and societies across many countries have codes of ethics that are very similar in content. Whilst the existence of a moral code does not necessarily imply that context will be properly addressed in all situations, it does provide a framework in which engineers should work and therefore has relevance to both the identification of context(s) and their resolutions in terms of subsequent engineering activities. The Code of Ethics for Engineers Ireland is not atypical and consists of four parts: (i) Relations with Colleagues, Clients, Employers and Society in general; (ii) Environmental and Social Obligations; (iii) Maintenance and Development of Professional Conduct and Standards; and (iv) Enforcement Procedures and Disciplinary Action (Engineers Ireland 2009). Within section (ii) it states that:

- Members shall have due regard to the effects of their work on the health and safety of individuals, and on the welfare of society and of its impacts on the natural environment.
- Members shall promote the principles and practices of sustainable development and the needs of present and future generations.
- Members shall strive to ensure that engineering projects for which they are responsible will, as far as is practicable, have minimal adverse effects on the environment, on the health and safety of the public and on social and cultural structures
- Members shall strive to accomplish the objectives of their work with the most efficient consumption of natural resources which is practicable economically, including the maximum reduction in energy usage, waste and pollution.
- Members shall promote the importance of social and environmental factors to professional colleagues, employers and clients with whom they share responsibility and collaborate with other professions to mitigate the adverse impacts of their common endeavours.
- Members shall foster environmental awareness within the profession and among the public.

Perhaps these six statements are open to some criticism, as in a lack of explicit reference to gender, race, religion etc., but it could be argued that such issues are already the subject of national equality legislation in most countries. Gender is mentioned as it can be an important context in major engineering projects as illustrated by Nina Laurie (Laurie 2011). And examples are not hard to find where race and equality in general are important issues: just think of the distribution and provision of hospitals, housing and education in developed countries let alone across the world. Nevertheless the above Code of Ethics sets a standard that is high and is framed in a manner that compliance is, in principle, measurable. And that perhaps is the key issue, namely, whilst engineers have collectively signed up to acting in an exemplary manner, is there sufficient adherence to its stated high ideals? This is somewhat akin to comparing the ideals of a Christian, say, with the actual behaviour of individuals of that religion. And similar statements could be made about the adherents to any belief system. On the general matter of compliance it is fair to say that transgressions brought before disciplinary panels are rare and more often than not are about matters related to 'Relations with Colleagues, Clients, Employers'. Finally, it is worth comparing the six items listed above with the eight habits of mind that Clark Miller and Sarah Pfatteicher promote as being appropriate for engineers and repeated in Engineering in Context (Miller and Pfatteicher 2008), (Fisher and Miller 2009).

1. Recognize that engineering work is a form of social engineering.
2. Develop a commitment to systematically inquire into the broad impact and import of engineering work.
3. Regularly seek out opportunities to learn new skills to successfully pursue such inquiries.
4. Recognize the obligation of engineers to work in partnership with those who will inhabit the technological worlds the engineers design and build.
5. Recognize that all design decisions involve the need to balance, choose, and evaluate interests, views and perspectives.
6. Look for ways to make those choices an explicit and integral part of the dialogues ta surrounds design decisions.
7. Develop a tolerance and appreciation for dissention, debate, and dialogue.
8. Involve the public more actively as participants in deliberations about the public good as embedded in technological systems.

What is worth noting is the similarity between the two lists and the sameness of tone, though the latter is more nuanced. What both lists imply is that engineers work in a socio-technical environment and not just a technical one and hence their interaction with engineers and non-engineers needs to be moderated: precisely the point of having a code of ethics. In a more general manner the social context of technology can be considered a process and hence similar in principle to the processes in engineering (Kroes and van de Poel 2009). This approach whilst not guaranteeing a successful resolution of contextual challenges at least incorporates context into a readily understood framework.

An Economic Perspective

Not surprisingly there are economic forces bearing on engineering education. On one hand some of the longest established and most confident engineering schools appear slow to adapt to changing circumstances where the rationale is one of taking a long term perspective whilst concentrating on universal basics. On the other hand younger engineering schools are often eager to adopt change and become early adopters in areas associated with technological shifts and new engineering paradigms. New contexts emerge and in a loosely coupled manner two-way interactions between society and academia develop – not always for good nor are they necessarily bad. As examples, most of the initial fears concerning nanotechnology appear to have abated whilst the technology supporting social media is lagging behind what could be required by way of either formal or informal regulation. What is clear is that context becomes an issue after the technology is introduced and as a result it is more difficult for engineers and indeed others to anticipate what will be unintended consequences. Perhaps this is always the case with new technologies (for example, consider the adverse effects following large irrigation and dam construction). Either way, context is at its most challenging when engineering takes the form of what Walter Vincenti referred to as *Radical Design* (Vincenti 1990).

There are other ways in which economics and associated policies influence engineering and engineering education. National research agendas and consequently funding are normally government led and have considerable power over what research takes place. In turn this has a trickle down impact on both graduate, first, education and then undergraduate education. In addition Andrew Jamison has referred to the greening of engineering and engineering education in which the borders between the academic and business worlds are increasingly transgressed (Jamison 2012). As a general observation the business or industry influence on education is significant across the world with contexts set by those external to the universities. In large parts of the former Soviet Union many universities were and still are vocationally based (mining, locomotive, chemical) where the context is effectively handed down. And finally and perhaps controversially, defense contracts have featured strongly in research funding for universities (not restricted to engineering) where again context is set by the funders. Unless a college has a strong resource base it is inevitable that the education it provides is in part determined by external economic factors.

Grand Challenges

The influential Finnish architect Eliel Saarinen held that one should 'always design a thing by considering it in its next larger context – a chair in a room, a room in a house, a house in an environment, an environment in a city plan.' Whilst this injunction has an obvious appeal – concentrate on one thing at a time – it is open to the criticism that proceeding from the bottom to the top does not necessarily yield a satisfactory overall outcome. But it must be admitted that much of engineering adheres to Saarinen's instruction. However one notable example of where engineering has taken a macro view is in the choice of the challenges for the twenty first Century as chosen by the US National Academy of Engineering (NAE 2013). There are 14 challenges and it is worth listing them here.

1. Make solar energy economical
2. Provide energy from fusion
3. Develop carbon sequestration methods
4. Manage the nitrogen cycle
5. Provide access to clean water
6. Restore and improve urban infrastructure
7. Advance health informatics
8. Engineer better medicines

9. Reverse-engineer the brain
10. Prevent nuclear terror
11. Secure cyberspace
12. Enhance virtual reality
13. Advance personalized learning
14. Engineer the tools of scientific discovery

Arguments can be made both for the rejection of some of these challenges and for the inclusion of others, nevertheless the list is evidence that one way or another context has been taken into account at a macro level. At least three of the challenges are directly concerned with climate change. Three straightforwardly are health related, and four others deal with the background in which we live and ideally prosper. But it is not the list of the challenges that is important since it is expected that other nations will have a different set of priorities with many Third World Countries having needs not considered important by the more developed countries. And in any case time will inevitably force the list to be revised. Instead it is, first, the process by which such challenges are identified and, second, the follow-through by which each challenge is addressed. Regarding the first, the National Academy of Engineering (NAE) did not rely solely on the engineering profession and the committee tasked with identifying the challenges consisted of 'a diverse group of people dedicated to improving quality of life around the globe'. In such an exercise it is the diversity of the group that brings some robustness to the process. Recognising the significance of NAE's Grand Challenges, the Royal Academy of Engineering (UK) held the inaugural Global Grand Challenges Summit in March 2013 involving over 450 leading engineers, artists, economists, designers, philosophers, scientists, politicians, industry leaders, educators and policy makers from across the globe (Global Grand Challenges Summit 2013). Again it is the diverse range of participants that justifies the hope that the challenges identified are indeed valid and deserving of sustained attention and effort. Regarding the second matter – the follow-through – this is more problematic and at root troubling. Significant resources and commitment need to be in place for a prolonged period of time if progress is to be made and this requires society mostly through governments to accept a responsibility that perhaps they are reluctant to accept. As an example the wrangling over various protocols and agreements together with lack of progress on climate change do not augur well for global success in meeting the grand challenges. And perhaps it is here that engineering has its biggest challenge, namely to canvas support at all levels, from corporations, foundations, various agencies and governments, to commit resources and effort to what are perceived to be the real and significant problems. Oddly enough the problem is still one of context but now it is essentially a socio-political one. In a partial conclusion, it could be argued that fundamental challenges with their associated contexts have not been ignored by engineering, but it is as yet unclear whether the profession is sufficiently persuasive to ensure the grand challenges are accepted, and acted upon, by society acting through governments and other agencies.

The Role of Professional Institutions and Academies of Engineering

Whilst undergraduate education is the main vehicle by which engineers are or can be sensitised to issues surrounding context it does not follow that exposure to this matter stops at graduation. In fact context becomes more important if not inevitable once an engineer commences practicing their profession. It follows that continuing professional development (CPD) plays a role in educating and re-educating engineers throughout their careers. In general institutions are well placed to identify topics that have particular importance and in many cases organise symposia, colloquia and conferences to address current and emerging subjects. An example would be the type of report that looks at the energy question where orthodox sources of power such as coal and nuclear are considered together with a range of alternative ones, and including material on sustainability, climate change, impact on economies, socio-political matters, local environmental conditions and other contexts. Other topics found on academy websites include reports on engineering the future of water, human enhancement and the future of work, energy storage, and the philosophy of engineering – in fact a rich and eclectic set of reports can easily be found each with their own relevant contexts.[1] In turn the material in such reports finds its way into textbooks intended for use in colleges and universities.

Legislation

Legislation as it impinges on engineering is essentially good practice that is encoded or framed in a way that forces the profession to comply. Health & Safety is one obvious area that has resulted in a raft of law setting out the conditions under which individuals must operate. Legislation of this type protects both the client and the engineer. Other types of legislation ensure that the greater interests of society are represented in the work of engineers and particularly at the early stages of a project. For example Environmental Impact Studies (EIS) are a mandatory part of any project such as building a new road or airport and take into account one set of contexts. Further, planning bodies then rule as to whether or not the relevant challenges for these contexts have been properly addressed. There are other less obvious mechanisms. For example in some jurisdictions professional engineers are regulated by an institution which is empowered through law to maintain a register of its members and who must act according to the bye-laws and regulations of that body. Ethical conduct expected of a member would normally be a strong feature of such bodies' laws. The systems in place may not be perfect but a general framework is in

[1] See for example http://www.raeng.org.uk/news/publications/list/mostrecent.htm, http://www.nae.edu/Publications/Reports.aspx and http://www.iae.ie/publications/

place through state legislation and the role of institutions in ensuring as far as practicable that engineers carry out their function in society in a responsible manner.

Dialogue Between Society and Engineering

The burden of identifying and then accounting in some reasonable manner for context should not be seen to rest solely on the shoulders of engineering. Society through groups and individuals has a role to play and not just the adversarial one that attracts the attention of the news media. Useful dialogue and the negotiation that needs to have formal support whereby those involved are properly informed. One such example is the Aarhus Convention supported by three ‘pillars’; namely, Access to Information, Public Participation in Decision-making, and Access to Justice, in Environmental Matters (www.unece.org). The underlying rationale, as the website makes clear, is that sustainable development is directly dependent on the meaningful engagement of civil society in decision-making. Whilst the primary concern is environmental, which in any case relates to much of what concerns society, the general approach is adoptable across the whole breadth of engineering as it is practiced.

Conclusion

No special claim can be made for engineering when it comes to the question of context in all its many facets. Without strong evidence to the contrary it can be assumed that engineering takes the issues surrounding context(s) no less seriously than other practical professions. But it can hardly be disputed that the effect on an ‘environment’ in the case of engineering endeavours where context has not been either understood or addressed can be of huge or even disastrous consequences. Even when well intentioned, engineering projects can be the victim of either unintended consequences or a lack of understanding of associated and previously known contexts. One area that has attracted adverse comment has been the building of dams and in general altering water courses. Too often such projects are undertaken by developed countries in developing countries and have resulted in a number of well publicized disasters. One feature of some of these failures has been a lack of real dialogue between project managers and local people who understand perfectly well their own surroundings. It has been pointed out by Peter McEvoy, Jane Grimson and William Grimson that engineers are well placed to contribute to what has been called ‘negotiated development’ for the simple reason that they are at the core of so many developments (McEvoy et al. 2012).

Civil engineering was so called to differentiate it from military engineering which for many centuries dealt with fortifications and weapons to breach enemies’ ramparts. The picture today is more complex with all branches of engineering deployed from time to time to support military engagements and interventions. The contexts surrounding military operations and peace are manifold and range from humanitarian to economic aspects. Engineering cannot be divorced from this complex background situation, with positions taken both in accord with international agreements and moral norms within any given country. For example, consider the use, production, transfer and stockpiling of cluster munitions (http://www.cluster-convention.org/). Engineers, scientists, medical doctors and other professionals all face the same or similar ethical questions about war, ones that have existed for thousands of years. Whether engineers are working within or outside the parameters set by society is a matter of debate but at the very least no individual can claim to be unaware of the various contexts associated with military operations and war.

Finally, engineering is an infuriating topic to some, for the field of endeavor that is engineering almost defies description. It involves mathematics, science, craft and various technologies in a mix that appears amoeba-like and having no clear boundaries. Carl Mitcham has claimed that engineering is philosophically weak when compared to other professions (Mitcham 2008). There is substance to this claim but in some respects it misses the point, for engineering to succeed it cannot afford the luxury of being soundly philosophically based as, say, mathematics. Its purpose lies elsewhere. It is the curse and blessing of engineering that it is both open-ended and forced to be a profession of everything (Williams 2002). As a result engineering is all too susceptible to failing to meet the heavy demands made of it, trying as it does to satisfy diverse and complex requirements. It would be ridiculous to claim that engineers have helped create the best of all possible worlds. But it is unthinkable that humanity could have developed to its current position or contemplated new developments without the direct involvement of engineering in one form or another. This chapter set out to show whether engineers are sufficiently context-aware and responsive. What the chapter shows is that through a multiple of means context is addressed at first during the educational formation of engineers and second throughout the professional life of an engineer. The means exist but whether the end-result is satisfactory is itself another and different question. Suffice to say that there is a realization within the engineering community that context is vitally important and deserves a rounded attention. And Samuel Florman's view that it is not the engineer's responsibility to impose their morals on their practice, considering that they are not responsible for the initial requirements, seems, at least to this author, nothing more than a convenient excuse for 'hand-washing' (Florman 1976).

References

Bucciarelli, L. (1994). *Designing engineers*. Cambridge, MA: MIT Press.

Christensen, S. H., Delahousse, B., & Meganck, M. (Eds.). (2009). *Engineering in context*. Aarhus: Academica.

Didier, C. (2012). Ex-students engaged in "engineers without borders". In S. H. Christensen, C. Mitcham, Li Bocong, & An Yanming (Eds.), *Engineering, development and philosophy: American, European and Chinese perspectives* (pp. 275–289). Dordrecht: Springer.

Engineers Ireland. (2009). *Code of ethics*. https://www.engineersireland.ie/about/code-of-ethics-and-bye-laws.aspx. Accessed 29 Aug 2013.

EUR-ACE (2008). http://www.enaee.eu/eur-ace-system. Accessed 29 Aug 2013.

European Network for Accreditation of Engineering Education (ENAEE) (2009). http://www.enaee.eu/. Accessed 29 Aug 2013.

Fisher, E., & Miller, C. (2009). Collaborative practices for contextualizing the engineering laboratory. In S. H. Christensen, B. Delahousse, & M. Meganck (Eds.), *Engineering in context*. Aarhus: Academica. Chapter 20.

Florman, S. (1976). *The existential pleasures of engineering*. New York: St Martin's Press.

Global Grand Challenges Summit. (2013). *Exploring collaborative approaches to tackling global grand challenges*. http://www.raeng.org.uk/international/global_grand_challenges_summit.htm. Accessed 29 Aug 2013.

Graham, R. (2010). *UK approaches to engineering project-based learning*. Bernard M. Gordon MIT Engineering Leadership Program. Available at: http://web.mit.edu/gordonelp/ukpjblwhitepaper2010. Pdf. Accessed 29 Aug 2013.

Grimson, W., & Murphy, M. (2013). A hippocratic oath for engineers. In D. Michelfelder, N. McCarthy, & D. Goldberg (Eds.), *Philosophy and engineering: Reflections on practice, principles, and process*. Dordrecht: Springer.

Grimson, W., Murphy, M., Christensen, S. H., Erno-Kjolhede, E. (2008). Philosophy matters in engineering studies. In *38th ASEE/IEEE frontiers in education conference*. New York: Saratoga Springs.

Herkert, J. (2009). Macroethics in engineering: The case of climate change. In S. H. Christensen, B. Delahousse, & M. Meganck (Eds.), *Engineering in context* (pp. 435–445). Aarhus: Academica.

Jamison, A. (2012). Turning engineering green: Sustainable development and engineering education. In S. H. Christensen, C. Mitcham, Li Bocong, & An Yanming (Eds.), *Engineering, development and philosophy: American, European and Chinese perspectives* (pp. 7–22). Dordrecht: Springer.

Koen, B. V. (2003). *Discussion of the method: Conducting the engineer's approach to problem solving (engineering & technology)*. Oxford University Press.

Krawczyk, E., & Murphy, M. (2012). The challenge of educating engineers for a close, crowed and creative world. In S. H. Christensen, C. Mitcham, Li Bocong, & An Yanming (Eds.), *Engineering, development and philosophy. American, European and Chinese perspectives* (pp. 109–122). Dordrecht: Springer.

Kroes, P., & van de Poel, I. (2009). Problematizing the notion of social context of technology. In S. H. Christensen, B. Delahousse, & M. Meganck (Eds.), *Engineering in context* (pp. 61–74). Aarhus: Academica.

Laurie, N. (2011). Gender water networks: Femininity and masculinity in water politics in Bolivia. *International Journal of Urban and Regional Research, 35*(1), 172–188.

Li Bocong. (2012). From a micro-macro framework to a micro-meso-macro framework. In S. H. Christensen, C. Mitcham, Li Bocong, & An Yanming (Eds.), *Engineering, development and philosophy. American, European and Chinese perspectives* (pp. 23–36). Dordrecht: Springer.

McEvoy, P., Grimson, J., & Grimson, W. (2012). Negotiated development: Rediscovering a global ethic. In S. H. Christensen, C. Mitcham, Li Bocong, & An Yanming (Eds.), *Engineering, development and philosophy. American, European and Chinese perspectives* (pp. 275–289). Dordrecht: Springer.

Miller, C., & Pfatteicher, S. (2008). Nanotechnology in society education: Cultivating the mental habits of social engineers and critical citizens. In A. Sweeney & S. Seal (Eds.), *Nanoscale science and engineering education*. Valencia, CA: American Scientific Publishers.

Mitcham, C. (1994). Engineering design research and social responsibility. In K. Shrader-Frechette (Ed.), *Ethics of scientific research* (pp. 153–163). London: Rowman & Littlefield.

Mitcham, C. (2008). The philosophical weakness of engineering as a profession. *Keynote talk at workshop on philosophy & engineering*. Royal Academy of Engineering, London.

NASA. (1995). *NASA systems engineering handbook*. SP-610S.

National Academy of Engineering (NAE). (2013). *NAE grand challenges for engineering*. Available at http://www.engineeringchallenges.org/. Accessed 29 Aug 2013.

Vincenti, W. (1990). *What engineers know and how they know it: Analytical studies from aeronautical history*. Baltimore: Johns Hopkins University Press.

White, L. (1967). Engineering and the making of a new humanism. *Journal of Engineering Education, 57*, 375–376.

Williams, R. (2002). *Retooling: A historian confronts technology change*. Cambridge, MA: MIT Press.

Wolff, R. P. (1992). *The ideal of the university*. Edison: Transaction Publishers.

William Grimson BAI, B.A., M.A.Sc., Chartered Engineer. An Electronic Engineering graduate of Trinity College, Dublin and the University of Toronto. He has recently retired but is still actively involved in re-thinking engineering education and the relevance of philosophy to engineering and engineering education. He first worked as a R&D engineer for Ferranti Ltd before joining the academic staff of the Dublin Institute of Technology (DIT) where he was at one time Head of the Department of Electrical Engineering. He ended his career as Registrar with overall responsibility for academic quality enhancement. His academic output was and remains eclectic ranging from publications in areas as diverse as plasma physics, clinical information systems, philosophy of engineering, and development issues. He is currently a member of the Executive of the Institution of Engineers of Ireland.

MIX
Papier aus verantwortungsvollen Quellen
Paper from responsible sources
FSC® C105338

If you have any concerns about our products,
you can contact us on
ProductSafety@springernature.com

In case Publisher is established outside the EU,
the EU authorized representative is:
Springer Nature Customer Service Center GmbH
Europaplatz 3, 69115 Heidelberg, Germany

Printed by Libri Plureos GmbH
in Hamburg, Germany